Biocatalysts in Biomass to Bioproducts

The Editors

Dr. K. Ramasamy is the Vice-Chancellor of Tamil Nadu Agricultural University, Coimbatore and Member of Planning Commission for Tamil Nadu State Government. He has completed his basic (1970) and masters' degrees (1974-Agricultural Microbiology) in Agriculture from Annamalai University. Later he studied M.S. in Fermentation Technology (1977) and Ph.D. in Industrial Microbiology (1980) at the Catholic University of Leuven, Belgium. In the same University, he gained his Post-Doctoral experience in Electron Microscopy and later on Gene Cloning from the Michigan State University, USA. He also had his specialized training in Applied Biology (1981) in UK.

He served in TNAU for more than 40 years in various departments and in various levels. He has also worked as Dean, School of Biosciences and Bioengineering, SRM University, Chennai and Vice-Chancellor, Karpagam University, Coimbatore. Currently as a VC of this great university, he made the TNAU to receive several **excellence awards** which includes **Best State Agricultural University Award, Overall Excellence Award (**by Federation of Indian Chamber of Commerce of India), **Global Quality Award,** *etc*. During his career, he obtained 21 awards, two medals, six fellowships (four national and two international) and organized 18 trainings and international workshops. He holds 120 research publications including 7 books and 23 bulletins.

Dr. S. Karthikeyan, Professor of Microbiology, Department of Bioenergy, TNAU, had served in this profession for 21 years. He had basics in agriculture, masters in microbiology and doctorate in environmental sciences from TNAU. He completed post doctoral training at Durban University of Technology, South Africa from 2008-2009. He holds 7 awards and had published 27 original research articles. He had published 7 book chapters, 34 seminar papers, 2 technical bulletins and 5 popular articles. He had handled several courses in under graduate, post graduate and doctorate level. His teaching excellency had won him the best teacher award in 2007. He had guided more than 2 doctoral and 7 masters' students in the field of agricultural microbiology. He had received 11 projects from different funding agency.

Dr. U. Sivakumar, presently working as Professor in Department of Agricultural Microbiology, TNAU started his career as Horticultural Officer (1991). He had his basics in Horticulture (1989), masters (1991) and doctorate (2003) in Agricultural Microbiology from TNAU. His post doctoral training was in Recombinant archaeal laccase production from University of Florida, USA during 2007-2008. He joined TNAU during 1995 and since then several technologies were perfected by him. He had published 36 research papers, 29 Seminar/Symposium Papers, 13 Booklets/Technical Manuals, 6 book chapters and 25 lecture notes. He won 5 awards and holds membership in 5 international and national bodies. He had received 11 research projects from several funding agency. Having 21years of experience he had handled many courses to UG, PG and PhD students, guided 5 M.Sc and 2 PhD students in the field of Microbiology.

Biocatalysts in Biomass to Bioproducts

— Editors —

K. Ramasamy

S. Karthikeyan

U. Sivakumar

2017

Daya Publishing House®

A Division of

Astral International Pvt. Ltd.

New Delhi – 110 002

ISBN 978-93-86071-69-9 (International Edition)

Published by   : **Daya Publishing House®**
                                     A Division of
                                     Astral International Pvt. Ltd.
                                     – ISO 9001:2015 Certified Company –
                                     4736/23, Ansari Road, Darya Ganj
                                     New Delhi-110 002
                                     Ph. 011-43549197, 23278134
                                     E-mail: info@astralint.com
                                     Website: www.astralint.com

Foreword

The need for various forms of energy to fuel the growth and development of any country need not be overstated. Much of the energy is derived from fossil fuels and hence the processes and products are engineered and designed around the utilisation of these fuels. As development has to keep pace with the population, the quantum of usage of these fuels has increased in geometric proportions. The negative aspect of using fossil fuels is the increasing trend towards emission of greenhouse gases and global warming. However, as the fuel availability declined the efficiency of utilisation increased, yet the rapid development and decline in fossil fuel reserves and the concern on global climate change has led to the search for alternative sustainable and renewable energy resources. Biomass is renewable and plays a unique role in carbon recycling. Producing fuels from biomass creates concern for use of land for fuel or food, the latter has to get priority to feed the millions. Hence the second generation technologies to produce fuels and chemicals from nonedible feed stocks such as agricultural residue, forest residue, municipal solid waste, industrial waste, and dedicated energy crops are gaining importance. In such approaches cost of production, the energy and material balance needs to be comprehensively addressed to prevent any lopsided development. Therefore, while developing second generation biofuels, several challenges in converting lignocellulosic biomass to fuels and chemicals using biochemical platforms need to be addressed in consonance with social issues. Hence this book is timely.

In this context, this book provides an overview of biomass to biofuel conversion technologies with the major challenges ahead that has to be addressed. Successful and economical biological conversion of lignocellulosics to fuels and chemicals using bio refinery approach depends on several key strategies in the production platform including biomass logistics. Development of energy efficient technologies (pretreatment, enzyme hydrolysis, and microbial fermentation), accompanied by coproducts and establishment of biofuel and biochemical standards are the need of the hour. Of these technologies, biocatalysts hold a significant and critical role for a successful bio refinery. The cutting edge technologies in biocatalysts *viz.*, protein engineering is one of the highlight in this book.

It must be emphasized that energy production as standalone technology can never work. There has to be a blend of energy strategies that works as a part of conglomerate with several network of consortiums that need to be evolved with decentralized approaches. Thus, apart from biofuels production from biomass, resource recovery of the non-cellulosic component of biomass such as lignin and hemicelluloses also find a major domain in this masterpiece. The emphasis on the generation of bio products such as acrylic acid, vanillin, poly hydroxyl butyric acid, ring opening products from lignin and furfural, xylitol, acetic acid from hemicellulosic component of the tropical biomass can make a breakthroughs in future research. Many technologies are available on the bench. The issue is how to put them together as per the requirement energy in the local ecosystem. I hope this masterpiece will be a treasure for entrepreneurs, researchers and students which gives a vivid understanding on Biomass conversion strategies and its challenges.

K. Gurumurthi
Ex-Chair of Excellence
Retd Director IFGTB
Formerly Senior Adviser DBT-GoI, New Delhi

Preface

Depleting fossil fuel reserves and adverse effects of fluctuating oil prices have renewed interest in alternative and sustainable sources of energy. Biofuels have emerged as a highly promising source of alternative energy, and have drawn global R&D for their production from biomass. Biofuel derived fromcorn, wheat, and sugar cane is the most commonly used first generation biofuel feed stock. This method poses a limit above which they cannot produce enough biofuel without threatening food supplies and biodiversity. With this limitation, second generation biofuels are manufactured from various types of lignocellulosic biomass. Bioethanol can be produced from biomass through chemical and biochemical approaches in which the later is cost-effective. Hence second generation biofuels promote sustainable energy policies that spur economic growth and environmental protection in a global context, particularly in terms of reducing greenhouse-gas emissions that contribute to climate change.

Microbes have been playing a key role by producing enzymes for bioconversion of various biomasses for biorefinery applications. This book on "Biocatalysts in Biomass to Bioproducts" summarizes state-of-the-art information on the logistics of biomass conversion to biofuels and related challenges to overcome in 13 chapters.

The chapter 1 brings out the availability of feed stock and the challenges faced in conversion of biomass to biofuel. Lignocellulose is a complex of lignin, hemicelluloses and cellulose. The lignin makes the biomass recalcitrant. So chapter 2 discuss on the various pre-treatments methods available for lignocellulosic biomass. Among the different pretreatment methods, enzymatic pretreatment using laccase is most effective and this multifunctional green catalyst is discussed in chapter 3. The cellulose portion is now accessible and upon the action of cellulase, the cellulose complex is broken down to simple sugars. Chapter 4 focuses on the fungal cellulase for biomass deconstruction and production of useful biofuel and bioproducts from them.

In biological conversion of biomass, the enzyme hydrolysis step has been identified as a major techno-economical bottleneck in the entire wood-to-ethanol bioconversion process. Cellulosome are a multienzyme complex that hydrolyze cellulose at a higher rate than fungal cellulases and was summarized in chapter 5. Enzymatic hydrolysis carried out at ≤ 50 °C exhibit slow rate of enzymatic hydrolysis. This has hastened the search for thermophillic enzyme and is detailed in chapter 6.

The use of chemical catalysts for biomass conversion is dealt in chapter 7. Research into the production of liquid transportation fuels from microscopic algae, or microalgae, is described in chapter 8. Glycosyl hydrolases and enzymatic hydrolysis of biomass and bioproduct development was elaborated in chapter 9 and 10 respectively. Chapter 11 focuses on development of enzymatic biorefinery for biofuel production from leather solid waste and sludges. In saccharification process, β-glucosidase is the rate limiting enzyme and its supplementation increases the hydrolysis rate. Hence chapter 12 highlights the functional role and involvement of BGL in cellulolysis. Indeed the last chapter 13 discuss about the synthesis and characterization of lignin adhesive.

The chapters have been contributed by the experts of the area across the globe and these chapters offer concise and clear information on the recent research developments in biomass utilization. Sustainable and affordable energy, underpinned by energy technology cooperation and innovation, is indispensable for economic growth and reduced carbon intensity. No doubt this book will provide readers with a comprehensive resource that connect theory to real-world implementation.

K. Ramasamy

S. Karthikeyan

U. Sivakumar

Contents

1

Dendro Biomass Resources for Solid and Liquid Fuel Generation: Opportunities and Challenges

K.T. Parthiban, N. Krishna Kumar, B. Palanikumaran and S. Umesh Kanna

Forest College and Research Institute,
Tamil Nadu Agricultural University,
Mettupalayam – 641 301

Introduction

Biomass contributes nearly 10-14 per cent of the global energy supply and is fourth largest after coal, petroleum and natural gas. The main utility of biomass has been directed towards direct combustion to generate steam based power in a decentralized approach. This biomass can very well be converted to liquid fuels like ethanol, butanol, biodiesel and biogas (Tuli, 2013). The demand for energy across the world is increasing steadily which resulted in fast depletion of petroleum resources and necessitated an increased interest in alternate fuels particularly on liquid transportation fuels (Wyman, 2007; Lynd, 2008). The predominant alternate fuels gaining significant attention across the world are biodiesel and bio petrol (Ethanol). The technologies for generating biodiesel from wide range of vegetable oils have been very well witnessed but for want of adequate and enough raw material resources, the achievement in biodiesel production is still dismally modest. Similarly bioethanol from ligno cellulosic biomass is considered as one of the alternate and easily adaptable fuel but due to availability of raw material and the associated cost effective processing technology limit the production and availability of bioethanol as well (Sukumaran *et al.*, 2009).

In India, the requirement of energy is steadily increasing due to progressive development in science and technology coupled with urbanization and the rural infrastructure development. It has been estimated that India has an installed capacity of 223 GW of energy and it will reach 400 GW in the year 2030. Similarly, 78 per cent of the current crude oil requirement is being imported which has taken heavy toll of the country's exchequer. These energy resources besides massive imports have contributed significantly towards accumulation of both gaseous and particulate pollutants thereby contaminate the environment. This has compelled development of alternate source of energy which are safe and environmentally viable. Under such circumstances, India needs to develop sustainable and decentralized energy planning process for which the role of biomass based energy generation both for liquid and solid energy generation is very significant. Though the current estimates indicated that biomass is available in surplus, but its quality and the availability in a decentralized farmlands and its value addition to other agricultural utility detract their availability to biofuel utility.

Hence the biomass derived from trees which are pronounced as dendro biomass will have an excellent role to play in biofuel generation due to their quality and their bulk availability. Considering the dendro biomass potential, the research group at Tamil Nadu Agricultural University has established a viable value chain approach to augment the Production to Consumption System (PCS) in Biofuel production process.

A. Opportunities and Significance

In India the biomass provides fuel for about 32 per cent of the total primary energy consumed and caters to almost 70 per cent of the population. Wood based dendro biomass has a significant and critical role not only in enhancing the biomass availability but also extending the scope for both solid and liquid fuel generation. The woody biomass are rich in carbon, hydrogen and oxygen and the wood resources have been classified as high cellulose, high lignin and low cellulose based on wood quality. Similarly the wood biomass resources are classified as dense wood, moderate and soft wood which may be suitable for meeting the demands of both solid and liquid fuel generation. The dendro biomass due to their high calorific value is directly deployed in combustion and gasification based biomass power generation. The potential species deployed in biomass power generation industry with their wood quality estimates are furnished in Table 1.1.

Similarly wood biomass also known for their lignocellulosic presence and the current research group has identified and developed improved varieties for high cellulose and low lignin in several wood species. The lignocellulosic species suitable for ethanol and butanol production are furnished in Table 1.2.

Table 1.1: Thermo Chemical Properties of Species Amenable for Biomass Power Generation

Sl.No.	Species	Specific Gravity	Volatile Matter (Per cent)	Ash Content (Per cent)	Fixed Carbon (Per cent)	Calorific Value (kcal kg⁻¹)	Fuel Value Index	Ash Deformation Temp. (°C)	Ash Fusion Temp. (°C)	Lignin Content (Per cent)	Heating Value (MJ kg⁻¹)
1.	Acacia holosericea	0.54	66.53	4.58*	21.25	3692	27.29	1043.33	1080.00	29.67	27.84
2.	Acacia auriculiformis	0.49	66.83	3.50	21.00	3557	30.23	1080.00	1140.00	30.00	28.45
3.	Anacardium occidentale	0.50	66.33	4.07	21.60	3441	28.16	1100.00	1180.00	30.00	28.38
4.	Bambusa vulgaris	0.54	66.21	3.71	22.92	4130	109.86	1130.00	1210.00	33.17	30.57
5.	Bambusa bambos	0.56	66.66	2.92	23.08	4150	115.51	1136.67*	1205.00	32.00	30.59
6.	Casuarina equisetifolia	0.58*	67.62	2.33	23.38*	4593*	251.65*	1156.67*	1230.33*	34.33*	31.35*
7.	Cassia siamea	0.48	65.87	4.47*	21.17	3462	25.15	1050.00	1110.00	28.00	28.18
8.	Caesalpinia pulcherimma	0.43	66.00	4.38	22.12	3479	23.32	1080.00	1120.00	28.17	28.21
9.	Dalbergia sissoo	0.58*	66.43	2.88	23.75*	4314*	236.81*	1136.67*	1230.00*	33.25*	30.81*
10.	Eucalyptus tereticornis	0.61*	67.00	2.50	23.85*	4614*	240.69*	1150.00*	1241.67*	33.67*	31.02*
11.	Gliricidia sepium	0.44	65.50	4.75*	20.58	3913	21.31	1083.33	1136.67	29.00	27.61
12.	Lannea coromandelica	0.45	66.33	4.37	21.30	3812	23.29	1060.00	1136.67	27.00	27.85
13.	Leucaena leucocephala	0.62*	68.08*	1.75	24.33*	4552*	265.68*	1173.33*	1273.33*	34.42*	31.50*
14.	Prosopis juliflora	0.78*	68.17*	1.48	24.88*	4860*	303.72*	1220.00*	1290.00*	35.17*	31.69*
15.	Simarouba glauca	0.55	65.83	4.97*	21.70	3781	31.48	1056.67	1196.67	29.33	28.02
	Mean	0.54	66.62	3.51	22.46	4023.33	121.62	1110.44	1178.36	30.61	29.47
	SEd	0.018	0.83	0.84	0.64	64.51	63.82	19.01	24.55	1.48	0.48
	CD (0.05)	0.036	1.19	0.91	0.750	131.75	85.34	25.83	49.13	2.57	1.15

Table 1.2: Wood Quality Analysis of Tree Species Amenable for Biofuel Production

Sl.No.	Species	Bulk Density (kg/m³)	Basic Density (kg/m³)	Holo Cellulose (Per cent)	Acid Insoluble Lignin (Per cent)	Pulp Yield (Per cent)
1.	Acrocarpus fraxinifolis	289.00	624.00	74.20	24.71	48.40
2.	Acacia auriculiformis	330.00	580.00	68.47	27.26	48.70
3.	Albizia falcataria	288.00	530.00	67.29	26.36	48.60
4.	Anthocephalus cadamba	180.00	385.00	68.82	26.33	48.50
5.	Bambusa balcooa	214.00	487.94	67.60	22.40	44.69
6.	Bambusa vulgaris	198.00	500.72	63.50	22.10	42.79
7.	Cassia siamea	250.00	530.00	67.73	28.31	45.50
8.	Chukrasia tabularis	257.33	467.11	73.68	26.14	46.50
9.	Dalbergia sissoo	286.00	610.00	69.35	25.62	49.40
10.	Erythrina spp.	220.00	430.00	72.00	24.00	42.00
11.	Lannea coramandalica	220.00	480.00	74.00	24.00	43.00
12.	Leucaena leucocephala	250.00	546.00	74.14	24.32	49.50
13.	Melia dubia	280.00	286.00	74.00	27.75	50.50
14.	Melia composita	260.00	510.00	78.00	22.0	48.50
15.	Populus deltoides	179.33	380.03	74.15	22.87	48.00
16.	Sweitenia macrophylla	270.25	520.00	69.54	26.39	41.89
17.	Thespesia populnea	253.00	464.77	71.91	26.35	48.00
18.	Gmelina arborea	258.00	486.00	69.16	26.67	49.10
19.	Casuarina	232.00	495.00	75.83	26.46	48.50
20.	Eucalyptus	240.00	540.00	74.36	26.73	44.20

B. Sources of Dendro biomass

Raw materials that can be used to produce biomass fuels are widely available across the country and come from a large number of different sources, and in a wide variety of forms. All of these forms can be used for fuel production purposes, however not all energy conversion technologies are suitable for all forms of biomass. The biomass sources from forestry sector alone are furnished.

a) Wood and Wood Residues

Wood consists of wood and other products such as bark and sawdust which have had no chemical treatments or finishes applied. Wood may be obtained from a number of sources which may influence its physical and chemical characteristics. The lignocellulosic wood is suitable for a range of energy applications. It can be burned for heat and/or power at a range of scales. New 'second generation' technologies are being developed which are capable of producing a range of liquid or gaseous transport bio fuels from ligno cellulosic wood.

i. Bark

Bark may be removed from saw logs and available as a residue from wood processing. Bark typically contains high levels of minerals and consequently is prone to give high levels of ash and slogging in combustion systems. It may, however, be a suitable fuel for generating process heat close to where it is produced, such as for firing drying kilns at a sawmill. Minerals will be retained in the ash and consequently this may be used as a soil fertilizer.

ii. Logs

Small Round Wood (SRW) may simply be cut into logs. This may be done in the plantation for ease of extraction and handling and to assist drying, and involve delimbing and cutting into logs of typically 2-3 m in length.

iii. Sawdust

Sawdust is typically available as a co-product of wood processing or manufacturing. It may be of high moisture content, *e.g.* from cutting green wood in sawmill, or very dry from furniture manufacturing. It may have a bulk density only 30 per cent of that of the solid wood and so, even if very dry, has a very low energy density. It does, however, present an extremely large surface area to volume ratio and is suitable for blowing into some combustion or gasification systems. Sawdust, especially dry sawdust, is particularly suitable for processing into pellets.

iv. Wood Chips

Although logs can be stored and transported conveniently when stacked, and the ease of air passage through a log pile allows good drying, they may not always be the most convenient form for automated handling and feeding. Also, the relatively small surface area to volume ratio is not ideal for efficient combustion or gasification. Wood chips can form a much more uniform fuel that can flow and can be fed to a boiler, gasifier or other conversion system as a steady flow using an auger feed or a conveyor. With a large surface area to volume ratio they can also be burned very efficiently. Wood chips may have a bulk energy density of about 50 per cent of that of the solid wood.

v. Wood Pellets and Briquettes

Wood pellets are made from dry sawdust compressed under high pressure and extruded through a die. They may include a low level of added binder, such as starch, but many use nothing other than steam. Wood pellets should be dry, clean, mechanically robust and have an ash content defined by the appropriate standard to which they have been made, which may also define other contaminants such as chlorine content. Briquettes are similar to wood pellets, but physically larger. Sizes vary but briquettes can vary in diameter from around 50 mm to 100 mm+. Briquettes are usually between 60 mm and 150 mm in length. They can offer a cleaner, more consistent alternative to firewood logs, offering higher energy density and steady combustion

vi. Wood from Plantations

This includes wood from private and state owned forest and plantations. As harvested, wood will be at a range of moisture content, and of a variety of physical shapes and sizes. In addition to harvesting, some level of pre-processing is likely to be required. Transport and storage will also be necessary.

Wood Processing Industry Co-products

The wood processing industries, such as sawmills and timber merchants are also a source of virgin wood in the form of off cuts, bark and sawdust. Many transport, pre-processing and storage issues may be similar to those for plantation products, however the material obtained is likely to be in different forms from plantation or arboricultural products.

There are likely to be a number of different output streams with different characteristics:

☆ Sawmill off cuts may include a high proportion of bark.

☆ Some material may have been kiln dried to extremely low moisture content making it potentially very suitable for wood pellet manufacturing or blending with material at higher moisture content.

☆ There may be sawdust at a range of moisture content from different stages of processing, and again very dry sawdust from kiln dried timber may be very suitable for wood pellet manufacture.

C. Challenges

i. Biomass Availability

In India, it has been estimated that 500 million tonnes of biomass is generated annually and of which 120 – 150 million tonnes is available for biomass utility. However several biomass industries are under threat due to non availability of biomass which indicated a wider gap between the demand and the actual viability. All agricultural residues *viz.*, forest and plant biomass are classified as lignocellulosic materials and are rich in carbon, hydrogen and oxygen. These biomass resources are abundantly available with low cost in India. However these resources are available in small and medium farm holdings coupled with diverse crop pattern which resulted in problems in the entire supply chain process. The biomass collection at the scale required and transportation of low bulk density biomass to a single large plant site would all make the feed stock availability not only difficult but also expensive for viability of large scale biomass biofuel refineries.

ii. Competition from other Industries

Though the biomass is available in plenty in the country but due to competition from other industries its conversion to biofuel industry is a major bottle neck. Because several times the biomass is converted into various value added products like cattle feed stock, biomass compost *etc.* which made them non available to the industries. The competition between industrial utility and cattle feed utilization lead to feed stock shortage. Hence there is a need to develop non food and feed quality raw material resources which may be amenable for biofuel generation. Under such circumstances wood based biomass resources will play an excellent role in meeting the biomass resources on a large scale due to their increased density, higher yield, non seasonal and their suitability as a ligno cellulosic material. The wood based ligno cellulosic biomass have to be characterized for their quality towards high cellulose and low lignin for liquid fuel generation and moderate cellulose with high lignin for solid fuel based power generation.

iii. Bioprocessing Technology

The woody biomass available can be used directly or woody biomass can be generated in the form of energy plantation through agro and farm forestry. However the major challenge in woody biomass is the technologies for pre treatment, fractionation of lignocellulosic biomass to cellulose, lignin and hemicellulose and their further conversion into sugars and finally into an ethanol coupled with the other value added products. The overall technology involved multiple steps which need to be standardized for economical viability and ecological sustainability.

iv. Lack of Organized Plantations

Establishment of exclusive biofuel plantation is yet another challenge faced by various stakeholders. The plantations available currently are sure that they are ligno cellulosic but they are used by multifarious industrial sector like pulp, paper, packing case, plywood, veneers etc. and limit the availability of biomass resources to biofuel industries. Hence development of exclusive biofuel plantations both for solid and liquid energy generation is very vital towards sustaining the process.

v. Non Availability of Quality Raw Material

The quality of currently available biomass is widely questioned due to multifarious feed stock. The quality in terms of cellulose, hemicellulose and lignin vary with species to species and increase or reduction of 1 per cent in the quality parameters will have a significant impact in the processing side. Hence development of quality raw material is very vital.

vi. Uncertain Supply Chain

One of the major challenges in biomass availability is the existence of unorganized and multipartite supply chain system. This unorganized supply chain system play a key role in biomass availability. The supply chain involves wide range of players with several middlemen and trader. Because of this multi-partite supply chain, variation in pricing system and marketing pattern delimit the availability of biomass resources for biofuel industries. This unorganized supply chain needs to be changed with organized institutional mechanism to create a platform for a sustained supply of dendro biomass resources. The alternate utility of biomass resources for other wood based industry is also a major problem in the supply chain process which needs to be addressed.

vii. Lack of Price Supportive System

For sustainability and commercial viability of dendro biomass resources, price supportive system is very critical both for promotion and commercial utilization. Unfortunately there is no price supportive mechanism for the ligno cellulosic biomass both for solid and liquid fuel generation. Due to lack of price supportive system and multi-partite supply chain, the price of ligno cellulosic biomass is highly variable and is the major challenge faced by the biofuel industries.

viii. Policy Support Mechanism

The promotion of biomass for biofuel utility demand a strong policy intervention towards incentives for biofuel plantation development programme coupled with assured and competitive price supportive system. This has to be addressed in detail inviting all stakeholders to draw a workable policy support.

D. Status of Dendro Biomass Feedstock Availability

The dendro biomass resources has a potential share in biofuel utility both for solid and liquid fuel generation. These dendro biomass resources can act as an excellent feed stock and with a planned resource planning, the availability can be matched with the demand for the raw material. For this purpose, it is estimated that one hectare of an energy plantation on an average is able to generate 100 tonnes of ligno cellulosic biomass suitable for biofuel utility. If this target is planned in a decentralized approach, the demand for raw material can easily be met through organized energy plantation development programme. The current research group has estimated the availability of nearly 74 million tonnes of lignocellulosic biomass which can be deployed for the biofuel industries.

Among 20 different tree species of ligno cellulosic biomass utility, only five species have been estimated for their biomass availability. These five species include Casuarina, Eucalyptus, Bamboos, Ailanthus and Melia and the first four species have been grouped under one category and it is estimated that an area of over 74000 ha is now available which has the biomass yield potential of over 74 million tonnes and indicated greater scope for their deployment in biofuel industry (Figures 1.1a and b; Plate 1.1). These biomass resources have excellent wood quality with high cellulose and low to moderate lignin content. With the development of suitable and cost effective processing technology, these biomass resources can be directly used for power generation through combustion technology and into ethanol through ligno cellulosic conversion technology.

Another major breakthrough in biomass improvement is the identification of *Melia dubia* as the potential lignocellulosic biomass species due to its higher cellulose and low lignin content. The development of high yielding and short rotation clone (Parthiban and Seenivasan, 2016) in the species has attracted its deployment in biofuel utility. This species is one of the fast growing tree amenable for harvest in 2 years and has a potential to yield the biomass volume of 100 tonnes/ha. This species has been planted in an area of over 900 ha and has a potential to yield 0.9 million tonnes of lignocellulosic biomass at present (Figures 1.2a and b). This biomass exhibited cellulosic pulp yield of over 50 per cent coupled with low lignin content

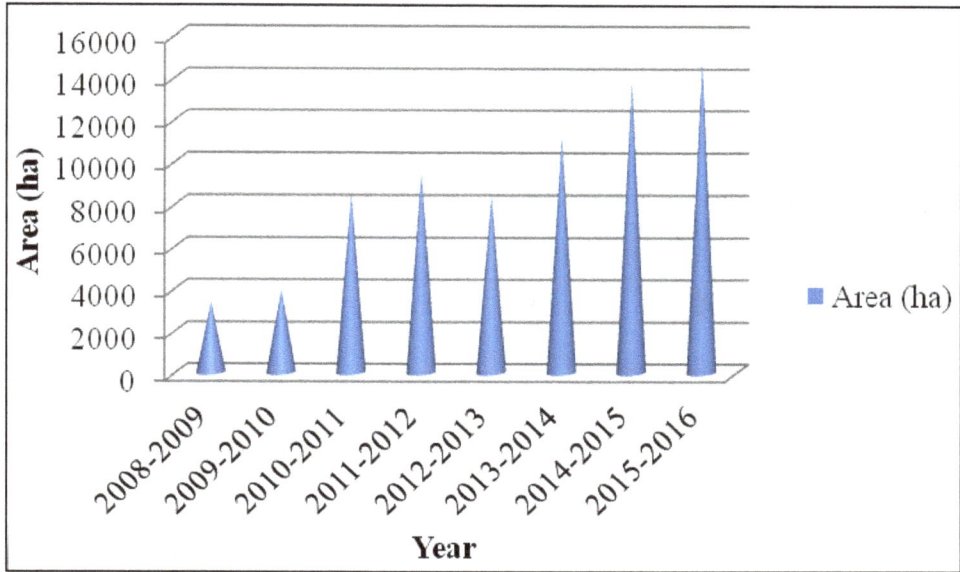

Figure 1.1a: Status of Lignocellulosic Biomass Plantations in Tamil Nadu.

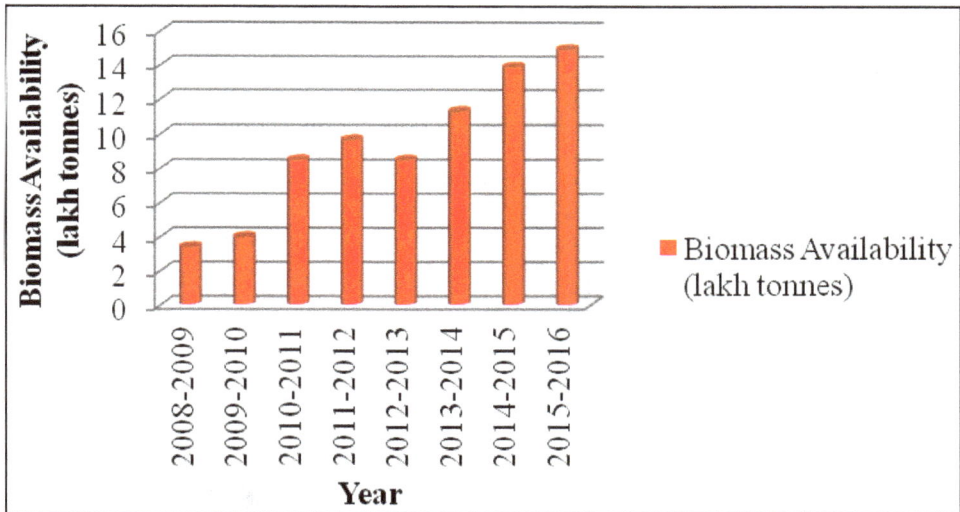

Figure 1.1b: Status of Lignocellulosic Biomass Availability in Tamil Nadu.

and extended greater scope for its adoption in liquid fuel generation particularly for ethanol production.

E. Strategies to Augment the Biomass Resources

Forests in general and forest biomass in particular have enormous role in biomass sector leading to the production of solid and liquid energy. The country's

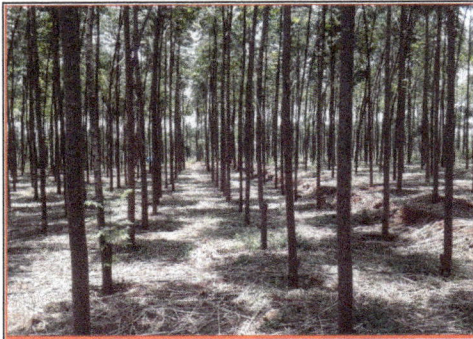

Melia dubia

Variety   : Melia MTP 26
Basic Density : 286.00 Kg/m³
Hollocellulose : 74.00 %
Lignin   : 27.74 %

Anthocephalus cadamba

Variety   : Kadam AC 13
Basic Density : 385.00 Kg/m³
Hollocellulose : 68.82 %
Lignin   : 26.33 %

Casuarina

Variety   : Casuarina MTP 2
Basic Density : 495.00 Kg/m³
Hollocellulose : 75.83 %
Lignin   : 26.46 %

Eucalyptus

Variety   : Eucalyptus EC 48
Basic Density : 540.00 Kg/m³
Hollocellulose : 74.36 %
Lignin   : 26.73 %

Plate 1.1: Lignocellulosic Biomass Plantations.

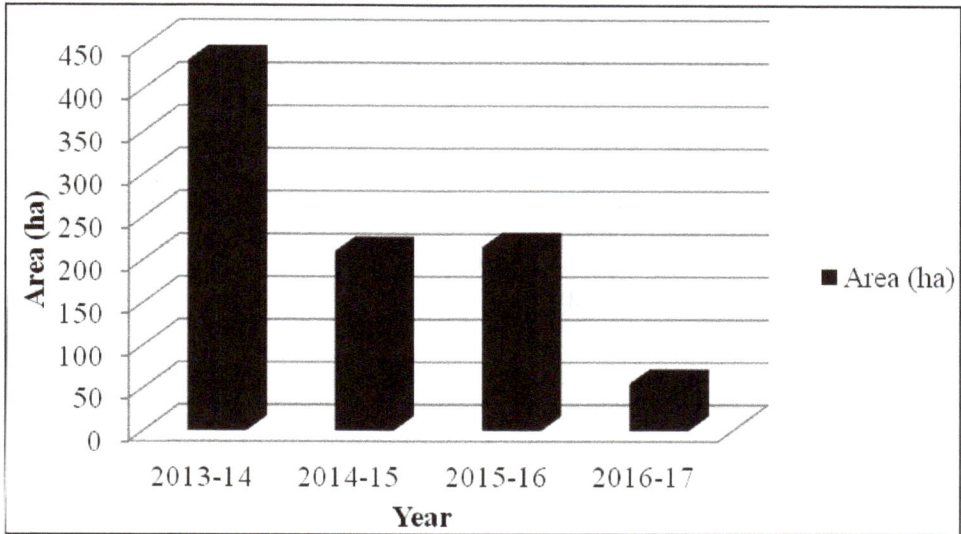

Figure 1.2a: Status of Melia Based Lignocellulosic Biomass Plantations in Tamil Nadu.

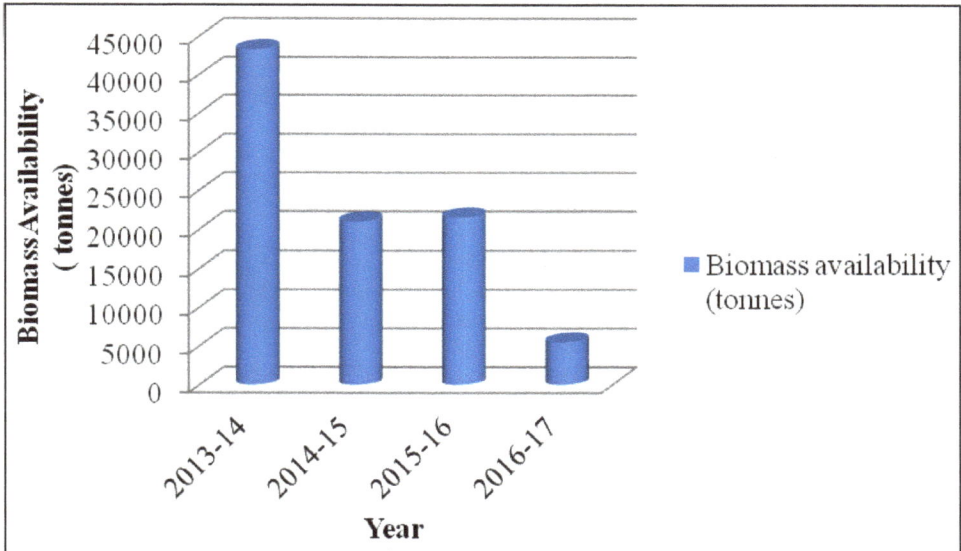

Figure 1.2b: Status of Melia Based Lignocellulosic Biomass Plantations in Tamil Nadu.

forest area is legally closed for any production oriented purposes which resulted in the wood based industries deriving raw material predominantly from natural forests till the recent past. These wood based industries have started generating their own raw material resources by establishing exclusive industrial wood plantations through a consortium mode. Similar approach can be established to augment the biomass resources to cater the feed stock needs of biomass and biofuel industries. The following strategies will help to improve the situation in biomass resource availability.

i. Development of Exclusive Biomass Energy Plantations

Since the government of India has mandated blending 20 per cent biofuels in transport sector from 2017 onwards and promotes decentralized energy planning process through biomass, it is very ideal to develop exclusive energy plantations suitable for liquid as well as solid energy generation. The Forest College and Research Institute of Tamil Nadu Agricultural University has identified the following tree species suitable for varied energy utility (Table 1.3).

Table 1.3: Potential Dendro Biomass Species for Biofuel Utility

Sl.No.	Energy Utility	Species	Product	Quality
1.	Biodiesel	*Jatropha curcas* *Pongamia pinnata* *Madhua latifolia* *Callophyllum inophyllum*	Seed Oil	☆ High oil content associated with the fatty acid profile ☆ 30 per cent and above
2.	Ethanol	*Melia dubia* *Thespesia populenea* *Eucalyptus* spp. *Casuarina* spp. Populars *Lannea coromandelica* Bamboos *Leucaena leucocephala*	Wood	☆ High cellulose and low lignin ☆ 48 per cent and above
3.	Power generation through gasification	*Acacia holloserecia* *Prosopis juliflora* *Gliricidia sepium*	Wood	☆ High calorific value ☆ Wood thickness of 1"–2"
4.	Power generation through combustion	*Prosopis juliflora* *Dalbergia sissoo* *Leucaena leucocephala* *Eucalyptus tereticornis* *Acacias* spp. Cassia spp.	Wood	☆ High calorific value 4000–5000 kg cal.

ii. Development of High Yielding Short Rotation Energy Plantations

One of the major strategies in promotion of biomass energy plantation is the development of High Yielding Short Rotation varieties preferably the clone based energy plantations. For this purpose, the Tamil Nadu Agricultural University has developed wide range of energy species which are amenable for high yielding and short rotation. The major species and the improved cultures within the species along with their potential yield are furnished in Table 1.4.

iii. Dendro Energy Models

The promotion of biomass resources for energy generation both for solid and liquid energy demands organized models for intensive promotion. The TNAU has developed various models which are discussed below.

Table 1.4: High Yielding Short Rotation Clones Suitable for Biofuel Utility

Energy Species	Variety	Spacing (in feet)	Density/ Acre	Period (In years)	Average Yield (Kgs)	Yield/ha (tonnes)
Eucalyptus	MTP 1	6x6 ft	1,200	3	40	36.00
Casuarina	MTP 2	5x5 ft	1,770	3	38	67.00
Subabul	FCRI LL15	4x4 ft	2,200	3	32	70.00
Dalbergia	MTPDS18	6x6 ft	1,200	3	40	48.00
Melia	Melia CL26	6x6 ft	1,200	2	50	60.00

a. High Density Short Rotation Model

This High Density Short Rotation (HDSR) Plantation models are vary from species to species and the general model adopted for various dendro energy species are depicted below in order to get sustained yield to meet the raw material requirement of biofuel industries (Figures 1.3 and 1.4).

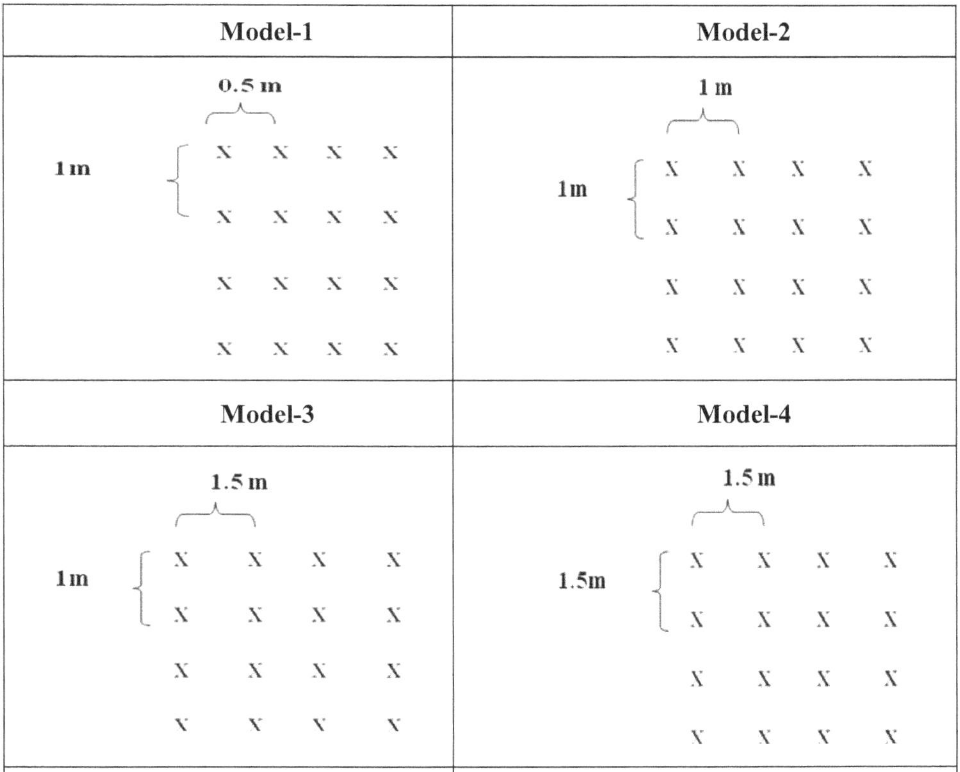

Model-1	Model-2
Model-3	Model-4
Spacing	1m x 0.5m, 1m x1m, 1m x1.5m, 1.5m x 5m
Number of plants ha[s1]	20 000, 10 000,6 666 and 4 444

**Figure 3: High Density Energy Models
(Durairasu and Parthibal *et al.*, 2013).**

Model-1	Model-2
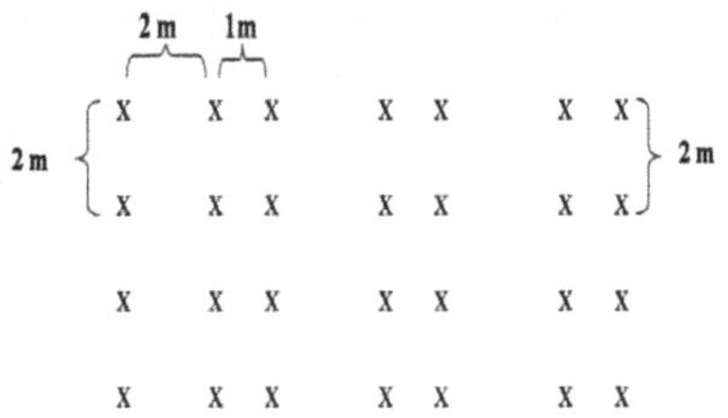	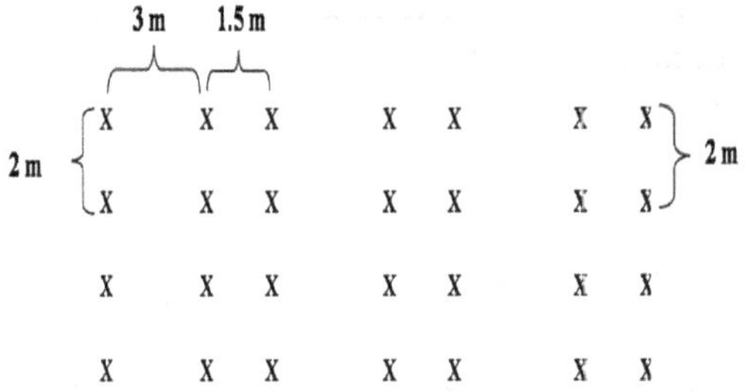
Spacing	2 m x 2 m and 1 m & 3 m x 1.5 m and 1.5
Number of plants ha^{-1}	2 900 & 2 250

Figure 1.4: Paired Row Energy Models.

b. Hybrid Tree Model

This model incorporates two different species in the same unit of land to auger the productivity and income by production of industrial wood for two different industries. Accordingly, Cauarina + Melia, Melia + Glyricidia, Melia – Subabul, Melia + Ceasalpinea etc., are recommended for hybrid tree model. Melia can be utilized for pulp and plywood industries and other trees can be profitably grown for biomass power.

c. Monoclonal Model

High yielding clones of various species like Casuarina (MTP 1 and 2), Eucalyptus (ITC 3, 7, 413 and TNAU 48), Melia (MTP 1, 2 and 7), Subabul (FCRI LL15), *Dalbergia sissoo* (MTPDS 18) and Glyricidia (FCRI 1-10) can be raised as high density monoclonal plantations. These plantations with intensive and precision silviculture management could be harvested on a sustainable basis either annually or once in two years depending on the species.

d. Polyclonal Model

Different clones in the same species can be planted as a mixed polyclonal model either with alternate rows/blocks of five to ten rows/paired row model etc. For this purpose, single species with multi clonal incorporation could be made for increased productivity and profitability. The medium to large land holdings are amenable for such poly clonal model.

e. Sporadic Model

This model incorporates sporadic distribution of trees. The small land holders cannot afford monocropping of tree crops and hence this model is suggested using sporadic distribution of biomass trees like Acacia, Melia, Eucalyptus and Casuarina.

f. Linear Model

This model incorporates linear planatation of single species of Eucalyptus/ Casuarina/Albizia/Subabul/Melia/Glyricidia/*Cassia siamea* etc., in the National and State Highways, along the railway lines and community lands. This will ensure availability of raw material on a large stretches of land which are currently unutilized.

g. Paired Row Model

This model ensures integration of single or two species in a paired row system. For every one pair an alternate single row can be established which will increase the productivity. This model is suitable for small and medium land holdings.

iv. Approaches for Augmenting Dendro Biomass Resources

The dendro biomass energy resources can be augmented by establishing adequate linkages between the stakeholders to complete the entire value chain process. For this purpose, the Tamil Nadu Agricultural University has conceived and implemented a consortia mode biomass energy plantation model through bi, tri and quad partite model which are indicated in Figures 1.5–1.7 and these model can be approached to generate adequate biomass resources both through captive as well as agroforestry based energy farming.

a) Bi-Partite Model

This model incorporates two stakeholders' *viz.*, biofuel industries and farmers (Figure 1.5). Under this system, the biofuel industry can supply quality planting materials of energy trees, site specific technical guidelines and periodical monitoring of various stages of tree growth by their scientific staff members. This besides, the biomass industry can assure a minimum support price of wood at farm site at the time of agreement. In case, if the market price exceeds the agreement price at the time of harvest or lower; in both the cases the biofuel industry should assure higher price to the grower. The biomass industry also assures transportation at its own cost from the farm land to the biofuel industry. The farmer assures to follow the

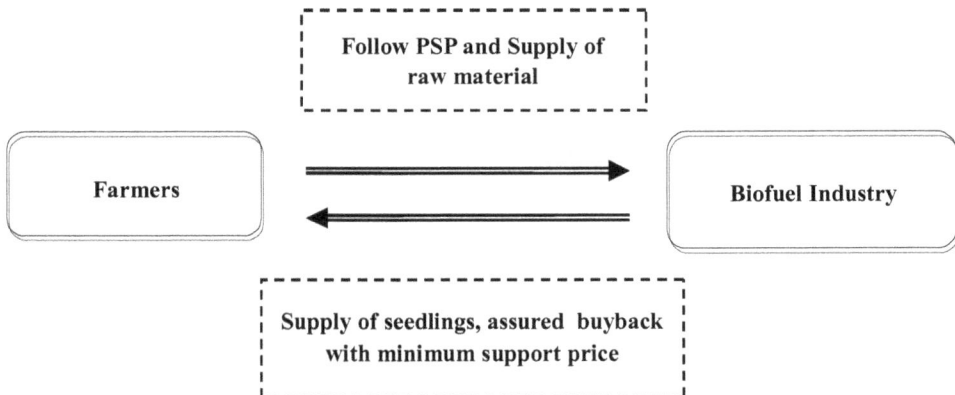

Figure 1.5: Bi-partite Model Energy Farming.

guidelines prescribed by biofuel industry and the growers are abide by the rules and purchase regulations of biofuel industry and supply the harvested produce to the firm as per the agreement.

b) Tri-partite Model

This model incorporates industry, growers and research institute. Under this system, the industry supplies quality planting material at subsidized rate and assures minimum support price or the prevailing market price whichever is higher. The farmer supplies the material to the contracting industry and the research institute advice site specific technology. The varieties already developed by the research institute are mass multiplied through its agri business incubators (ABI) distributed across the state and supplied to the contracting farmers at subsidized and affordable prices (Figure 1.6).

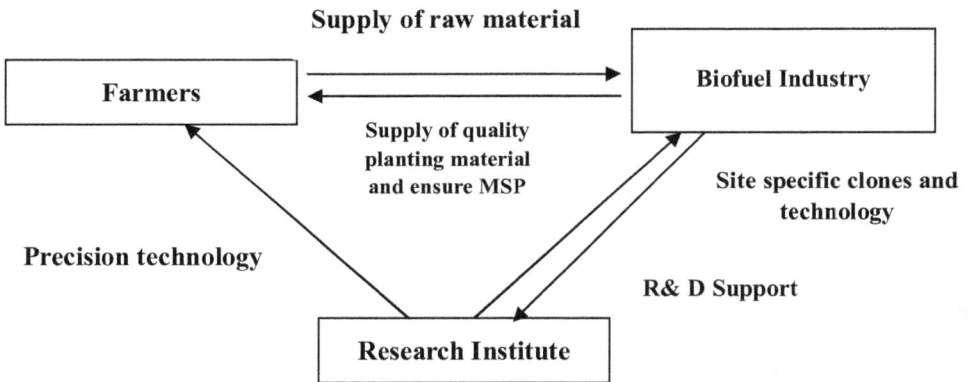

Figure 1.6: Tri-partite Model Energy Farming.

c) Quad-partite Model

This system is similar to tri-partite model barring the involvement of financial institution (Figure 1.7). Financial institutions provide credit facilities to the growers. For credit facilities, a simple interest rate at 9 per cent is followed and the repayment starts after felling. The research institute particularly Forest College and Research Institute (TNAU) play a significant role for technological advancements through varietal development and also to advice site specific precision silvicultural technology to the growers. A pre and post-plantation scientific advice helps to develop human resources through on and off institute mode to farmers and plantation staff of the industries.

The industry mass multiply the potential genetic materials identified by the research institute in a decentralized manner and supply them at subsidized costs. The industry also facilitates felling and transport at their own costs, which resulted in strong linkage between industry and the farmers. The industry helps to repay the credit amount after felling of farm grown raw materials there by help the financial institutions for timely repayment, which resulted in strong institutional mechanism

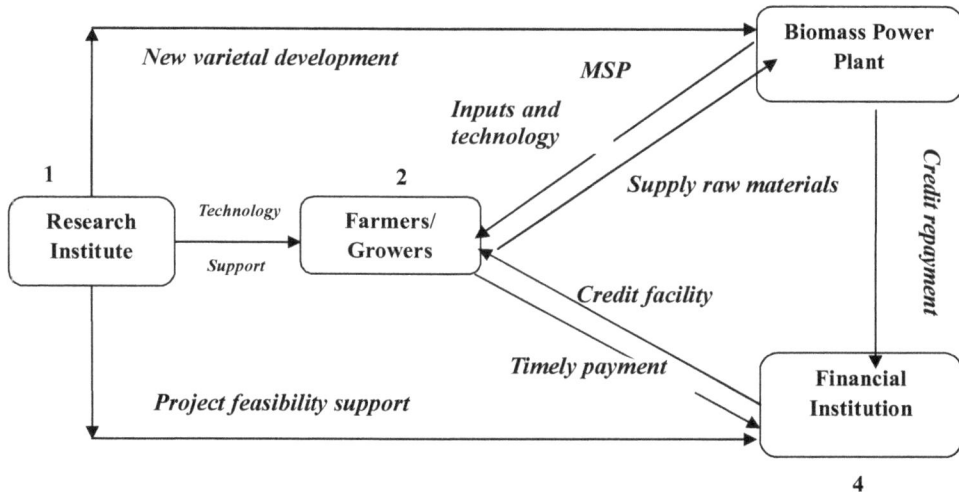

Figure 1.7: Quad-Partite Model Energy Farming.

for sustainability of the contract dendro energy farming system in the state. These models were successfully adopted in other industrial wood plantation programmes (Parthiban and Govinda Rao 2008; Parthiban *et al.*, 2010 and Parthiban, 2016) and extend scope for its application in biofuel utility.

Consortia Mode Energy Farming Methods

The following contract energy farming methods designed by Forest College and Research Institute of Tamil Nadu Agricultural University can be implemented in the farm lands in association with biofuel industries to generate solid and liquid fuels.

a) Farm Forestry Model

In this method, the biofuel industry can supply the quality planting materials of site specific variety identified by the research institute to the interested farmers on a subsidized rate. The farmers in turn develop his own plantation and obtain the needed technological support from the research institute. If needed, the farmers can get credit and insurance facility from the financial institutions and final felling and transportation by the biomass industry. An agreement can be made before the plantation establishment. In this method, the farmers grow only trees without any intercrop and mostly practiced in dry land condition.

b) Agroforestry Model

In this model the farmers can raise energy plantation as a major tree crop and grow suitable intercrops and this method can be practiced mostly in garden land conditions. The other conditions are similar to that of Farm Forestry model.

c) Captive Model

In this method, the industry can develop their own plantation through land lease or benefit share model. The large land holders and the lands available with government and private sector which are unutilized for a longer period can be utilized. A minimum land size requirement of 10 ha. for 1 cluster unit is prescribed. Once the land is identified, the land holder can opt either land lease model or benefit share model.

i. Land Lease Model

In this method, the land owner and the industry will have an agreement for lease amount and the lease period. Once the lease agreement is signed, the industry establishes their captive plantation and the period of lease is for one rotation and tenable further based on mutual consultations.

ii. Benefit Share Model

Under this method, the land owner agrees to share the benefits at the time of harvest. Accordingly, an agreement can be made wherein the industry establishes the plantation at its own cost and the benefit is shared at 70 (industry): 30 (land owner) for dry lands and 60 (industry): 40 (land owner) for garden lands. The benefit share is worked at the total yield of the produce.

F. Summary and Conclusion

Energy is a critical and significant input for urbanization, industrialization and the associated socio-economic development. The energy strategy of India aims at providing energy to all through decentralized approach with a primary focus on providing clean and green source of energy. Accordingly energy generation has been attempted both for solid and liquid fuel generation using gasification, combustion and biofuel based processing technology. Wide range of biofuel industries have been established across the country with the available estimates of biomass availability from agriculture and allied sectors. However several of these industries are under threat due to non availability of adequate quantity and quality raw material resources. The existing agriculture based lignocellulosic biomass are deployed for wide range of value addition industries and hence their availability is dismally modest. Under such circumstances, the Tamil Nadu Agricultural University has developed a value chain model and demonstrated successfully in association with few industries. However for biofuel industry, this model needs to be demonstrated on a large scale in association with both liquid and solid based biofuel industries. The high yielding short rotation clones identified by TNAU have the great potential for biofuel utility and the current estimate indicated in this paper has extended greater scope for its adoption in biofuel sector.

Acknowledgement

The authors are thankful to the members of Consortium of Industrial Agroforestry for sharing the database on plantation development. The authors are also thankful to Mrs. R.D.Sharmila for her patience in assisting towards collection, processing the data and executing the manuscript.

References

Durairasu, P and K.T.Parthiban, 2013.Dendrobiomass Based Power Generation- A PCS (Production to Consumption System) Approach in Tamil Nadu, India. *Indian Journal of Forestry*, Vol.36(1) : 31-40.

Lynd, L.R. Laser, M.S., Bransby, D., Dale, B.E., Davision, B., Hamilton, R. Himmet, M., Keller, M. McMillan, J.D., Sheehan, J., Wyman, C.E., 2008. How Biotech can transform biofuels, *Nat. Biotechnol.* 26, 169-172.

Parthiban K.T., and M.Govinda Rao. 2008. Pulpwood based Industrial Agroforestry in Tamil Nadu-A case study. *Indian Forester*. 134(2):155-163.

Parthiban K.T, R. Seenivasan, and M.Govinda Rao. 2010.Contract Tree Farming in Tamil Nadu – A Successful Industrial Farm Forestry Model. *Indian Forester*. 136 (2) pp:187-197.

Parthiban, K.T. 2016. Industrial Agroforestry : A Successful Value Chain Model in Tamil Nadu, India. In: Agroforestry Research Developments. Nova Science Publishers, Inc., New York. pp.523-537.

K.T.Parthiban and R.Seenivasan. 2016. Forestry Technologies – A Complete Value Chain Approach. Scientific Publishers, Jodhpur. p. 629.

Sukumaran, R.K., Singhania, R.R., Mathew, G.M., Pandey, A. 2009. Cellulase production using biomass feed stock and its application in lignocelluloses saccharification for biotehanol production. *Renew energy*. 34 (12): 421 – 424.

Tuli, D.K. 2013. Biofuels can solve oil crisis only if we overcome technological challenges. *Energy Next*, Vol.3 (7) : 18-21.

Wyman, C.E. 2007. What is (and is not) vital to advancing cellulosic ethanol. *Trends in biotechnol*. 25, 153-157.

2

Biomass Pre-Processing Technologies for Biofuel Production

Desikan Ramesh and Subburamu Karthikeyan

Department of Bioenergy,
Agricultural Engineering College and Research Institute
Tamil Nadu Agricultural University, Coimbatore – 641 003

1. Introduction

Searching of new transportation fuels is gaining momentum in worldwide due to fast shrinking of limited fossil fuels reserves and ever rising oil prices, their impact on environment and global climate change. Biofuels are one such alternative and can be produced from bioresources via thermochemical or biochemical route either as liquid or gas or solid form. Biodiesel and bioethanol are recommended as substitute to replace or blending for diesel and petrol. Bioalcohols including bioethanol and biobutanol are considered as next-generation transportation fuels and significant quantities are currently being produced and utilized by different countries. The usage of biofuels in transportation sector in European Union will be enhanced to 10 per cent in 2020 to reduce considerable GHG emissions (Soimakallio and Koponen, 2011). However biofuels are critiqued because of their low production efficiency and moreover biofuels are low value commodities. But integrating biofuel production systems to other industries for use of the by-products of biofuel process is one promising option. There are many by-products that are highly valued in other industries such as glycerol. At present At present, first generation bioethanol is mostly produced from cereals, corn or grain and sugarcane or sweet sorghum. However, the use of these agrarian crops for biofuel production is unsustainable in the immediate future as it is conflicts with food and feed production and perhaps very expensive. The abundant and inexpensive lignocellulosic biomass, such as agro residues, forestry wastes and municipal solid wastes extant a sustainable source for the production of biofuels and other high valuable biomaterials (designated 2nd generation biofuels). These facts have motivated extensive research toward making an efficient conversion of lignocellulosic materials into sugar monomers for subsequent fermentation to bioproducts. The complex structure of native lignocellulosic biomass makes it however, difficult for microorganisms to access; it is mainly composed of cellulose, hemicelluloses, and lignin. Therefore, production of liquid biofuels from lignocellulosic biomass creates technical challenges, such as the need for pretreatment to make sugars available for the subsequent fermentation steps. Therefore, pretreatment is a crucial process to break the lignocellulosic structure and to make it accessible for hydrolyzing enzymes to release the monomeric sugars which can be finally converted into alcohol and other valued chemical feestocks (Georgieva *et al.*, 2008).

2. Biofuel Feedstocks and its Cohort

Generally, the biofuels are primarily classified into first and second-generation biofuels. The feedstocks for biofuels produced by first, second, third and fourth generation are food crops (starch, sugarcane, sugar beet, and oils), lignocellulosic biomass feedstocks, and microalgae (Chaturvedi and Verma, 2013). According to recent estimates, Global biofuel production is 127.7 billion litres, with contribution of ethanol (74 per cent), biodiesel (23 per cent) and hydrotreated vegetable oil (HVO) (3 per cent). The top biofuel producing countries are the United States, Brazil, Germany, China, and Argentina (Anonymous, 2015). The annual bioethanol required for 5 per cent blending in petrol in India was 3.45 billion liters (Saxena *et al.*, 2009). In order to meet the huge quantity of ethanol requirement, several biofeedstocks were tried for bioethanol production. First generation ethanol biofuel is produced from food feedstocks, such as starch and sugarcane and finally, they ended with food versus fuel controversy due to (i) consumer demand and (ii) competition between land requirement for food and fuel crops. To solve this issue, the second generation technologies are targeted for non-food biomass feedstocks for production of biofuel and multiple bioproducts. Second generation non-food biomass feedstocks includes different biomass including wood crops, grasses and agro crop residues. Among the non-food biomass feedstocks, lignocellulosic biomass (LCB) feedstocks has added advantages i) it offers platform for multiple products, i) low cost and iii) huge availability.

The major compositions of LCB feedstocks are cellulose (30–60 per cent), hemicellulose (20-40 per cent) and lignin (10-30 per cent) (Dietmar Peters, 2007) and it may vary with varieties, crops, seasons and regions. The composition of different sources of LCB feedstocks are furnished in Table 2.1. For LCB feedstocks, there is no room for conflict on food security versus energy security. In order to meet out the increasing biofuel demand, high yielding new breeding/genetically modified varieties with lesser lignin content may be developed.

Table 2.1: Composition of different Lignocellulosic Biomass Feedstocks

Biomass Type	Cellulose (Per cent)	Hemicellulose (Per cent)	Lignin (Per cent)	Reference
Energy crops, dedicated plants, and aquatic plants				
Switch grass	5–20	30–50	10–40	McKendry (2002)
Miscanthus	38.2–40	18–24.3	24.1–25	Brosse *et al.* (2010)
Grass Esparto	33–38	27–32	17–19	Sanchez (2009)
Elephant grass	22	24	23.9	
Bermuda grass	25	35.7	6.4	Prasad *et al.* (2007)
Grasses (general)	25–40	25–50	10–30	Saini *et al.* (2015)
Alfalfa stalks	48.5	6.5	16.6	Chandel *et al.* (2007)
Sugarcane whole	25	17	12	Saxena *et al.* (2009)
Napier grass	32	20	9	
S32 ryegrass (early leaf)	21.3–26.7	15.8–25.7	2.7–7.3	Sanchez (2009)
Orchard grass	32	40	4.7	
Water hyacinth	18.2–18.4	48.7–49.2	3.5–3.55	Kumar *et al.* (2009), Nigam (2002)
Municipal solid wastes (MSW) and Industrial wastes				
General MSW	33–49	9–16	10–14	Li *et al.* (2012), Saxena (2009)
Kraft paper	57.3	9.9	20.8	Schmitt *et al.* (2012)
High–grade paper	87.4	8.4	2.3	
Mixed or low grade paper	42.3	9.4	15	
Food waste	55.4	7.2	11.4	
Office paper	68.6	12.4	11.3	Mosier *et al.* (2005)
Newspaper	40–55	25–40	18–30	Howard *et al.* (2003)
Waste papers from chemical pulps	60–70	10–20	5–10	Prasad *et al.* (2007)
Sorted refuse	60	20	20	
Primary wastewater solids	8–15	0	24-29	Sanchez (2009)
Agricultural residues and agro-wastes				
Leaves and grass	15.3	10.5	43.8	Schmitt *et al.* (2012)
Solid cattle manure	16–4.7	1.4–1.33	2.7–5.7	Sanchez (2009)
Coffee husk	43	7	9	Gouvea *et al.* (2009)
Nutshells	25–30	25–30	30–40	Howard *et al.* (2003)
Corn cob	42–45	35–39	14–15	Prasad *et al.* (2007), Kuhad and Singh (1993)
Cotton seed hairs	80–90	5–20	0	Prasad *et al.* (2007)
Corn stover	38–40	24–26	7–19	Saini *et al.* (2015), Zhu *et al.* (2005)
Corn fiber	14.28	16.8	8.4	Mosier *et al.* (2005)

Contd...

Table 2.1–*Contd...*

Biomass Type	Cellulose (Per cent)	Hemicellulose (Per cent)	Lignin (Per cent)	Reference
Coir	36–43	0.15–0.25	41–45	Saini *et al.* (2015)
Sugarcane bagasse	42–48	19–25	20–42	Saini *et al.* (2015), Kim and Day (2011)
Rice straw	28–36	23–28	12–14	Saini *et al.* (2015)
Wheat straw	33–38	26–32	17–19	
Barley straw	31–45	27–38	14–19	
Sweet sorghum bagasse	34–45	18–27	14–21	
Banana waste	13.2	14.8	14	Medina de Perez and Ruiz Colorado (2006)
Sponge gourd fibers	66.6	17.4	15.5	Guimaraes *et al.* (2009)
Pineapple leaf fiber	70–82	18	5–1	Reddy and Yang (2005)
Oat straw	31–37	27–38	16–19	Sanchez (2009)
Ryestraw	33–35	27–30	16–19	
Bamboo	26–43	15–26	21–31	
Bast fiber Seed flax	47	25	23	Sanchez (2009)
Bast fiber Kenaf	31–39	22–23	15–19	
Bast fiber Jute	45–53	18–21	21–26	
Leaf Fiber Abaca (Manila)	60.8	17.3	8.8	
Leaf Fiber Sisal (agave)	43–56	21–24	7–9	
Leaf FiberHenequen	77.6	4–8	13.1	
Coffee pulp	35	46.3	18.8	
Banana waste	13.2	14.8	14	
Rice husk	25–35	18–21	26–31	Ludueña *et al.* (2011)
Woods				
Softwood	27–30	35–40	25–30	McKendry (2002)
Softwood bark	18–38	15–33	30–60	Saini *et al.* (2015)
Softwood stem	45–50	25–35	25–35	Sanchez (2009)
Pine	44–46.4	8.8–26	29–29.4	Mosier *et al.* (2005), Olsson and Hahn-Hagerdal (1996)
Hardwood	20–25	45–50	20–25	McKendry (2002)
Hardwood bark	22–40	20–38	30–55	Saini *et al.* (2015)
Hardwood stem	40–55	24–40	18–25	Sanchez (2009)
Poplar	47.6–49.9	27.4–28.7	18.1–19.2	Mosier *et al.* (2005), Olsson and Hahn-Hagerdal (1996)

Generally, the moisture content of harvested lignocellulosic biomass is high. In order to attain low moisture of biomass before pre-processing, drying can be done by different methods. The limiting factor for the use of second generation feedstock for biofuel production is the non-availability of low-cost production and processing technologies that efficiently convert biomass into liquid fuel. Consequently, current production economics are more favourable for conversion of first generation feedstock into biofuels.

For economically feasible bioethanol production, we should utilize the both cellulose and hemicellulosic sugars of LCB feedstocks. The recovery of fermentable sugars from LCB is very difficult due to recalcitrant nature of lignin structure. Therefore, the pretreatment of LCB is an inevitable step to remove the lignin and disrupt crystalline structure of cellulose for easy accessibility to hydrolytic enzymes. Though consolidated bioprocessing is one other option available, the lower yields of ethanol and the longer duration of production process limits it from adoption. The selection and design of downstream processing equipment for LCB into fuels fully depends on pretreatment method and its product output. As first generation biofuels are produced from the sugars, starch and vegetable oils, advanced biofuels such as biobutanol can be produced from lignocellulosic biomass, because of huge potential and low cost of biomass.

3. Biomass Preprocessing

3.1. Densification and Handling

In general, LCB feedstock has low bulk density and energy density. Biomass densification involves in collection and packing of harvested LCB feedstocks from the field and can be achieved through baling, bundling, module building, boxing, packing, and pelletizing. Baling is the first step after collection of harvested biomass, which may square or round bales used for LCB feedstocks for easy transport and storage. Other process like pelletization is also used for densification of LCB feedstocks. The main purpose of pelletization is to make the uneven size of feedstocks into a uniform granular sizes and it also enhance energy density. The ranges of diameter and bulk density of commercially available pellets are 4.5 to 6.5 mm and 450 to 600 kg DM m^{-3}, respectively. The particle size required by the pelletizing machinery ranges from 3 to 8 mm which is why the process is energy intensive and further predisposition is that there is high biomass loss in this process.

3.2. Storage

The proper logistics should be provided for transporting of biomass materials from field to biofuel industry by any mode of truck, rail and pipe conveyer. The biomass can be stored in indoor or outdoor depending biomass types. Outdoor storage may not be recommended for all LCB feedstocks due to fast degrading nature. Also it will affect the further pre-processing operations like grinding and handling. In order to avoid spoilage of biomass due to higher moisture content, the moisture content of biomass used for storage should be less than 20 per cent depending on method used.

3.3. Drying

The harvested fresh LCB feedstocks usually have higher moisture content.

However, in order to avoid rotting, poor quality, low fermentable sugar yield, microbial activity and mass loss in the biomass stock, the excess moisture present in the biomass should be removed. The drying processes used to dry the harvested biomass are sun drying, solar drying and mechanical drying systems. Drying at higher temperature would affect the sugar recovery from the plant biomass materials. The biomass should be dried at ambient conditions. For bulk drying of harvested biomass at ambient conditions can be achieved by solar tunnel dryer with control mechanism or mechanical dryers.

4. Biomass Pretreatment

The key aim of biomass pretreatment is increasing the biomass surface area and to break the lignin barrier so as to expose cellulose and hemicellulose and in the event facilitating decrease in crystallinity of the cellulose. Pretreatment is among the most cost incurring process in biochemical conversion of LCB accounting to about 40-45 per cent of the total processing cost (Wyman *et al.*, 2005; Percival Zhang *et al.*, 2009). Hence cost effective pretreatment of LCB is the foremost task of LCB based biofuels.

The criteria considered for selection of biomass pretreatment method are higher sugar yield, more lignin reduction with minimum energy input, low cost, less inhibitors formation and lesser environmental impact. Techno-economically viable pretreatment technology will offer for the commercialization of low cost bioethanol production from LCB feedstocks. The design and selection of suitable downstream processes for biofuel production from LCB feedstocks completely depends on products produced from biomass pretreatment method. This necessitates the careful selection of a suitable pretreatment process and the optimal treatment parameters for effective biofuel production process.

Lignocellulosic biomass comprises about 50 percent of world biomass with annual production of 1 x10^{10} MT worldwide (Sanchez and Cardona, 2008) and 686 MT in India (Hiloidhari *et al.*, 2014). Figure 2.1 represents the outline of different biofuels produced from lignocellulosic biomass (LCB) feedstocks. Pretreatment is unavoidable step for lignocellulosic biomass used for lignin removal and to release the sugars for fermentation. Pretreatment of lignocellulosic biomass is one of the expensive processes and it shows about 18 per cent of the total production cost (Lynd *et al.*, 1996; Mosier *et al.*, 2005; Yang and Wyman, 2008). Though various pretreatment technologies have been studied extensively, no single method can be solely effective to process various biomass for cellulosic ethanol process.

The major hurdles in commercialization of lignocellulosic based biofuels are high energy intensive process and cost involved in different unit operations. The major criteria considered for process design of the pretreatment used for lignocellulosic biomass are (i) higher yield of total releasing sugars without degradation/loss of holocellulose, (ii) minimize inhibitors formation, (iii) low processing cost and (iv) easy to upscale.

An efficient pretreatment is that it should avoid size reduction, should preserve hemicellulose fractions, allow no or lesser formation of inhibitor compounds, minimize energy requirements and low cost. In addition to these

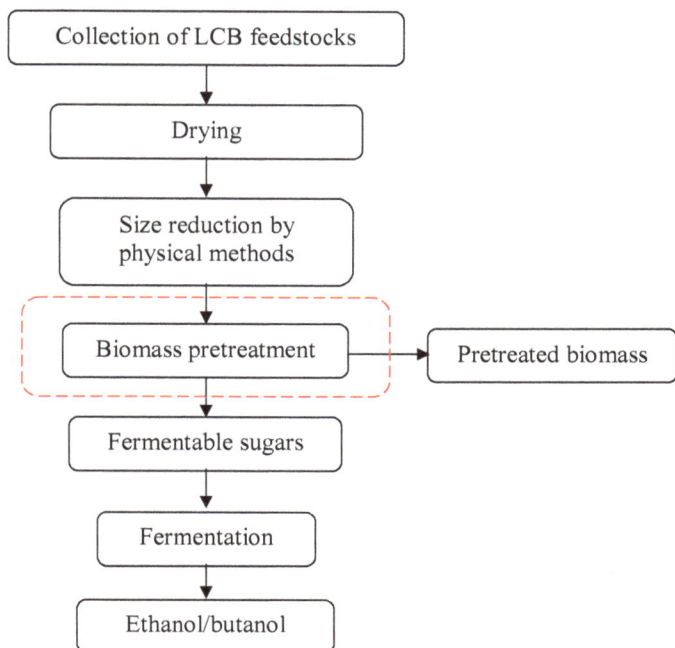

Figure 2.1: Process Flow Diagram of Biofuel Production of Lignocellulosic Biomass Feedstocks by Biochemical Route (- - - Line shows energy intensive process).

several other parameters like recovery of high value added co-products like lignin derived chemicals and protein; catalyst used for pretreatment and its recovery or recycle options; and waste treatment also needs consideration. A comprehensive consideration of all these criteria is essential and additionally the pretreatment should be evaluated on the basis of ease of operation and the cost of downstreaming processes.

5. Pretreatment Strategies

The main classification of pretreatment methods may be individual (physical, chemical, biological) or combined methods (*e.g.*, liquid hot water, ammonia fiber/ freeze explosion etc). Combine methods of pretreatment is generally more efficient and effective method for lignin and hemicellulose removal than that of individual methods.

5.1. Physical Methods

5.1.1. Size Reduction by Mechanical Comminution

In order to increase the surface area, lower the cellulose crystallinity and particle size, and improve the digestibility of biomass feedstocks, different mechanical size reduction machineries can be employed (Sun and Cheng 2002, Palmowski and Muller 1999). The size reduction can be achieved by (i) impact, (ii) attrition, (iii) shear and (iv) compression. The machineries used to make a fine powder of LCB feedstocks are chopper, chipper, shredder, grinder and ball mills. The size of the biomass particles produced in the chopper/chipper and milling/grinding are 10 to 20 and 0.2 to 2 mm respectively (Agbor *et al.*, 2011). The main aim of size reduction by mechanical method is to produce fine biomass particles for enhanced specific surface area, reduce of cellulose crystallinity index and degree of polymerization. The selection of these machineries for size reduction depends on the type and nature of LCB feedstocks, moisture content and particle size used in the pretreatment process. This method can consume more energy for its operation.

5.1.2. Extrusion

The LCB slurry fed in to screw speed and barrel of extrusion system would under mixing and shearing actions and it causes heat, which is sufficient for modifying the structure of LCB. This would lead to cutting of the lengthy fibers into smaller pieces, which will help for enhancing the accessibility of enzymes (Karunanithy *et al.*, 2008). Combination of extrusion with chemicals can improve the delignification process.

5.1.3. Steam Explosion

Autohydrolysis or uncatalyzed steam explosion is one of the common pretreatment method for LCB. In this the LCB particles are rapidly heated by pressure saturated steam to promote the hydrolysis of hemicellulose. This process is terminated by swift release of pressure, which causes the biomass to undergo an explosive deconstruction. During the pretreatment, the hemicellulose is often hydrolyzed by organic acids such as acetate and other acids formed from acetyl or other functional groups, released from biomass. In addition, the water itself, by virtue of high temperature possesses certain acid properties, which further catalyze hemicellulose hydrolysis. Therefore, the degradations of sugars might happen during uncatalyzed steam-explosion due to acidic conditions. The major factors for steam explosion are treatment time, temperature, particle size and moisture content. Steam explosion is typically conducted at a temperature of 160-270! for several seconds to a few minutes before pretreated contents are discharged into a vessel for cooling. Lower temperature and longer residence time are more favourable and the use of small micron size particles in steam explosion would not be favorable in optimizing the effectiveness of the process for improved economics. The major physicochemical changes of LCB during the uncatalyzed steam-explosion are attributed to the hemicellulose removal and lignin modification. These changes improve the digestibility of biomass to enzymes. In pretreated poplar chips, enzymatic digestibility achieved 90 per cent, compared to only 15 per cent hydrolysis of untreated chips (Grous *et al.*, 1986). In addition, the rapid thermal expansion opens up the biomass particle structure leading to the reduction of particle size and increased pore volume and effectively improve the digestibility of the pretreated cellulose residue.

5.1.4. Hot Water Pretreatment

In hot water pretreatment, pressure is utilized to maintain water in the liquid state at elevated temperatures. This have the potential to enhance cellulose digestibility, sugar extraction and pentose recovery, with the advantage of producing hydrolysates that contain little or no inhibitor of sugar fermentation. It has been shown to remove up to 80 per cent hemicellulose and to enhance the enzymatic digestibility of pretreated sugarcane bagasse (Laser *et al.*, 2002).

5.1.5. High Energy Radiation

Digestibility of cellulosic biomass has been enhanced by the use of high energy radiation methods, including ultrasound, electron beam, pulsed electrical field and microwave heating (Saha *et al.*, 2008). The action mode behind the high energy radiation could be one or more changes of features of cellulosic biomass, including increase of specific surface area, decrease of the degrees of polymerization and crystallinity of cellulose, hydrolysis of hemicellulose and partial depolymerization of lignin. However, these high energy radiation methods are usually slow, energy intensive and expensive and hence lack commercial appeal. During the microwave treatment process, the microwave generated from the equipment will produce heat. Exposing of LCB in such environment would lead to disrupt the structure of LCB feedstocks, which will helps in accessibility of cellulose and hemicellulose by hydrolytic enzymes (Sarkar *et al.*, 2012 and Chaturvedi and Verma, 2013).

5.2. Chemical Methods

Chemical pretreatments were originally developed and have been extensively used in the paper industry for delignification of cellulosic materials to produce high quality paper products. The possibility of developing effective and inexpensive pretreatment techniques by modifying the pulping processes has been considered. Chemical pretreatments that have been studied to date have had the primary goal of improving the biodegradability of cellulose by removing lignin and/or hemicellulose and to a lesser degree decreasing the degree of polymerization (DP) and crystallinity of the cellulose component.

5.2.1. Acid and Alkali

Chemical pretreatment is the most studied pretreatment technique among pretreatment categories. The catalyst like acids (*e.g.*, H_2SO_4 (sulphuric acid), hydrochloric, nitric acids) or alkali (*e.g.*, calcium, sodium, potassium and ammonia hydroxide) may be employed in the pretreatment process. In case of acid pretreatment, dilute or concentrated form of acids can be employed. Acid pretreatment method was derived from the concentrated acid hydrolysis such as concentrated H_2SO_4 and HCl hydrolysis, which had been a major technology for hydrolyzing lignocellulosic biomass for fermentable sugar production. Even though it is powerful and effective for cellulose hydrolysis, concentrated acid is toxic, corrosive and hazardous and requires reactors that need expensive construction material resistant to corrosion. Additionally, the concentrated acid must be recovered and recycled after hydrolysis to render the process economically feasible. Alternatively, dilute acid pretreatment has been applied to a wide range of feedstocks, including softwood, hardwood, herbaceous crops, agricultural residues and municipal solid waste. Mostly alkali pretreatment are often preferred due to effective lignin removal, lesser loss of carbohydrates and lesser corrosion problems to equipment (Karunanity and Muthukumarappan, 2011). It employs various bases, including sodium hydroxide, calcium hydroxide (lime), potassium hydroxide, aqueous ammonia, ammonia hydroxide and sodium hydroxide in combination with hydrogen peroxide or others. Alkaline pretreatment is basically a delignification process, in which a significant amount of hemicellulose is solubilized as well. Solvation and saponification reactions occurred during alkali pretreatment are main cause for swelling of the lignocellulose structure and reducing the degree of polymerization, which makes more accessibility for enzymatic and microbial degradation.

5.2.2. Ionic Liquids

The selective ionic liquids can be used to dissolve the lignin or cellulose present in the LCB feedstocks and then, adding of water to precipitate and washing to recover the soluble lignin. Ionic liquids shows promise as efficient and green novel cellulose solvents. They can dissolve large amounts of cellulose at considerable mild conditions and feasibility of recovering nearly 100 per cent of the used ionic liquids to their initial purity makes them attractive. Upon interaction of the cellulose-OH and ionic liquids, the hydrogen bonds are broken, resulting in opening of the hydrogen bonds between molecular chains of the cellulose. In this case, there is no modification in the structure of lignin and hemicellulose (Wyman *et al.*, 2009) and process can take place at low temperature (90 to 130 °C) and ambient pressure. Further they have several advantages over regular volatile organic solvents of biodegradability, low toxicity, broad selection of anion and cation combinations, low hydrophobicity, low viscosity, enhanced electrochemical stability, thermal stability, high reaction rates, low volatility with potentially minimal environmental impact and non-flammable property. Deterrents of this method are higher cost of ionic solvents and presence of solvent act as inhibitor for cellulase enzyme.

5.2.3. Oxidative Delignification

In case of oxidative delignification, process, oxidizing agents with aromatic rings (*e.g.*, ozone, hydrogen peroxide, peracetic acid, or oxygen) are employed to pretreat the LCB feedstocks. Oxidizing agent can convert the lignin compound into carboxylic acids, whereas hemicellulose may be partly degraded. Ozonolysis oxidations are effective in removing the lignin compound and other compounds structure (hemicellulose and cellulose) are not altered (Zhi *et al.*, 2012).

5.2.4. Organosolv

The organosolv process is a delignification process, with varying simultaneous hemicellulose solubilization. In this process, an organic or aqueous organic solvent mixture with or without an acid or alkali catalysts is used to break the internal lignin and hemicellulose bonds. The organic solvents used in the organosolv process include methanol, ethanol, acetone, ethylene glycol, triethylene glycol, tetrahydrofurfuryl alcohol, glycerol, aqueous phenol, aqueous n-butanol. The LCB feedstocks can be treated with the combination of organic solvent/mixtures with water and yields higher quality lignin (Marzieh Badiei *et al.*, 2014). Organic solvents are costly and their use requires high-pressure equipment due to their high volatility. The used solvents should be recovered and recycled to reduce the operation costs. Removal of solvents from the pretreated biomass is necessary because the residual solvents may be inhibitors to enzymatic hydrolysis and fermentation. The main advantage of organosolv over other chemical pretreatments is that relatively pure, low molecular weight lignin is recovered as a by-product. However, the limitation of method is more cost involved for solvent removal and reuse.

5.3. Biological Methods

The biological pretreatment can be carried out with help of different lignin degrading microorganisms like fungi (white-rot fungi, brown-rot fungi, soft-rot fungi) or bacteria. Fungi have distinct degradation characteristics on lignocellulosic biomass. In general, brown and soft rots mainly attack cellulose while imparting minor modifications to lignin, and white-rot fungi are more actively degrade the lignin component. The examples of delignifying white-rot fungi are *Phanerochaete chrysosporium, Phlebia radiata, Dichmitus squalens, Rigidosporus lignosus,* and *Jungua separabilima* (Hatakka *et al.*, 1993). The added advantages of method are no chemicals employed, eco-friendly, low cost and operated at low temperature and pressures. The disadvantages of this method are slower reaction rate, consumption of hemicellulose and cellulose (eg., brown and soft rots first targeting cellulose) and unwanted byproducts/inhibitors formation due to lignin derivatives.

The added advantages of biological methods are eco-friendly, low energy required and mostly ambient conditions sufficient for the pretreatment process. The deterrents of this method are larger space required, slow process and monitoring required. Some of the microorganisms are capable to break the barrier made lignin in LCB feedstocks by lignin-degrading enzyme activities. Based on lignin degradation capability, the microorganisms are selected and used in the biological pretreatment methods and different microorganisms used for this purpose are fungi, bacteria and actinobacteria. Most of them are collected from forest leaf litter *ie.*, natural environment and they have to be subjected to optimize the conditions for higher delignification in the pretreatment process. Further, the future research on lignin degrading microorganisms may be developing genetically engineered microorganisms with higher specific activity (Bon and Ferrara, 2007). The lignin present in the LCB feedstocks is removed by fungi in an anaerobic environment with help of lignases (Howard *et al.*, 2003). The main families of lignolytic enzymes widely used in lignin degradation are (i) phenol oxidase (*e.g.*, laccase) and (ii) peroxidases (*e.g.*, lignin peroxidase, LiP and manganese peroxidase, MnP) (Krause *et al.*, 2003; Malherbe and Cloete, 2003).

Generally, laccases are blue copper oxidase enzyme family, similar to other phenol-oxidizing enzymes which preferably polymerize lignin by coupling of the phenoxy radicals produced from oxidation of lignin phenolic groups (Kunamneni *et al.*, 2007). The list of microorganisms used for lignin degradation is furnished in the Table 2.3.

Table 2.3: List of Lignin Degrading Microorganisms Used for Biomass Pretreatment

Sl.No.	Name of the Micro-organisms	Substrate	Remarks	Reference
1.	Phanerochaete chrysosporium	Polymeric dyes	Rate of decolourization rangedfrom 0.16 to 0.62 absorbance ratio per hour	Glenn and Gold (1983)
2.	Merulius tremellosus	Aspen wood	Cellulose digestibilityincreased from 18 to 53 per cent	Reid ID (1985)
3.	Fungal isolate, RCK-1	Wheat straw	Volumetric productivityincreased from 0.30 g/LH to0.54 g/LH	Kuhar et al., 2008
4.	Coriolus versicolor B1	Bamboo residues	Maximum saccharification rate was 37.0 per cent, reducing sugar yield of 223.2 mg/g ofbamboo residues (2.34 timesthat of raw material)	Zhang et al., 2007
5.	Irpex lacteus	Cotton stalks	Xylan loss was 7.84 per cent with rawcorn stalks and 21.86–51.37 per cent; glucan digestibility showed a 14 per cent increase	Yu et al., 2010
6.	Ceriporiopsis subvermispora	Cotton stak	Glucose yields ranged from 57.67 to 66.61 per cent	Wan and Li, 2010
7.	Lactobacillus fermentum	Sugar beety pulp	improved enzymatic saccharification by as much as 35 per cent	Cheng et al., 2011,
8.	Ceriporiopsis subvermispora	Softwood chips	Energy savings of around 30 per cent in subsequent pulping	Canam et al., 2013
9.	Echinodontium taxodii	Water hyacinth	Not available	Ma et al., 2010
10.	Stereum hirsutum	Softwood	Attributed to an increase in the pore size of the substrate	Lee et al., 2007

5.4. Other Imminent Methods

5.4.1. Hydrodynamic Cavitation based Acid/Alkali Pretreatment

Cavitation can be defined as the phenomena of the formation, growth and subsequent collapse of microbubbles or cavities occurring in an extremely small interval of time (milliseconds) releasing large magnitudes of energy. The effects of the cavitation phenomena can generate very high temperatures (*ie.,*1000 to 5000 K) and pressures (100 to 5000 bar) (Gogate and Pandit, 2001) at millions of places in the reactor. Hydrodynamic cavitation reactor consists of circulation tank, orifice plates, flanges, pump, electrical motor, gate valves and accessories. The inlet pipe of the pump is connected to circulation tank. The pipe line is connected to delivery pipe of the pump. The biomass slurry can be prepared by mixing of catalyst, powdered LCB sample and water. For closed loop circulation, the LCB slurry can be supplied from circulation tank to orifice plate via pump and then returned back to circulation tank. In order to improve the effectiveness of delignification process, acid/alkali/ biological catalysts may be used.

5.4.2. Downflow Liquid Contact Reactor for Biomass Pretreatment

Downflow liquid contact reactor can be used to pretreat LCB feedstocks. In this reactor, there is no mechanical device employed for mixing of biomass slurry and reactants. The components of this reactor are similar to hydrodynamic cavitation reactor. Instead of orifice plate, reactor column is employed in this reactor and there is a provision for air entry to the reactor. Oxygen present in the air is acting as oxidizing agent to convert the lignin into carboxylic acids. The biomass slurry is prepared by mixing of catalyst, powdered biomass sample and water. In order to make a closed loop circulation, the biomass slurry is supplied from circulation tank to reactor column with the help of pump and returned to circulation tank. Due to this arrangement, the effective mixing of reactants and air takes place to help delignification process. The energy consumed for this kind of reactor is very low as compared that of other pretreatment methods.

5.5. Combined Pretreatment Methods

In order to improve the efficiency and performance of the pretreatment process, the combinational of physical, chemical or biological methods are to be employed. The summary of different biomass pretreatment methods and inhibitors formation are listed in the Table 2.2. The examples for physicochemical and thermochemical pretreatment process are: steam explosion, liquid hot water with acid/alkali pretreatment, hydrodynamic cavitation with catalyst, hydrothermal pretreatment, ammonia fiber/freeze explosion, CO_2 explosion etc.

6. Validation of Biomass Deconstruction

6.1. Key Indicators for Measuring Effectiveness of Biomass Pretreatment

The effectiveness of pretreatment are measured in terms of cellulose crystallinity, reduction in lignin and hemicellulose content by proximate compositional analyses,

Table 2.2: Summary of Biomass Pretreatment Methods and Inhibitors (Modified from Hendricks and Zeeman 2009)

Sl.No.	Mode	Types	Effectiveness in Removal of		Inhibitors*
			Hemicellulose	Lignin	
1.	Physical methods	Grinding	–	–	- -
		Milling	–	–	- -
		Extrusion	+	+	- -
2.	Chemical methods	Alkaline	+	+	–
		Acid	+	+	+
		Ionic liquid	–	+	–
		Oxidative delignification	–	+	–
		Organosolv process	+	+	–
3.	Biological method	Bacteria	+	+	- -
		Fungi	–	+	–
4.	Thermo-physical methods	Steam explosion	+	+	+
		Liquid hot water	+	+/-	–
		Ultrasound	+	+	–
		Hydrodynamic cavitation	+	+	- -
5.	Thermochemical methods	Hydrothermal	+	+	NA
		Ammonia Fiber/ Freeze Explosion	+	–	–
		CO_2 explosion	–	+	NA
6.	Other methods	Microwave pretreatment	+	–	–
		Wet oxidation	+	+	–

* May vary with reaction conditions used (acids, furans, phenols, acetic acid, 5-HMF, solvent).

+: Major effect; –: Minor effect, - -: Nil effect; NA: Not available.

structural disruption by scanning electron microscopy (SEM), changes in the functional groups and newer products formation using Fourier transform infrared spectroscopy (FTIR), the morphological changes and thermal behaviour in thermal gravimetric analyser (TGA).

6.2.1. Cellulose Crystallinity

Generally, the cellulose in LCB feedstocks contains $2/3^{rd}$ portion belongs to crystalline and remaining part in amorphous forms (Chang and Holtz/apple 2000). XRD analysis helps the detection of the amorphous part of the lignocellulosic biomass, as well as the modification of the crystalline structure of the cellulose after pretreatment. The crystallinity index (CI) is obtained from the ratio of the maximum peak intensity 002 (I002, $2\theta = 22.0$) and minimal depression (Iam $2\theta = 16.5$) between peaks 001 and 002 (Segal *et al.*, 1959; Rodrigues *et al.*, 2007).

$$CrI = \frac{I_{002} - I_{am}}{I_{002}} \times 100$$

wherein,

I_{002} is the diffraction intensity at 2=22.5 which represents both the crystalline and amorphous regions and I_{am} is the diffraction intensity at 2=18.5, which represents the amorphous regions. Decrease in crystallinity index may increase in digestibility of LCB and vice versa (Kumar and Wyman, 2009).

6.2.2. Surface Area

Surface area accessibility is one of key factor affecting enzymatic hydrolysis of LCB. That is one good reason why preprocessing procedures necessitates particle size reduction. Internal and external are the two types of surface area in the LCB feedstocks, which may be related to particle size, shape and capillary structure of cellulosic fibers respectively. The surface of LCB can be determined by morphological structural analysis with SEM. Pinholes or holes appeared in the surface is the indicator for lignin and hemicellulose removal, which is directly related to digestibility of LCB.

6.2.3. Lignin Content

Lignin in the LCB feedstocks is act as protective shield to cellulose and hemicellulose compounds. If the lignin compound altered/modified/removed, enzymes can easily attack/access the cellulose and hemicellulose of LCB. Lignin removal is key parameter considered for enhancing the enzymatic hydrolysis process. From SEM image analysis, we can easily find the pores appeared in the surface to know about lignin removal.

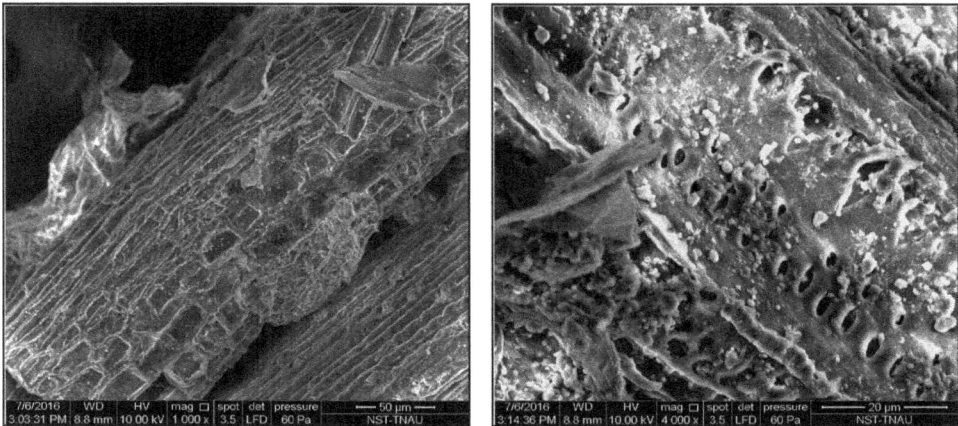

Figure 2.2: Raw and Lime Treated Biomass Samples.

6.2.4. Hemicellulose Content

Hemicellulose is another protective shield to enclose the cellulose content, which will prevent the accessibility of enzymes. The existence of hemicelluloses hinders the accessibility of enzymes to cellulose and consequently results to low enzymatic hydrolysis of LCB. The removal of hemicellulose moiety from biomass increases the pore size and the contact of the lignocellulosic materials. From SEM image analysis, it is evident of the pores on the surface and that is indicative of removal of hemicellulose in LCB feedstocks.

6.2.5. Functional Group Analysis

Effect of pretreatment process on biochemical composition of biomass can be verified by observing the changes in the raw and pretreated biomass. The FTIR spectra of the both the samples can be obtained from FT-IR equipment by recording the absorbance spectra.

Table 2.4: Assignment of Functional Groups and their Corresponding Polymers in LCB Biomass

Wavenumber	Functional Groups	Name of the Polymer
3340	O-H stretch	Lignin
1634	Aromatic ring vibration	Lignin
1509	C=C	Lignin
1422	CH_2	Lignin
1321	C-O-CH vibration	Lignin
1247	C-O stretching	Syringyl units
1157	C-O-C asymmetrical stretching	Hemicellulose (Xylose)
1031	C-O,C=C,C-C-O stretching	Cellulose, Hemicellulose and lignin

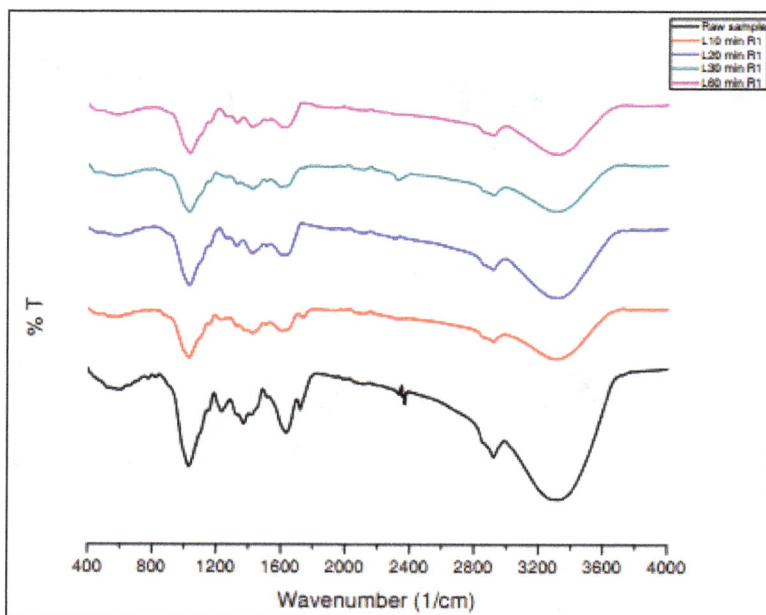

Figure 2.3: FTIR Spectra for Raw and Lime Treated Biomass Samples.

6.2.6. Thermogravimetric Analysis

Thermal behaviour studies of raw and pretreated biomass sample in a thermogravimetric analyzer is convenient to see the structural changes due to pretreatment effect. Generally, the temperature of two distinct DTG peaks of pretreated samples is shifted to higher temperatures than that of raw samples, which represents the removal of lignin and hemicellulose content.

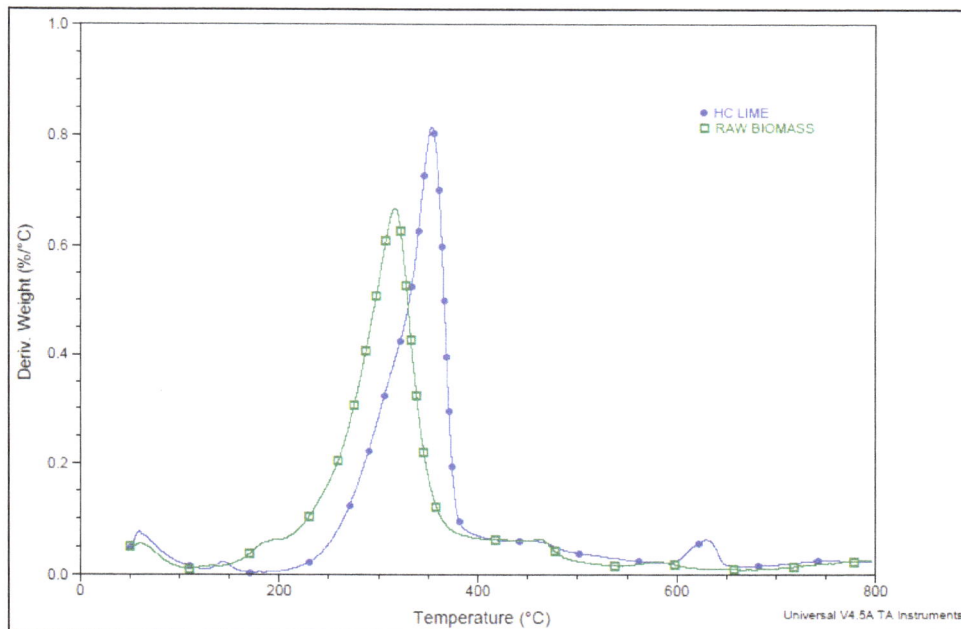

Figure 2.4: DTG Curves for Raw and Pretreated Biomass Samples.

7. Conclusion and Future Prospects

Although extensive research has been done to develop numerous effective pretreatment techniques on various lignocellulosic biomass feedstocks, none has been commercialized for cellulosic ethanol production due economic feasibility. Development of cost effective and energy efficient technologies for different unit operations of downstream processes *viz.*, pretreatment, fermentation and distillation process are overcome the hurdles for production of biofuels from lignocellulosic feedstocks. The major factors affecting breaking the barrier to overcome recalcitrance of lignocellulosic biomass for easy accessibility are lignin and hemicellulose content, crystallinity of cellulose, particle size and its surface area, and selection of pretreatment type. Biomass pretreatment is one of the important operations in bioprocessing of LCB based biofuel production. An effective pretreatment should focus on reduction of water and chemical use; recovery of carbohydrates and value added by-products to improve economic feasibility; a clean delignification system without any fermentation inhibitors. In this context, and in order to make the process technologically and economically feasible biorefinery approach of effective utilization of bioresources for both biofuel and chemicals production have to be addressed. A crucial step in making this system an improved operational biobased refinery is to establish integrated production processes capable of efficiently converting a broad range of LCB into inexpensive high-value products such as pharma, food and feed, bioplastics and polymers, bulk chemicals and biofuels. As an integrated biorefinery, it should employ novel technologies and diverse LCB which will need significant investments in research, development, and deployment projects to reduce costs and thus, improve competitiveness with fossil products. Furthermore, it must also optimize the use of LCB raw materials to create a product mix that can match to market demand and compete with the current products from fossil oil.

References

Agbor *et al.*, 2011. Biomass pretreatment: fundamentals towards application. *Biotechnology Advances*, 29:675-685.

Anonymous, 2015. Renewables 2015 Global Status Report. REN21.

Bon EPS and Ferrara MA (2007) Bioethanol production via enzymatic hydrolysis of cellulosic biomass. In: The Role of Agricultural Biotechnologies for Production of Bioenergy in Developing Countries. FAO.

Brosse, N., Hage, R.E., Sannigrahi, P., Ragauskas, A. 2010. Dilute sulphuric acid and ethanol organosolv pretreatment of Miscanthus Giganteus. *Cellul Chem Technol.*, 44:71–8.

Canam T, Town J, Iroba K, Tabil L, and Dumonceaux T. 2013 Pretreatment of Lignocellulosic Biomass Using Microorganisms: Approaches, Advantages, and Limitations, from Sustainable Degradation of Lignocellulosic Biomass—Techniques, Applications and Commercialization, Anuj Chandel and Silvio Silvério da Silva (Ed.), Rijeka, Croatia: InTech,.

Chandel, A. K., Chan, E., Rudravaram, R., Narasu, M. L., Rao, L. V., Ravindra, P. 2007. Economics and environmental impact of bioethanol production technologies: an appraisal. *Biotechnol. Mol. BiolRev.*, 2:14–32.

Chaturvedi V, Verma P. 2013. An overview of key pretreatment processes employed for bioconversion of lignocellulosic biomass into biofuels and value added products. 3 *Biotech.*, 3:415–431.

Cheng YS, Zhang R, Jenkins B, VanderGheynst JS. Effects of ensilage on storage and enzymatic degradability of sugar beet pulp. *Bioresour Technol.* 2011;102(2):1489-95.

Dietmar Peters.2007. Raw materials. *Adv Biochem Engin/Biotechnol.* 105: 1–30.

Eriksson K-EL. 2000. Lignocellulose, lignin, ligninases. In: Schaechter M, editor. Encyclopedia of microbiology, vol 3. San Diego: Academic press, pp. 39–48.

Georgieva, T. I., Hou, X., Hilstrøm, T., Ahring, B. K., 2008. Enzymatic hydrolysis and ethanol fermentation of high dry matter wet-exploded wheat straw at low enzyme loading. *Appl. Biochem. Biotechnol.* 148, 35-44.

Glenn KJ, Gold HM. 1983. Decolorization of several polymeric dyes by the lignin-degrading basidiomycete Phanerochaete chrysosporium. *Appl Environ Microbiol* 45:1741–1747.

Gogate, P. R., Pandit, A. B. 2001. Hydrodynamic cavitation reactors: A state of the art review. *Reviews in Chemical Engineering*, 17:1–85.

Gouvea, B., Torres, C., Franca, A., Oliveira, L., Oliveira, E. 2009. Feasibility of ethanol production from coffee husks.*Biotechnol.Lett.*, 31:1315–9.

Grous W R, Converse A O, Grethlein H E. Effect of steam explosion pretreatment on pore size and enzymatic hydrolysis of poplar. *Enzyme Microb Technol*, 1986; 8: 274-280.

Guimaraes, J. L., Frollini, E., Da Silva, C. G., Wypych, F., Satyanarayana, K. G. 2009. Characterization of banana, sugarcane bagasse and sponge gourd fibers of Brazil. *Industrial Crops and Products*, 30(3): 407-415.

Harmsen P, Hujigen W, Bermudez L, bakker R, Literature review of physical and chemical pretreatment processes for lignocellulosic biomass. Wageningen UR Food and Biobased Research, September 2010.

Hatakka A. 1994. Lignin-modifying enzymes from selected white-rot fungi: production and role from in lignin degradation. *FEMS Microbiol Rev* 13:125–135.

Hiloidhari, M,Dhiman Das, D C Baruah. 2014. Bioenergy potential from crop residue biomass in India. *Renewable and Sustainable Energy Reviews* 32: 504-512.

Howard RL, Abotsi E, Jansen van Rensberg EL, Howard S. 2003. Lignocellulose biotechnology: issues of bioconversion and enzyme production. *Afr J Biotechnol.*2:602–619.

Howard, R. L., Abotsi, E., Van Rensburg, E. J., Howard, S. 2003. Lignocellulose biotechnology: issues of bioconversion and enzyme production. *African Journal of Biotechnology*, 2(12):602-619.

Karunanithy C, Muthukumarappan K, Julson JL. 2008. Influence of high shear bioreactor parameters on carbohydrate release from different biomasses. American Society of Agricultural and Biological Engineers Annual International Meeting 2008. St. Joseph, Mich.

Karunanity C and Muthukumarappan K. 2011. Optimization of alkali, big bluestem particle size and extruder parameters for maximum enzymatic sugar recovery using response surface. *Bioresources*, 6:762-790.

Kim, M., Day, D. F. 2011. Composition of sugar cane, energy cane, and sweet sorghum suitable for ethanol production at Louisiana sugar mills. *Journal of Industrial Microbiology and Biotechnology*, 38(7):803-807.

Krause DO, Denman SE, Mackie RI. 2003. Opportunities to improve fibre degradation in the rumen: microbiology, ecology, and genomics. *FEMS Microbiol Rev.*;797:1–31. [PubMed].

Kuhad, R. C., Singh, A. 1993. Lignocellulose biotechnology: current and future prospects. *Crit. Rev. Biotechnol.*,13:151–72.

Kumar, A., Singh, L., Ghosh, S. 2009. Bioconversion of lignocellulosic fraction of water- hyacinth (*Eichhornia crassipes*) hemicellulose acid hydrolysate to ethanol by *Pichia stipitis*. *Bioresour. Technol.*, 100:3293–7.

Kunamneni A, Ballesteros A, Plou FJ, Alcalde M. 2007. Fungal laccase—a versatile enzyme for biotechnological applications. In: Mendez-Vilas A, editor. Communicating current research and educational topics and trends in applied microbiology. Badajoz: Formatex Research Center, pp. 223–245.

Laser M, Schulman D, Allen S G. 2002. A comparison of liquid hot water and steam pretreatments of sugar cane bagasse for conversion to ethanol. *Bioresour Technol*, 81: 3344.

Lee JW, Gwak KS, Park JY, Park MJ, Choi DH, Kwon M, *et al.*, 2007. Biological pretreatment of softwood Pinus densiflora by three white rot fungi. *J Microbiol.*;45(6):485-91.

Li, S., Zhang, X., Andresen, J. M. 2012. Production of fermentable sugars from enzymatic hydrolysis of pretreated municipal solid waste after autoclave process.*Fuel* 92:84–8.

Ludueña, L., Fasce, D., Alvarez, V. A., Stefani, P. M. 2011. Nanocellulose from ricehusk following alkaline treatment to removesilica. *Bioresources*, 6:1440–53.

Lynd L, Elamder R, Wyman C. 1996. Likely features and costs of mature biomass ethanol technology. *Applied Biochemistry and Biotechnology* 57-58: 741-761.

Ma F, Yang N, Xu C, Yu H, Wu J, Zhang X. 2010, Combination of biological pretreatment with mild acid pretreatment for enzymatic hydrolysis and ethanol production fromwater hyacinth. *Bioresour Technol.* 2010;101(24):9600-4.

Malherbe S, Cloete TE. 2003. Lignocellulosic biodegradation: fundamentals and applications: a review. *Environ Sci Biotechnol.* 2003;1:105–114.

Malherbe, S., Cloete, T. E. 2003. Lignocellulosic biodegradation: fundamentals and applications: a review. *Environ SciBiotechnol*, 1:105–114.

Marzieh Badiei, Nilofar Asim, Jamilah M Jahim and Kamaruzzaman Sopian. 2014. Comparison of Chemical Pretreatment Methods for Cellulosic Biomass. *APCBEE Procedia* 9 : 170 – 174.

McKendry, P. Energy production from biomass (part1): overview of biomass. 2002. *Bioresour Technol.*, 83:37–46.

Medina de Perez, V. I., Ruiz Colorado, A. A. 2006. Ethanol production of Banana Shell and Cassava Starch. *Dyna.*, 73:21–7.

Mosier, N., Wyman, C., Dale, B., Elander, R., Lee, Y., Holtzapple, M. *et al.*, 2005. Features of promising technologies for pretreatment of lignocellulosic biomass. *Bioresour. Technol.*, 96:673–86.

Nair S, LM, Kuhad RC. 2008. Pretreatment of lignocellulosic material with fungi capable of higher lignin degradation and lower carbohydrate degradation improves substrate acid hydrolysis and eventual conversion to ethanol. *Can J Microbiol* 54:305–313.

Nigam, J. 2002. Bioconversion of water-hyacinth (*Eichhornia crassipes*) hemicellulose acid hydrolysate to motor fuel ethanol by xylose – fermenting yeast. *J Biotechnol.*, 97:107–16.

Olsson, L., Hahn-Hägerdal, B. 1996. Fermentation of lignocellulosic hydrolysates for ethanol production. *Enzym. Microb. Technol.*, 18:312–31.

Percival Zhang, Y. H., Berson, E., Sarkanen, S., Dale, B.E. 2009. Pretreatment and biomass recalcitrance: Fundamentals and Progress. *Appl. Biochem. Biotechnol.* 153, 80-83.

Prasad, S., Singh, A., Joshi, H. 2007. Ethanol as an alternative fuel from agricultural, industrial and urban residues. *Resour. Conserv. Recycl.* 50:1–39.

Reddy, N., Yang, Y. 2005. Biofibers from agricultural byproducts for industrial applications. *Trends Biotechnol.*, 23:22–7.

Reid ID. 1985. Biological delignification of aspen wood by solidstate fermentation with the white-rot fungus, Merulius tremellosus. *Appl Env Microbiol* 50:133-139.

Saha B C, Biswas A, Cotta M A. Microwave pretreatment enzymatic saccharification and fermentation of wheat straw to ethanol. *J Biobased Mater Bioenergy*, 2008; 2: 210-217.

Saini, J. K., Saini, R., Tewari, L. 2015. Lignocellulosic agriculture wastes as biomass feedstocks for second-generation bioethanol production: concepts and recent developments. *Biotech.*, 5:337–53.

Sanchez OJ, Cardona CA. 2008. Trends in biological production of fuel ethanol from different feedstocks. *Bioresour. Technol.* 99: 5270-5295.

Sánchez, C. 2009. Lignocellulosic residues: biodegradation and bioconversion by fungi. *Biotechnol Adv.*, 27:185–94.

Sarkar N, Ghosh SK, Bannerjee S, Aikat K. 2012. Bioethanol production from agricultural wastes: an overview. *Renew Energy* 37(1):19–27.

Saxena RC, Adhikari DK, Goyal HB. 2009. Biomass-based energy fuel through biochemicalroutes: a review. *Renew Sustain Energy Rev* 13:167–178.

Saxena, R., Adhikari, D., Goyal, H. 2009. Biomass-based energy fuel through biochemical routes:a review.*Renew. Sustain. Energy. Rev.*, 13:167–78.

Schmitt, E., Bura, R., Gustafson, R., Cooper, J., Vajzovic, A. 2012. Converting lignocellulosic solid waste into ethanol for the State of Washington: an investigation of treatment technologies and environmental impacts. *Bioresour. Technol.*,104:400–9.

Soimakallio S, Koponen K. 2011. How to ensure greenhouse gas emission reductions by increasing the use of biofuels? – Suitability of the European Union sustainability criteria. *Biomass and Bioenergy*, 35, 3504–3513.

Wan C, Li Y. 2010. Microbial pretreatment of corn stover with Ceriporiopsis subvermispora for enzymatic hydrolysis and ethanol production. *Bioresour Technol* 101:6398–6403.

Wyman CE, Dale BE, Elander RT, Holtzapple M, Ladisch MR, Lee YY. 2005. Comparative sugar recovery data from laboratory scale application of leading pretreatment technologies to corn stover. *Bioresour Technol* 96, 2026-2032.

Wyman C E, Dale BE, Elander RT. 2009. Comparative sugar recovery and fermentation data following pretreatment of poplar wood by leading technologies. *Biotech progress*, 25: 333–339.

Yang B, Wyman CE. 2008. Pretreatment: the key to unlocking low-cost cellulosic ethanol. Biofuels, *Bioproducts and Biorefining*, 2: 26-40.

Yu H, Du W, Zhang J, Ma F, Zhang X, Zhong W. 2010. Fungal treatment of cornstalks enhances the delignification and xylan loss during mild alkaline pretreatment and enzymatic digestibility of glucan. *Bioresour Technol* 101:6728–6734.

Zhang X, Xu C, Wang H. 2007. Pretreatment of bamboo residues with Coriolus versicolor for enzymatic hydrolysis. *J Biosci Bioeng* 104:149–151.

Zhi S, Yu X, Wang X, Lu X. 2012. Enzymatic hydrolysis of cellulose after pretreated by ionic liquids/; focus on one-pot process. *Energy Procedia*, 14: 1741-47.

Zhu, Y., Lee, Y. Y., Elander, R. T. 2005. Optimization of dilute acid pretreatment of corn stover using a high-solids percolation reactor. *Applied Biochemistry and Biotechnology*, 121-124, 1045-1054.

3

Tuning Up of Laccases as Multifunctional Green Catalysts

Sujatha Kandasamy[1], Sivakumar Uthandi[2],
Kalaichelvan Gurumurthy[3]and Julie A. Maupin-Furlow[4]

[1]Molecular Microbiology Lab, Department of Food science and Technology,
Pondicherry University, Pondicherry – 605 014
[2]Biocatalysts Lab., Department of Agricultural Microbiology,
Department of Agricultural Microbiology, Tamil Nadu Agricultural University,
Coimbatore – 641 003
[3]Department of Microbiology and Cell Science,
University of Florida, Gainesville, Florida 32611-7200

1. Introduction

Laccases (benzenediol oxygen oxidoreductase, EC 1.10.3.2) are glycosylated blue multicopper oxidases that oxidize phenolic and non-phenolic lignin substrates with a concomitant reduction of oxygen to water through a radical-catalyzed reaction mechanism. Some biological functions of laccase include biosynthesis and degradation of lignin, pigmentation, plant pathogenesis, fungal morphogenesis, sporulation, iron metabolism, resistance against abiotic stress and browning of kernel in plants(Christopher *et al.*, 2014). Even though, laccases are known to exist in plants, fungi, prokaryotes, arthropods and archaea, predominant studies are on enzymes mainly from fungal origin, including ascomycetes, deuteromycetes and basidiomycetes (Senthivelan *et al.*, 2016).

Due to broad substrate specificity, low catalytic requirements and oxidative versatility laccases are considered for use as green catalysts in several industries such as food and beverages (baking and processing),textiles (bleaching and dye decolourization), paper and pulp (delignification), bioremediation of soil and industrial effluents, cosmetics, biosensors and nano biotechnology (Virk *et al.*, 2012; Ba *et al.*, 2013; Christopher *et al.*, 2014). This chapter is focused on the sources, enzymatic properties, production and varied applications of laccases.

2. Sources

2.1. Plant Laccases

The first laccase was discovered by Yoshida in 1883 from the sap of Japanese lacquer tree *Rhus vernicifera*. Consequently, laccases from other plants such as lacquer, pine, prune, mung bean, mango, peach, tobacco, ryegrass, corn and sycamore are partially characterized (Morozova *et al.*, 2007a). Multiple laccase isoforms from poplar (*Populus trichocarpa*) and loblolly pine (*Pinus taeda*) are identified (Lafayette *et al.*, 1999; Sato *et al.*, 2001). Further, plant laccases are involved inlignin polymerization, wound healing and iron oxidation (Wang *et al.*, 2015). During the past ten years, over expression as well as down regulation of laccase genes has been used successfully to engineer transgenic plants with low-lignin content for effective conversion of plant biomass (Dwivedi *et al.*, 2011).

2.2. Bacterial Laccases

Bacterial laccase was first reported from the plant root associated bacterium *Azospirillum lipoferum* which was found to be involved in melanin formation (Givaudan *et al.*, 1993). Later laccases were reported from several bacterial species including *Bacillus subtilis, Bordetella campestris, Pseudomonas syringae, P. aeruginosa, Bordetella compestris, Caulobacter crescentus, Escherichia coli, Mycobacterium tuberculosum, Stenotrophomonas maltophilia* and *Yersinia pestis* (Sharma and Kuhad, 2008; Galai *et al.*, 2012). Thermostable laccases from *Bacillus subtilis* and *B. licheniformis* were noted for producing brown spore pigment which protects the endospore against damage from UV and hydrogen peroxide (Enguita *et al.*, 2003; Koschorreck *et al.*, 2009). Bacterial laccases are also noted for manganese oxidation, detoxification of phenols and antibiotics synthesis (Chandra and Chowdhary, 2015). Compared to fungi, bacterial laccases are mostly intracellular with lower redox potential, higher thermal and pH stability, as well as greater resistance towards chloride and copper ions (Bugg *et al.*, 2011). Laccases from immobilized bacterial spores are identified for almost all industrial processes (Chandra and Chowdhary, 2015). Until now, the purification and complete characterization of bacterial laccases are primarily from *A. lipoferum, Marinomonas mediterranea* and *B. subtilis* (Christopher *et al.*, 2014).

Several laccases from the genus *Streptomyces* have been identified and characterized such as *Streptomyces cyaneus, S. coelicolor, S. ipomea, S. lavendulae, S.psammoticus, S. violaceusniger, S. lividans, S. viridosporus* and *S. bikiniensis*. All the laccases of *Streptomyces* are reported to play major roles in morphogenesis, sporulation, pigmentation and lignocellulose degradation (Martins *et al.*, 2015; Devi *et al.*, 2016).

2.3. Archaeal Laccases

Laccase from extremophiles like *Aquifex aeolicus* and *Pyrobaculum aerophilum* have also been characterized (Alcalde, 2007).Recently, a highly thermostable and salt/solvent tolerant laccase, LccA was identified from a novel halophilic archaeon, *Haloferax volcanii*. LccA oxidizes a wide variety of phenolic compounds and is the only archaeal laccase that has been described experimentally to date (Uthandi *et al.*, 2010).

2.4. Fungal Laccases

Among fungal species, laccase producers are widely demonstrated among ascomycetes, deuteromycetes and basidiomycetes, none has been revealed in lower fungi, *i.e.* zygomycetes and chytridiomycetes. Among laccase producing fungi, the white rot fungi (WRB)are the predominant group noted for their high level secretion and ability to oxidize various recalcitrant aromatic compounds (Baldrian, 2006). Purified fungal laccases are reported from *Trametes versicolor, T. hirsute, T. ochracea, T. villosa, T. gallica, Trichoderma reesei, T. harzianum, Anthracophyllum discolor, Cerrena maxima, C. unicolor, Phlebia radiata, Pycnoporus sanguineus, Coriolopsis polyzona, Lentinus tigrinus, Pleurotus eryngii, Melanocarpus albomyces, Magnaporthe grisea* and *Xylaria polymorpha* (Piscitelli *et al.*, 2016).

More than 100 laccases are now purified and characterized from various fungi and the number is still increasing. Owing to their higher redox potential (+ 800 mV) and extracellular nature, fungal laccases are functional in a wide variety of biotechnological applications especially in lignin degradation (Upadhyay *et al.*, 2016). Furthermore, fungal laccases are noted for biological functions in spore pigmentation, rhizomorph production, fruiting body development, fungal morphogenesis, detoxification, resistance towards abiotic stress, sporulation and pathogenicity (Fang *et al.*, 2010). Recently, *Cladosporium tenuissimum* a psychrotolerant fungus newly isolated from the cold deserts of the Indian Himalayas is reported to produce a laccase with enhanced cold stability (Dhakar and Pandey, 2015). In addition, a newly isolated WRB, *Hexagonia hirta* MSF2 was reported for its high-level secretion of a laccase (LccH) (Table 3.1), that displays thermal and solvent stability (Sujatha *et al.*, 2016).

Yeasts are a physiologically specific group of ascomycetes and basidiomycetes. To date, only one type of yeast, *Cryptococcus neoformans,*is reported to oxidizephenols and aminophenols (Viswanath *et al.*, 2014).

2.5. Insects Laccases

Laccases are detected in insects of different genera including *Bombyx, Calliphora, Diploptera, Drosophila, Lucilia, Manduca, Musca, Oryctes, Papilio, Phormia, Rhodnius, Sarcophaga, Schistocerca* and *Tenebrio* (Arora and Sharma, 2010). In *Tribolium castaneum,* two laccase isoforms are identified that are needed for larva, pupa and adult cuticle tanning (Arakane *et al.*, 2005). The insect laccase is characterized by having an extended amino-terminal sequence with a unique domain consisting of several conserved cysteine, aromatic and charged residues that are associated with cuticle sclerotization (Sharma and Kuhad, 2008).

3. Structure

Laccases are extra- or intracellular multicopper glycoproteins with molecular mass ranging from 50 to 130 kDa, of which 10-50 per cent might be attributed to glycosylation.

The glycan group, which contributes to the stability of the enzyme is composed of mannose, N-acetylglucosamine and galactose in fungal laccases (Baldrian, 2006). Overall laccase structure is made up of three equal sized cupredoxin-like domains (A, B and C) which have significant catalytic activity. A cleft between domains B and C serves as a site of substrate-binding, while the mononuclear and trinuclear copper (Cu) centres are located in domain C and at the edge between domains A and C, respectively (Dwivedi *et al.*, 2011; Piscitelli *et al.*, 2016).

The active site of the laccase monomer contains four Cu atoms, categorized based on spectroscopic features as a Type 1 (T1) Cu and a trinuclear cluster (T2/T3) consisting of one type 2 (T2) and two type 3 (T3) Cu's (Figure 3.1). Substrate oxidation occurs at the T1 Cu while oxygen reduction takes place at the T2 and T3 Cu's. The T1 Cu is in a paramagnetic blue form with a strong absorption at 610 nm due to a covalent Cu-cysteine bond. T1 Cu is trigonally coordinated with two histidines, one cysteine and one variable axial ligand (Piontek *et al.*, 2002).Variations on the axial ligand include phenylalanine and leucine in the fungal laccases of *Trametes versicolor*and*Myceliophora thermophila*, respectively and methionine in the CotA of *Bacillus subtilis* (Shleev *et al.*, 2005).

Type 2 (T2) is a paramagnetic non-blue Cu coordinated by two histidines and a water molecule, with no absorption in the visible region, but detectable electron paramagnetic resonance (EPR) signals. The type 3 (T3) Cu's are in a diamagnetic spin-coupled Cu-Cupair coordinated by six histidines with anti-ferromagnetic coupling and a hydroxyl bridge between the Cu-Cu pair that shows a weak UV absorbance at 330 nm (Dwivedi *et al.*, 2011).

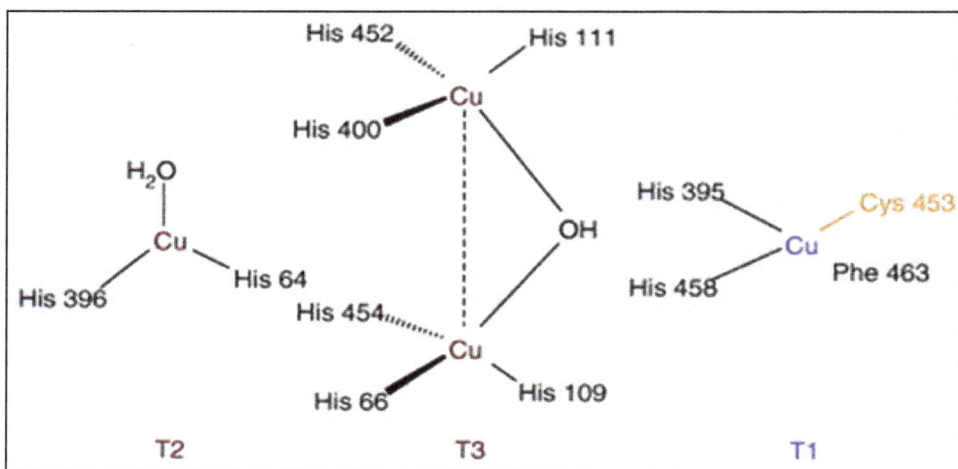

Figure 3.1: Model Structure of Laccase from *Trametes versicolor* (*Source*: Adapted from Riva *et al.*, 2006).

The 3D structures of fungal laccases from *Coprinus cinerius, Melanocarpus albomyces, Trametes versicolor, Pycnoporus cinnabarinus*and *Rigidoporus lignosus* are resolved to harbor three sequentially arranged cupredoxin domains (Senthivelan *et al.*, 2016). Even though, structurally different from fungal laccases, the Cot A laccase of*Bacillus subtilis* endospores alsocomprises three cupredoxin-like domains (Enguita *et al.*, 2003). In contrast, laccases from the bacteria *Streptomyces grizeus* (Endo *et al.*, 2002) and *S.coelicolor* (Skalov'a *et al.*, 2007) contain only two cupredoxin domains.

Some laccase variants lack T1 Cu and are often referred to as yellow and white laccases, characterized by the absence of absorption spectrum around 600 nm. White laccase from *Phellinus ribis* comprises one Mn atom per molecule, while *Polyporus ostreatus* holds only one Cu, one Fe and two Zn atoms per molecule (Palmeiri *et al.*, 1993).

4. Redox Properties and Laccase-Mediator System (LMS)

The redox potential (E°) of the T1 Cu is one of the most important features of laccases that measures the tendency of a chemical species to get reduced for acquiring electrons. Although, having EPR parameters with strong similarities, the E° of the T1 Cu sites can differbroadly between laccases depending upon the sources (Tadesse *et al.*, 2008). The ability of laccase to oxidize a potential substrate or redox mediator is determined based on the E° difference between the substrate and T1 Cu (Desai and Nityanand, 2011). Thus laccases are categorized by this E° difference ensure overall enzyme performance in a particular application.

High rates of substrate oxidation are facilitated by E° that are high for the laccase T1 Cusite or low for the substrate. Depending on the E° of the T1 Cu site, laccases are classified into high (730–780 mV), mid (470–710 mV) and low (340–490 mV) E° laccases. Indeed, fungal laccases (mainly white rot fungi) exhibit T1 Cu sites of high E° (800 mV), while plant and bacterial laccases exhibit lower E° (400 mV) (Mate and Alcalde, 2015). The E° is directly associated to the oxidation-reduction reactions catalyzed by the laccase. The type of amino acid in the axial ligand, distance between the ligand and copper atom, hydrogen bonding between the amino acid, electrostatic interactions and the hydrophobic effect (see below) are the major determinants that affect the E° of the T1 Cu (Li *et al.*, 2004).

The hydrophobic effect is due to the relative stabilization of a less charged Cu (I) state. Generally, a more hydrophobic surrounding of the T1 Cuis associated with a higher E° because of the more energetically favourable Cu (I) state that increases the electron susceptibility of the T1 Cu (Cambria *et al.*, 2012). Structural analysis of laccase reveals that the axial ligand in T1 Cuoccupied by non-ligating hydrophobic residues strongly influences the E°. A high E° (780 mV) is observed in the*Trametes versicolor*laccase when the axial ligand is phenylalanine (Figure 3.2 A), while a mid E° (710 mV) is observed in the *Myceliophora thermophila* laccase which has aleucine and low E° (60 mV) is identified in the CotA laccase from *Bacillus subtilis* which contains a methionine (Figure 3.2 B) in the axial ligand (Shleev *et al.*, 2005). Replacement of methionine by a leucine or phenylalanine in the axial ligand of CotA laccase by site-directed mutagenesis increased the E° approximately 100 mV compared to

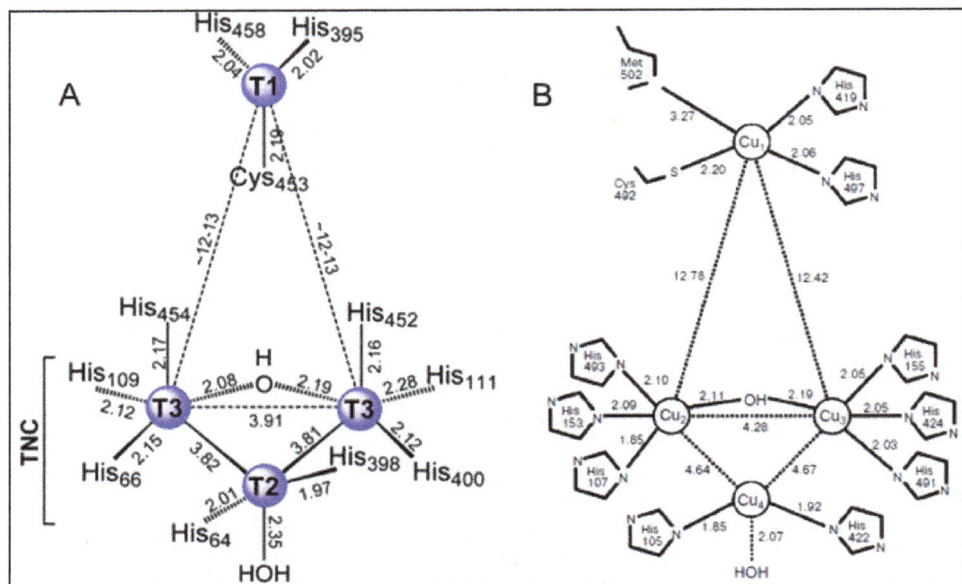

Figure 3.2: Schematic Representation of the Active Site of Lacase Showing Coordinating Residues and Interatomic Distances (A).

A: *Trametes versicolor laccase*; B: *Cot A laccase* from *Bacillus subtilis* (Adapted from Cannatelli and Ragauskas, 2016; Chandra and Chowdhary, 2015).

wild-type (Melo *et al.*, 2007). In contrast, mutation of phenylalanine to methionine in *Trametes villosa* significantly lowered the E° (Kumar *et al.*, 2003).

Increase in distance between Cu–N bond also affects the E° due to unavailability of a free electron pair of nitrogen. In the case of high E° laccase from *T. versicolor*, the Cu ion is closely located to ligand His458 (Piontek *et al.*, 2002), whereas in the mid E° laccase from *Coprinopsis cinerea* the Cu ion lies far from this ligand due to the hydrogen bonding between Glu460 and Ser113 leading to a shift in the whole helix (Ducros *et al.*, 1998). Addition of hydrogen bonding in the backbone of the T1 Cu coordinating cysteine increased the E° in the *Alcaligenes faecalis* and *Paracoccus denitrificans* laccases (Machczynski *et al.*, 2002). Simple electrostatic repulsion between T1 Cuand charged dipolesfavour the Cu(I) state by destabilizing the Cu(II). Inthe *Pseudomonas aeruginosa* laccase, when methionine was replaced by a positively charged lysine the E° was raised by 60 mV (Battistuzzi *et al.*, 2001).

4.1. Substrates

Laccases are remarkably non-specific due to their unique substrate specificity and ability to oxidize a wide range of natural substrates (phenolic compounds, aromatic amines, polyaromatic hydrocarbons and phenylpropanoids). This ability is usually attributed to the high E° variation in various laccase enzymes. Laccase substrates that are ortho-substituted (*o*-phenylenediamine, guaiacol, gallic acid, dihydroxyphenylalanine, caffeic acid, pyrogallol) tend to beoxidized at a higher rate than *para* (*p*-cresol, *p*-phenylenediamine) and *meta*-substituted compounds (*e.g. m*-phenylenediamine, orcinol, resorcinol). Laccases that are of broad substrate specificities are useful for a diversity of applications in bioremediation, organic synthesis and biosensors (Chandra and Chowdhary, 2015).

Due to their low E°, phenolic compounds are typical substrates for laccases. However, some nonphenolic and phenolic compounds are not readily oxidized due to a high E° (Couto and Toca-Herrera, 2006). Use of redox mediators in a so-called laccase mediator system (LMS) overcomes this hindrance thereby enabling substrate with high E° to be oxidized. Redox mediators are small molecules with a higher E° than laccase that act as electron shuttles to speed up the reaction rate in the oxidation of lignin polymers (Husain and Husain, 2008).

In LMS, the mediator is oxidized by laccase and the oxidized mediator further oxidizes the substrate (Figure 3.3). Use of redox mediators increases the substrate range of laccases and at present nearly 100 natural and artificial substrates have been identified that include unusual substrates such as Mn^{2+} and certain lipids (Strong and Claus, 2011).

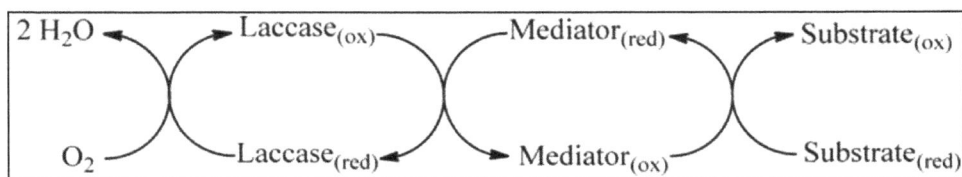

Figure 3.3: Laccase-mediator System (LMS)
(*Source*: **Adapted from Morozova *et al.*, 2007**).

Laccase mediators may be either natural or synthetic. Some common natural mediators (Figure 3.4) are 3-hydroxyanthranilic acid(HAA), acetosyringone, vanillin, acetovanillone, methylvanillate, p-coumaric acid, ferulic acid, sinapic acid, 2,4,6-trimethylphenol and syringaldehyde which are highly efficient in paper and pulp delignification (Canas and Camarero, 2010). Though nontoxic and cheap compared to synthetic ones, the generation of unstable phenoxy radicals during oxidation make natural mediators less effective. Synthetic mediators (Figure 3.5) include 2,2'-azino-bis-(3-ethylbenzothiazoline-6-sulphonic acid) (ABTS), N-hydroxy type mediators such as violuric acid (VLA), N-acetyl-N-phenylhydroxylamine (NHA), N-hydroxybenzotriazole (HBT) and N-hydroxyphthalimide (HPI) and TEMPO (2,2,6,6-tetramethyl-1-piperidinyloxy radical).

Depending upon the substrate, the oxidation mechanism (Figure 3.6) occurs by an electron transfer (ET) pathway (ABTS) or a radical hydrogen atom transfer (HAT) route (N-OH type) or an ionic pathway (TEMPO) (Coniglio *et al.*, 2008). ABTS, syringaldazine and 2,6-DMP are commonly used for monitoring laccase activity due to formation of intensely colored oxidation products (Witayakran and Ragauskas, 2009).

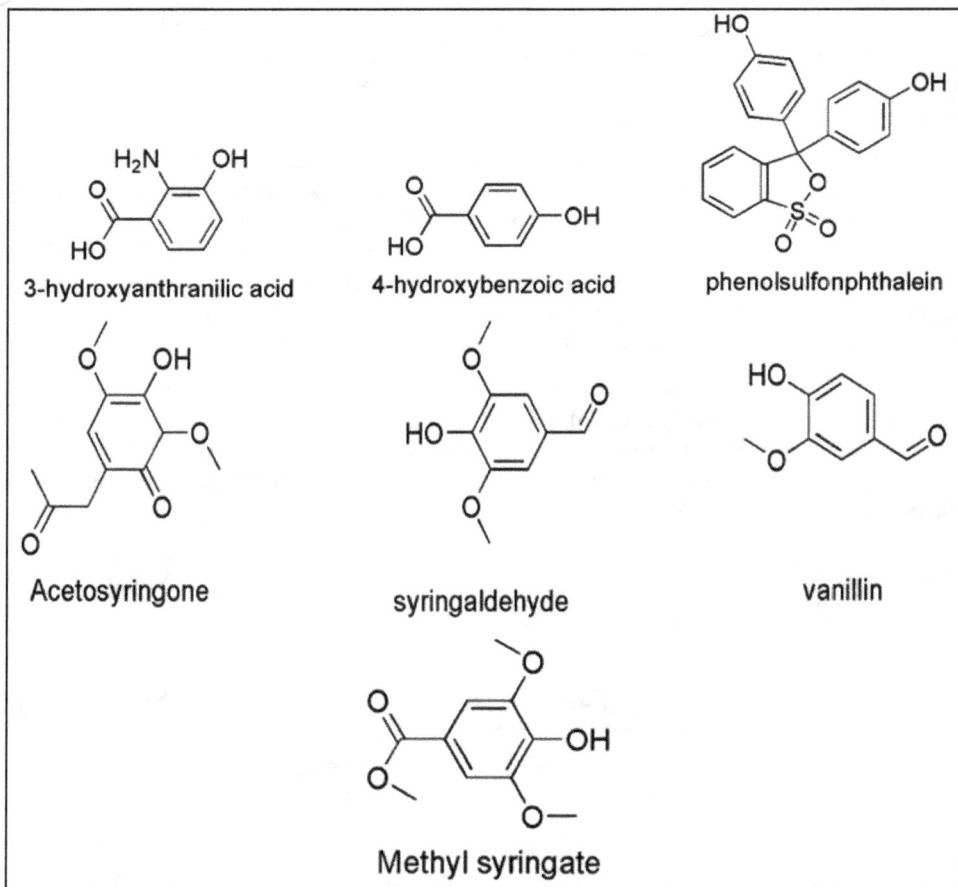

Figure 3.4: Chemical Structure of Natural Laccase Mediators.

Moreover, laccases are described in the oxidization of inorganic ions or metal complexes. The laccase CueO from *E. coli* oxidize Cu(I) to the less toxic Cu(II), while McoA from *Aquifex aeolicus* oxidizes both Cu(I) and Fe(II) and the laccase from *Trametes hirsuta* oxidizes Mn(II) (Martins *et al.*, 2015).

4.2. Inhibitors

Substances that inhibit laccase (inhibitors) are also of great consequence in industrial application. Inhibition of enzyme activity occurs by a number of mechanism including binding of Type 2 or 3 Cu that result in amino acid modification, prevention of electron transfer from the T1 to the T2/T3 site (azide, cyanide and fluoride), conformational changes by replacing or chelating Cu centres (dithiothreitol, fatty acids, glutathione, hydroxyglycine, metal ionsand sulfhydryl reagents,) and selective Curemoval (EDTA, dimethyl glyoxime, N,N'-dimethyldithiocarbamate, NTA) (Witayakran and Ragauskas, 2009; Dwivedi *et al.*, 2011). Copper-chelating agents

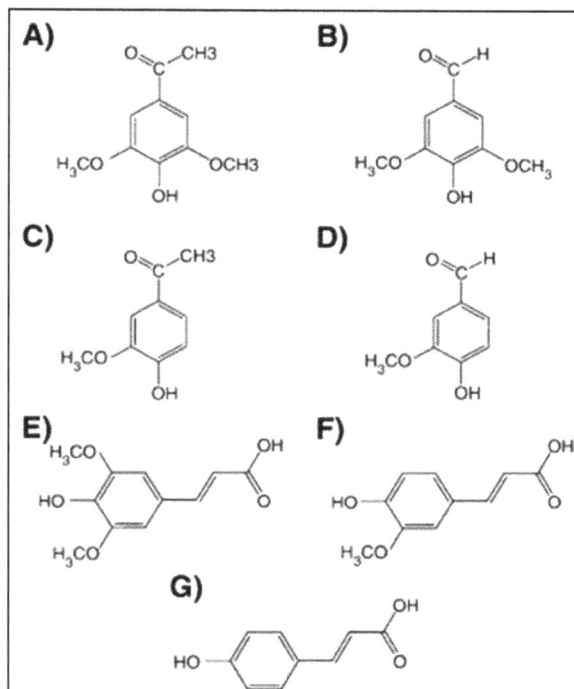

Figure 3.5: Chemical Structure of Synthetic Laccase Mediators.
(A) Acetosyringone; (B) Syringaldehyde; (C) Vanillin, (D) Acetovanillone; (E) Sinapic acid; (F) Ferulic acid and (G) p-coumaric acid.

such as sulphamic acid, oxalic acid, hydroxylammonium chloride, malonic acid, citric acid and EDTA inhibit the laccase activity of *T. versicolor* (Lorenzo *et al.*, 2005).

Although, laccase has an intrinsic sensitivity to halides and alkaline conditions, the polyphenol oxidase from *Marinomonas mediterranea* show activity above a neutral pH and tolerance towards 1 M chloride (Jimenez-Juarez *et al.*, 2005).

Rosconi *et al.* (2005) reported inhibition of laccase activity against 1mM NaCN, NaF (electron flow inhibitors) and 100 mM NaCl, that limitsthe laccase utility in various industrial processes.

5. Catalytic Mechanism

The laccase reaction mechanism is a reduction of mono-electron that generates a free radical through oxidation of four substrate molecules (Figures 3.7 and 3.8). Binding of substrate molecules to the enzyme surface and oxidization at T1 Cu centre is the initial step of the catalytic mechanism. In the substrate binding site, the Näl of a highly conserved T1 Cu coordinating His residue is assumed as the first electron acceptor from the substrate (Mate and Alcalde, 2016).

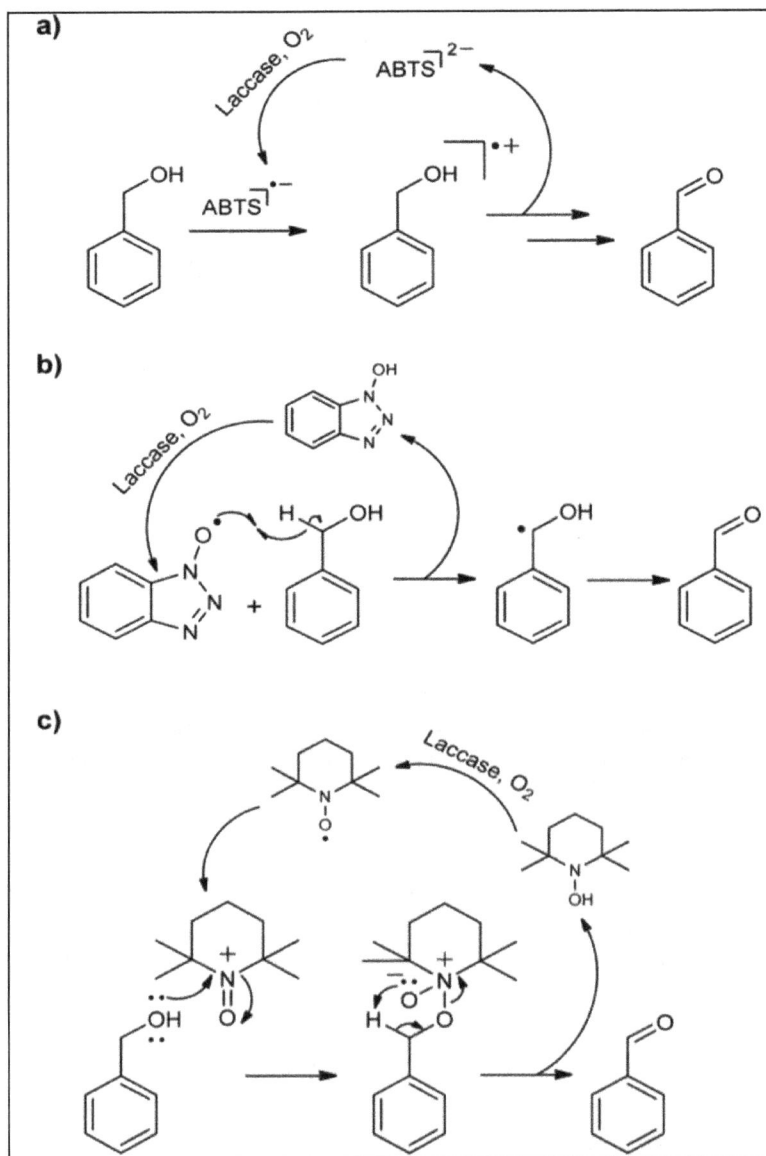

Figure 3.6: LMS Routes.
(a) Electron transfer (ET) mechanism; (b) Hydrogen atom transfer (HAT) mechanism; (c) Ionic mechanism. (*Source*: Adapted from Mate and Alcalde (2016).

Initially, substrates are converted into a free unstable radical that undergoes a second oxidation either by enzymatic or non-enzymatic (hydration, deprotonation or polymerization) reactions (Reece and Nocera, 2009). Then electron transfer occurs internally from T1 to the T2 and T3 Cu trinuclear cluster (TNC) via a cysteine-

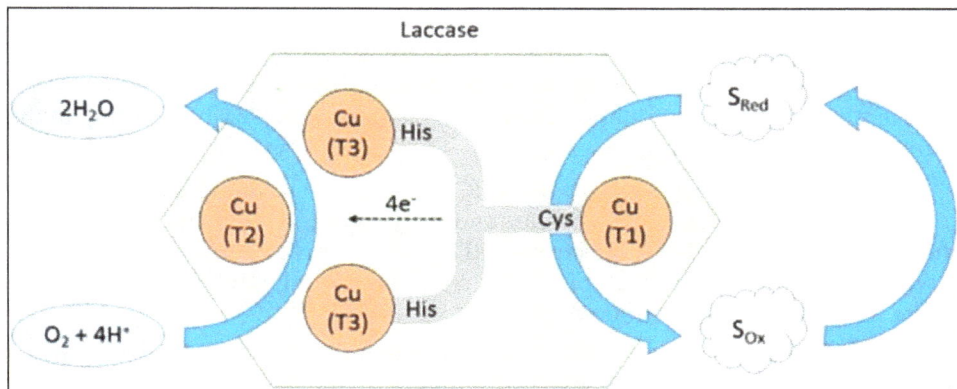

Figure 3.7: A Simplified Reaction Mechanism of Laccase Oxidation of Suitable Substrate (Adapted from Rodriguez-Delgado *et al.* (2015).

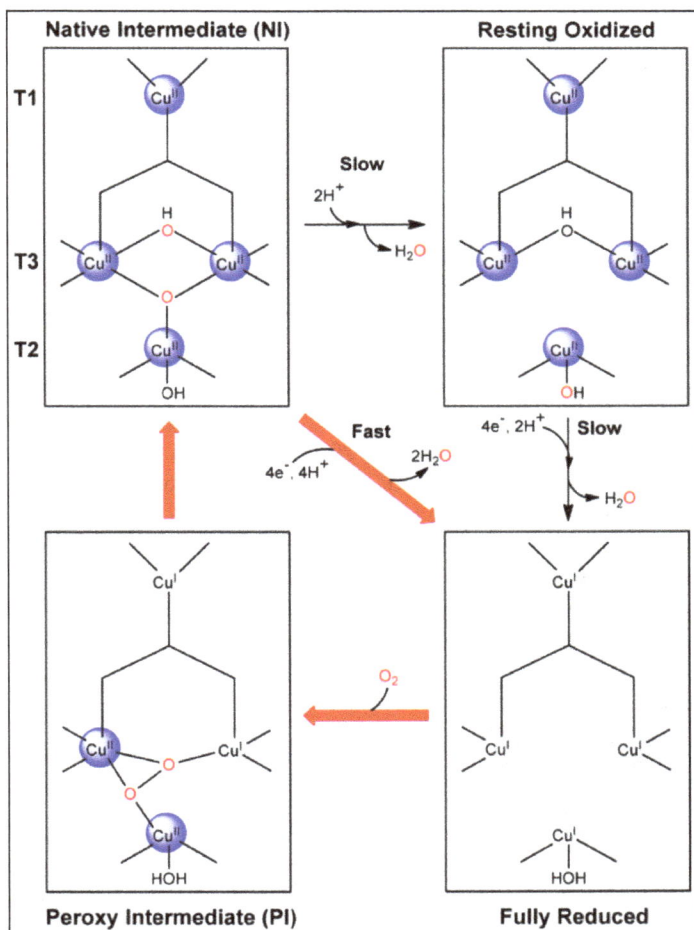

Figure 3.8: Laccase Catalytic Cycle.

histidine (superexchange) pathway. Laccase serves as a battery and electrons released from individual oxidation reactions are stored for reduction of oxygen (Solomon *et al.*, 2008). At the fully reduced state, all four Cu(II) ions are converted to a Cu(I) state with concomitant reduction of oxygen at the T2 Cu, resulting in the formation of two water molecules.

Initially, the oxygen binds to the TNC and reduction of two-electron results in a peroxy intermediate (Witayakran and Ragauskas, 2009). This intermediate undergoes reduction with two electrons and leads to a native intermediate formation in which all Cu ions of the TNC bind with a β-3-oxo ligand, while both T3 Cu bind with a ì2-oxo ligand (Mate and Alcalde, 2015).

Acidic residues near the TNC cluster protonate the oxygen molecule in the native intermediate resulting in two molecules of water (Bento *et al.*, 2010; Galli *et al.*, 2013). Consequently, the catalytic cycle of laccases is completed by oxidation of the four substrate molecules again by the Cu centres. In absence of substrates, the native enzyme intermediate undergoes a slow decay to the oxidized resting state in which the T3 Cu pair is hydroxyl bridged (Figure 3.8). Heterologously expressed and isolated laccases are usually under oxidized resting state (Jones and Solomon, 2015).

6. Optimization of Laccase Production/Activity

Several factors are known to influence laccase production and or activity such as pH, temperature, carbon and nitrogen source, inducers and type of cultivation (submerged or solid state).

6.1. pH

Growth and laccase production by the organism are greatly influenced by pH of the cultivation medium. Several reports describe a pH between 4 and 6 is suitable for laccase production (Vasconcelos *et al.*, 2000). Maximum laccase titres and biomass are observed when medium is at pH 6.0 for *Fomes sclerodermeus* and at pH 7.0 for *T. modesta* (Nyanhongo *et al.*, 2002).Optimum pH for laccase production from bacteria is normally higher than fungi that prefer acidic pH, while plant laccases prefer neutral range due to intracellular physiological properties. Due to their function at low pH, fungal laccases have widespread applications ranging from medicinal sector to pulp and paper industry (Christopher *et al.*, 2014). In contrast, bacterial laccases are generally more stable at high pH and temperature. Due to the vast environmental flexibility and biochemical adaptability, prokaryotes are more studied as hosts for laccase production in industrial and medical science applications (Chandra and Chowdhary, 2015).

The optimum pH value for laccase activity differs accordingly to the substrate. Optimum pH is more acidic (pH 3.0-5.0) when using ABTS as a substrate, while syringaldazine works under near neutral to alkaline conditions (Sharma and Kuhad, 2008). Optimal oxidation of ABTS and SGZ by an archaeal laccase *H. volcanii* is observed at pH 6 and pH 8.4, respectively (Uthandi *et al.*, 2010). With pH, mostly a bell-shaped curve is derived with laccase activity. Fungal laccases mostly function at acidic and near neutral pH, but lose activities under alkaline conditions due to a reduction of ionization potential and inhibition of -OH binding with T2/T3 Cusite (Zumarraga *et al.*, 2008).

6.2. Temperature

Laccase production can be affected by temperature. For example, fungi grown under light conditions produce laccase optimally at 25°C, while 30°C is best when grown in the dark (Cordi *et al.*, 2007).

Most laccases display a temperature optimum of 25–30°C with minimal thermostability; however, some are found to be thermostable and thermoactive. Laccase from *P. ostreatus* is active at 40 to 60 °C, with an optimum at 50 °C; the activity remains stable at 40 °C for more than 4 hours (Palmieri *et al.*, 1993). Similarly, the laccase from *T. modesta* is fully active at 50 °C and stable at 40 °C; however, at 60 °C, its half-life is reduced within 2 hours (Nyanhongo *et al.*, 2002). Preincubation at 40 and 50 °C prominently improved the thermoactivity of a laccase purified from Marasmius quercophilus (Farnet *et al.*, 2000).

Laccases from *Rigidoporus lignosus* (Cambria *et al.*, 2012) and *H.hirta* (Sujatha *et al.*, 2016) display optimal activity at 40 °C, while *Haloferax volcanii* shows at 45°C (Uthandi *et al.*, 2010). Differences in laccase activity under different temperatures might be due to the number of disulphide bonds and thermal dissociation of the molecule and release of Cu ions at elevated temperature (Ramirez *et al.*, 2012).

6.3. Carbon and Nitrogen

In laccase production, glucose, mannose, maltose, fructose and lactose are commonly used as carbon sources for the microbial host. Studies relating to higher laccase production mostly use glucose as a carbon source (Table 3.1) rather than complex sources. Laccase production is improved when glucose is as carbon source for *Hexagonia hirta, Coriolus versicolor, Trametes versicolor, Trametes gallica, Pleurotus florida, Lentinula edodes* and *Grifola frondosa* (Viswanath *et al.*, 2014; Sujatha *et al.*, 2015). Utilization of other carbon sources for enhanced laccase production includes starch by *Cerrena* sp WR1 (Revankar and Lele, 2006), mannitol by *Ganoderma lucidum* andmalt extract by *Phlebia floridensis, P. brevispora, P. radiata, and P. fascicularia* (Viswanath *et al.*, 2014).

Yeast extract, peptone, urea, $(NH_4)_2SO_4$ and $NaNO_3$ are generally used as nitrogen sources for laccase production. In certain strains, laccase production is initiated under depletion of nitrogen while in other strains productivity remains unaffected. Defining an optimal carbon-to-nitrogen ratio can also increase laccase activity (Brijwani *et al.*, 2010). Alternative nitrogen sources such as ammonium sulphate (*C. unicolor*), mixtures of amino acids (glutamic acid and glycine) and yeast extract (*S. psammoticus*) are also reported to improve higher laccase production (Shekher *et al.*, 2011; Muthukumarasamy and Murugan, 2014).

6.4. Inducers

Use of an appropriate inducer can greatly improve laccase production which is a prerequisite for effective large/industrial scale synthesis. Addition of inducers such as copper sulphate, anisidine, ethanol, ferulic acid, guaiacol, gallic acid, veratryl

Table 3.1: Comparison of Laccase Production by White Rot Fungi and Recombinant Microbes[a] (Adapted from Sujatha et al., 2015)

Organism	Major Components in the Medium	Inducer	Laccase Yield ($U.ml^{-1}$)	Reference
Trametes pubescens	Glucose 40 g/L and Peptone 10 g/L	2 mM $CuSO_4$	330 (ABTS)	(Galhaup et al., 2002)
		1 mM Gallic acid	350 (ABTS)	
		1 mM 2,5 Xylidine	275 (ABTS)	
Pycnoporus cinnabarinus	Maltose 20g/L	Ethanol 35 g/L	266 (ABTS)	(Lomascolo et al., 2003)
Pleurotus ostreatus	Glucose 10 g/L, Peptone 0.5 g/L and Vit. B150 mg/L	1 mM ABTS	400 (ABTS)	(Hou et al., 2004)
Ganoderma sp.	Glycerol 40 g/L,	0.85 mM veratryl alcohol	240 (ABTS)	(Teerapatsakul et al., 2007)
Lentinus strigosus	Glucose 20 g/L,	2mM $CuSO_4$ and 2,6–dimethyl phenol	186 (ABTS)	(Myasoedova et al., 2008)
WR1	Glucose20 g/L	—	124 (ABTS)	(Revankar and Lele, 2006)
	Starch 20 g/L	—	288 (ABTS)	
	Starch 20 g/L	2 mM $CuSO_4$	410 (ABTS)	
	Starch 20 g/L	0.8 mM Xylidine	692 (ABTS)	
Pleurotus ostreatus	Glucose 10.5 g/L and Yeast extract 5 g/L	0.25g/L $CuSO_4$	150 (DMP)	(Tlecuitl-Beristain et al., 2008)
Cerrena unicolor	Ethanol production residue (40 g/L),	0.5 mM TNT	165 (ABTS)	(Elisashvili et al., 2010)
Cerrena sp.	4 per cent PDB and5 per cent soytone	0.4 M $CuSO_4$ and 2 mM 2,5 Xylidine	202 (ABTS)	(Chen et al., 2012)
Shiraia sp Super H-168 Recombinant laccases	Starch 20 g/L and Yeast Extract 4 g/L,	0.6 mM $CuSO_4$	101 (ABTS)	(Yang et al., 2013)
Schizophyllum commune in Aspergillus sojae	Glucose 53 g/L, Tannic acid 25g/L	0.005 per cent $CuSO_4$	770 (ADPB)	(Hatamoto et al., 1999)
Trametes versicolor in Pichia pastoris	YNS 13.4 g/L and Biotin 400 µg/L	0.1 mM $CuSO_4$ and 0.5 per cent Methanol	140 (ABTS)	(Hong et al., 2002)

Contd...

Table 3.1–*Contd*...

Organism	Major Components in the Medium	Inducer	Laccase Yield $(U.ml^{-1})$	Reference
Trametes sp. 420 in *Pichia pastoris*	Methanol 0.5 per cent	0.3 mM $CuSO_4$ and 0.6 per cent Alanines	83 (ABTS)	(Hong *et al.*, 2007)
Overproduction strain of *Haloferax volcanii*	YPC	100 µM $CuSO_4$	2.84(SGZ)	(Uthandi *et al.*, 2010)
Hexagonia hirta MSF2	PDB	1 mM $CuSO_4$	217 (ABTS)	Sujatha *et al.*, 2016
	PDB	500 µM $CuSO_4$	1260 (ABTS)	
	Glucose 10.5 g/L and yeast extract 5 g/L)	500 µM $CuSO_4$	1940 (ABTS)	
	Glucose 10.5 g/L and yeast extract 5 g/L)	500 µM CuSO4 + 1 mM 2,5 xylidine	5670 (ABTS)	

a: Substrate used to determine laccase assay is given in parenthesis.

Abbreviations: PDB: Potato dextrose broth; YPC: Yeast-peptone-casamino acids; YN: Yeast nitrogen base with ammonium; GY: Glucose 10.5 g/L and yeast extract 5 g/L.

alcohol, ethanol, 2,5-xylidine,syringaldazine, vanillin and lignosulfonates could significantly enhance laccase production (Christopher *et al.*, 2014).

The most effective and common inducersused in laccase production are copper sulphate ($CuSO_4$) and 2,5-xylidine (Table 3.1). Copper plays a key role as a metal activator in the active site of laccase and induces both transcription and activity. Laccase activity is described to improve at Cu^{2+} concentrations of 0.5 and 3.5 mM with higher concentrations inhibiting cell growth and laccase production by generating hydroxyl radicals that directly damage DNA, proteins and membrane lipids (Elisashvili *et al.*, 2010).

In proteobacteria laccase production is increased 13-fold by addition of $CuSO_4$ after an onset of growth (Muthukumarasamy and Murugan, 2014). Different fungal strains such as*Phanerochaete chrysosporium,Ganoderma applanatum, Eriophora* sp, *Pycnoporus sanguineus, Coriolus versicolor* increase laccaseproductionbyCuSO$_4$ induction (Shekher *et al.*, 2011).Xylidine stimulates laccase production in *Fomes annosus, Pholiota mutabilis, Pleurotus ostreatus* and *Trametes versicolor* (Brijwani *et al.*, 2010). A combination of 500 μM $CuSO_4$ and 1 mM of 2,5-xylidine enhances laccase production of *H.hirta*to a maximum of 5670 U.ml^{-1} (Sujatha *et al.*, 2016).

7. Industrial Application

7.1. Food industries

Laccase-assisted homo- and heterocoupling of proteins and carbohydrates affords an opportunity in developing value-added products by improving functional and rheological properties of food biopolymers. Laccases can be exploited in a number of food industries to enhance the colour, appearance and texture of food or beverages leading to improvement in quality, shelf life, new functionalities and cost reduction. Laccases are widely used in the stabilization of beer and wines, processing of fruit juices, improvement of food sensory parameters and generation of sugar beet gelatin (Osma *et al.*, 2010).

Laccases are useful in clarification of fruit juice, beer and wine through removal of phenolic compounds liable for browning, haze formation and turbidity (Rodríguez-Delgado *et al.*, 2015). To enhance the storage life of beer, the excess $O_2$is removed during beer processing through the addition of laccases (Alcalde, 2007).

Laccase-assisted cross-linking of the feruloylated arabinoxylan fraction of flour improves dough strength and stability along with reduced stickiness (Labat *et al.*, 2000).

Laccase from *T. hirsute* undergoes polymerization to improve the dough resistance and decreases the dough extensibility in both flour and gluten (Selinheimo *et al.*, 2006).

7.2. Textile

Laccase is increasingly useful in commercial textile applications to enhance the whiteness in conventional bleaching of cotton and biostoning. Oxidative hetero-coupling of dye precursors by laccases yield better and more efficient heteropolymeric dyes that can be used in fabric dyeing of cotton and wool. Laccase may be included in a cleansing formulation to eliminate the odour on fabrics, including cloth, sofa surface, and curtain, or in a detergent to eliminate the odour generated during cloth washing (Tzanov *et al.*, 2013).

Around the world, there are more than 19 commercial laccase-based products for denim bleaching marketed by at least 14 companies. Studies on synthesis of C-N heteropolymeric dyes using alkaline laccases from *M. thermophile* and synthesis of substituted phenoxazinones and phenazines dyes using laccase *B.subtilis* CotA and *T. versicolor* laccase is also described (Mate and Alcalde, 2016).

7.3. Paper and Pulp

Depolymerization of lignin and delignification of wood pulp by laccases potentially degrade toxic phenols produced during lignin degradation. Laccase-mediated grafting of phenolic compounds is employed to improve hydrophobicity of paper (Elegir *et al.*, 2008). Biobleaching of paper pulp using a laccase mediator system (LMS) acts as an alternative in reducing the usage of less toxic chlorinated phenols such as chlorine (C), chlorine dioxide (D), and sodium hypochlorite (H) (Canas and Camarero, 2010). Reduction in kappa number and enhancement of paper making qualities of pulp using laccases are some other advantages in the paper and pulp industry (Upadhyay *et al.*, 2016).

Biobleaching of pulp using laccases from bacteria (*Streptomyces cyaneus, Pseudomonas stutzeri)* is also studied using ABTS and HOBT as mediators (Virk *et al.*, 2012). An effective reduction in lignin content of eucalyptus wood is also noted from laccase of *Cryptococcus albidus* that helps for biopulping in the pulp and paper industry (Singhal *et al.*, 2013).

Enzymatic delignification of pulp using fungal laccases is reported with *T. versicolor, P. chrysosporium, Ceriporiopsis subvermispora, Coriolus versicolor, Pleurotus eryngii* and *Pycnoporus cinnabarinus* (Singh *et al.*, 2010). Delignification using native laccase of *H. hirta* achieves a maximum lignin reduction of 28.6 and 16.5 per cent in wood and corncob, respectively. Comparable is the lignin reduction levels using recombinant *Myceliophtora thermophila* laccase in eucalyptus (28.4 per cent) and commercial laccase from *T. villosa*in elephant grass (20.7 per cent) and eucalyptus (17.5 per cent) at 50 U.g^{-1} of dry biomass (Sujatha *et al.*, 2016).

ABTS, violuric acid (VLA) and N-hydroxybenzotriazole (HBT) are proved to be efficient mediators in the LMS delignification of kraft pulp as they achieve higher delignification rate at reduced time intervals (Cannatelli and Ragauskas, 2016). Using ABTS in LMS delignification, *Sclerotium* sp. attain 82 per cent lignin reduction in wheat straw, while a lignin reduction of 50 and 54.2 per cent in wood pulp and hardwood chips is observed using HBT as a mediator with laccase from *T. villosa* and *T. versicolor* (Plácido and Capareda, 2015).

7.4. Pharmaceutical Industry

Most of the products of laccases are antimicrobial, detoxifying, or active personal-care agents. Laccases are involved in synthesis of antioxidants, anaesthetics, anti-cancer drugs, anti-inflammatory, antibiotics, antiviral derivatives and sedatives.

Actinocin is the first anti-cancer chemical prepared using laccase which block the transcription of DNA in tumor cells (Imran *et al.*, 2012). Laccases are also reported for significant HIV-1 reverse transcriptase inhibitory activity and fighting against aceruloplasminemia (Kunanmeni *et al.*, 2008).

A novel application field for laccases is in cosmetics. Laccases are effective replacers for H_2O_2 as an oxidizing agent in the formulation of hair dyes because of less irritation and easier handling than current hair dyes. Novel laccases from *Thermobifida fusca* and *Flammulina velutipes* oxidize the dye intermediates widely used in hair colouring (Chen *et al.*, 2013). Laccases can also find use in deodorants, toothpaste, mouth wash, detergent, soap, and diapers (Upadhyay *et al.*, 2016).

7.5. Bioremediation and Biodegradation

In the presence of a synthetic mediator, laccase from *Pleurotus ostreatus* degrades polycyclic aromatic hydrocarbons (PAHs) like anthracene, fluoranthene, fluorene, perylene, phenanthrene and pyrene. Rather than substrate oxidation, laccases immobilize soil pollutants by coupling thereby lowering the biological availability and toxicity to soil humic substances(Pozdnyakova *et al.*, 2006). Some of the laccase immobilized xenobiotics reported include 3,4-dichloroaniline, 2,4,6-trinitrotoluene or chlorinated phenols (Baldrian, 2006).Laccase is also shown to be responsible for the biodegradation of dichlorodiphenyltrichloroethane (DDT) and 2,4-dichlorophenol (Upadhyay *et al.*, 2016).

Laccases degrade different dyes such as triarylmethane, indigoid, azo and athraquinone used in dyeing textiles. Decolorization of textile effluents with immobilized laccase might be used for dyeing and acceptable colour difference measurements (Soares*et al.*, 2011). Immobilized laccase of *T. hirsute* enhances thermal stability and tolerance towards halides,copper chelators and dyeing additives. Treatment of dyes with an immobilized laccase from *Pseudomonas putida* can reduce toxicity up to 80 per cent (Abadulla *et al.*, 2000).

Kalme *et al.* (2009) disclosed a process of laccase decolorization of azo dyes through a highly non-specific free radical mechanism that avoids direct cleavage of azo bonds and formation of toxic aromatic amines. DeniLite® is the first commercial laccase from Novozymes (Denmark) used in textile industry for indigo decolorization in denim finishing. In 2001, Zytex (India) developed a formulation for dye decolorization based on a LMS (Couto and Herrera, 2006).

Isoxaflutole is an herbicide that is activated in soils and plants to diketonitrile (active). Using ABTS as a redox mediator, laccases from *P. chrysosporium and T. versicolor* are found to to transform diketonitrile into nontoxic product for bioremediation (Mougin *et al*, 2000).

7.6. Biofuel, Biosensing and Immunological Biosensors

Laccases can be used as biosensors or bioreporters, where an electrode records the oxygen consumption when laccases reduce O_2 to H_2O during analyte oxidation. In food industry, laccase-based sensors are used to detect polyphenols in fruit juices, wine and teas and to quantify fungal contamination in grape musts (Rodríguez-

Delgado *et al.,* 2015). Kulys and Vidziunaite, (2003) employed recombinant fungal laccases from *Polyporus pinsitus* and *Myceliophthora thermophila* as biosensors for measuring phenolic compounds during continuous flow-through in alarm systems.

In biomedicine, laccase-based biosensors are established to detect bilirubin, insulin, morphine and codeine. Covalent conjugation of laccase to an antibody or antigen is used as a reporter in various assays of immunochemical (ELISA, Western blotting), histochemical, cytochemical or nucleic acid-detection. Laccase-based miniature biological fuel cells are of particular interest for many medical applications calling for a power source implanted in a human body (Mate and Alcalde, 2016).

8. Conclusion and Future Directions

Laccase is an ideal green catalyst that has application in several industries such as food, textile pulp and paper, pharma, cosmetic, paint or furniture. To fully understand the potential of laccases to compete in the biotechnology race, several hurdles must still be overcome. Even though a variety of laccases have been purified and characterized, not all types of laccase activities have been defined at the molecular level, thus, impeding progress in the conversion of certain phenolic and non-phenolic compounds that are substrates to only a subset of laccases. Isoenzymes produced by the same strain puts an additional barrier to complete study of these unusual laccases. Although recombinant laccases are successfully employed in various industrial applications for their use in generation of high value products, application of laccases on a large industrial scale for bulk products is still not possible. Efficient use of LMS under large scale is also limited by cost and their inhibitory potential. In this aspect, studies on employing LMS based on natural mediators obtained under lignin combustion will be noteworthy. Designing laccases through genetic and protein engineering to boost enzyme kinetics, substrate binding and stability with further enlarge the portfolio of effective laccase alternatives and expand their use in a wide range of biotechnology applications, from delignification to organics synthesis and beyond.

References

Abadulla, E., Tzanov, T., Costa, S., Robra, K. H., Cavaco-Paulo, A., and Gübitz, G. M. (2000). Decolorization and detoxification of textile dyes with a laccase from *Trametes hirsuta*. *Applied and environmental microbiology*, 66(8), 3357-3362.

Alcalde, M. (2007). Laccases: biological functions, molecular structure and industrial applications. In: *Industrial enzymes* (pp. 461-476). Springer Netherlands.

Alexandre, G., and Zhulin, I. B. (2000). Laccases are widespread in bacteria. *Trends in biotechnology*, 18(2), 41-42.

Arakane, Y., Muthukrishnan, S., Beeman, R. W., Kanost, M. R., and Kramer, K. J. (2005). Laccase 2 is the phenoloxidase gene required for beetle cuticle tanning. Proceedings of the National Academy of Sciences of the United States of America, 102(32), 11337-11342.

Arora, D.S., and Sharma, R.K. (2010). Ligninolytic fungal laccases and their biotechnological applications. *Applied biochemistry and biotechnology*, 160(6), 1760-1788.

Ba, S., Arsenault, A., Hassani, T., Jones, J. P., and Cabana, H. (2013). Laccase immobilization and insolubilization: from fundamentals to applications for the elimination of emerging contaminants in wastewater treatment. *Critical reviews in biotechnology*, 33(4), 404-418.

Babu, P.R., Pinnamaneni, R. and Koona, S. (2012). Occurrences, physical and biochemical properties of laccase. *Universal J Environ Res Technol*, 2, 1-13.

Baldrian, P. (2006). Fungal laccases–occurrence and properties. *FEMS microbiology reviews*, 30(2), 215-242.

Battistuzzi, G., Borsari, M., Canters, G. W., de Waal, E., Loschi, L., Warmerdam, G., and Sola, M. (2001). Enthalpic and entropic contributions to the mutational changes in the reduction potential of azurin. *Biochemistry*, 40(23), 6707-6712.

Bento, I., Silva, C. S., Chen, Z., Martins, L. O., Lindley, P. F., and Soares, C. M. (2010). Mechanisms underlying dioxygen reduction in laccases. Structural and modelling studies focusing on proton transfer. *BMC structural biology*, 10(1), 1.

Bertrand, T., Jolivalt, C., Briozzo, P., Caminade, E., Joly, N., Madzak, C. and Mougin, C. (2002). Crystal structure of a four-copper laccase complexed with an arylamine: insights into substrate recognition and correlation with kinetics. *Biochemistry*. 41, 7325-7333.

Brijwani, K., Rigdon, A., and Vadlani, P. V. (2010). Fungal laccases: production, function, and applications in food processing. *Enzyme Research*, 2010.

Bugg, T. D., Ahmad, M., Hardiman, E. M., and Singh, R. (2011). The emerging role for bacteria in lignin degradation and bio-product formation. *Current opinion in biotechnology*, 22(3), 394-400.

Camarero, S., Ibarra, D., Martínez, M. J., and Martínez, Á. T. (2005). Lignin-derived compounds as efficient laccase mediators for decolorization of different types of recalcitrant dyes. *Applied and environmental microbiology*, 71(4), 1775-1784.

Cambria, M. T., Gullotto, D., Garavaglia, S., and Cambria, A. (2012). In silico study of structural determinants modulating the redox potential of *Rigidoporus lignosus* and other fungal laccases. *Journal of Biomolecular Structure and Dynamics*, 30(1), 89-101.

Canas, A.I., and Camarero, S. (2010). Laccases and their natural mediators: biotechnological tools for sustainable eco-friendly processes. *Biotechnology advances*, 28(6), 694-705.

Cannatelli, M. D., and Ragauskas, A. J. (2016). Two decades of laccases: Advancing sustainability in the chemical industry. *The Chemical Record*.

Chandra, R., and Chowdhary, P. (2015). Properties of bacterial laccases and their application in bioremediation of industrial wastes. *Environmental Science: Processes and Impacts*, 17(2), 326-342.

Chen, C.Y., Huang, Y.C., Wei, C.M., Meng, M., Liu, W.H., and Yang, C.H. (2013) Properties of the newly isolated extracellular thermo-alkali-stable laccase from thermophilic actinomycetes *Thermobifida fusca* and its application in dye intermediates oxidation. *AMB Express* 3: 49.

Chen, S. C., Wu, P. H., Su, Y. C., Wen, T. N., Wei, Y. S., Wang, N. C., and Shyur, L. F. (2012). Biochemical characterization of a novel laccase from the basidiomycete fungus *Cerrena* sp. WR1. Protein Engineering Design and Selection, 25(11), 761-769.

Christopher, L. P., Yao, B., and Ji, Y. (2014). Lignin biodegradation with laccase-mediator systems. *Frontiers in Energy Research*, 2, 12.

Christopher, L. P., Yao, B., and Ji, Y. (2014). Lignin biodegradation with laccase-mediator systems. *Frontiers in Energy Research*, 2, 12.

Coniglio, A., Galli, C., Gentili, P., and Vadala, R. (2008). Oxidation of amides by laccase-generated aminoxyl radicals. *Journal of Molecular Catalysis B: Enzymatic*, 50(1), 40-49.

Cordi, L., Minussi, R. C., Freire, R. S., and Durán, N. (2007). Fungal laccase: copper induction, semi-purification, immobilization, phenolic effluent treatment and electrochemical measurement. *African Journal of Biotechnology*, 6(10).

Couto, S. R., and Herrera, J. L. T. (2006). Industrial and biotechnological applications of laccases: a review. *Biotechnology advances*, 24(5), 500-513.

Desai, S. S., and Nityanand, C. (2011). Microbial laccases and their applications: a review. *Asian J Biotechnol*, 3(2), 98-124.

Devi, P., Kandasamy, S., Chendrayan, K., and Uthandi, S. Laccase producing *Streptomyces bikiniensis* CSC12 isolated from composts. *J Microbiol Biotech Food Sci*, 6 (2), 794-798

Dhakar, K., and Pandey, A. (2015). Extracellular laccase from a newly isolated psychrotolerant strain of *Cladosporium tenuissimum* (NFCCI 2608). Proceedings of the National Academy of Sciences, India Section B: Biological Sciences, 1-6.

Ducros, V., Brzozowski, A.M., Wilson, K.S., Brown, S.H., Østergaard, P., Schneider, P., Yaver, D.S., Pedersen, A.H. and Davies, G.J. (1998). Crystal structure of the type-2 Cu depleted laccase from *Coprinus cinereus* at 2.2 Å resolution. *Nature Structural and Molecular Biology*, 5(4), 310-316.

Dwivedi, U. N., Singh, P., Pandey, V. P., and Kumar, A. (2011). Structure–function relationship among bacterial, fungal and plant laccases. *Journal of Molecular Catalysis B: Enzymatic*, 68(2), 117-128.

Elegir, G., Kindl, A., Sadocco, P., and Orlandi, M. (2008). Development of antimicrobial cellulose packaging through laccase-mediated grafting of phenolic compounds. *Enzyme and Microbial Technology*, 43(2), 84-92.

Elisashvili, V., Kachlishvili, E., Khardziani, T., and Agathos, S. N. (2010). Effect of aromatic compounds on the production of laccase and manganese peroxidase by white-rot basidiomycetes. *Journal of industrial microbiology and biotechnology*, 37(10), 1091-1096.

Endo, K., Hosono, K., Beppu, T., and Ueda, K. (2002). A novel extracytoplasmic phenol oxidase of *Streptomyces*: its possible involvement in the onset of morphogenesis. *Microbiology*, 148(6), 1767-1776.

Enguita, F. J., Marçal, D., Martins, L. O., Grenha, R., Henriques, A. O., Lindley, P. F., and Carrondo, M. A. (2004). Substrate and dioxygen binding to the endospore coat laccase from *Bacillus subtilis*. *Journal of Biological Chemistry*, 279(22), 23472-23476.

Fang, W., Fernandes, É. K., Roberts, D. W., Bidochka, M. J., and Leger, R. J. S. (2010). A laccase exclusively expressed by *Metarhizium anisopliae* during isotropic growth is involved in pigmentation, tolerance to abiotic stresses and virulence. *Fungal Genetics and Biology*, 47(7), 602-607.

Farnet, A. M., Criquet, S., Tagger, S., Gil, G., and Petit, J. L. (2000). Purification, partial characterization, and reactivity with aromatic compounds of two laccases from *Marasmius quercophilus* strain 17. *Canadian journal of microbiology*, 46(3), 189-194.

Fernandes, A. T., Soares, C. M., Pereira, M. M., Huber, R., Grass, G., and Martins, L. O. (2007). A robust metallooxidase from the hyperthermophilic bacterium *Aquifex aeolicus*. *FEBS Journal*, 274(11), 2683-2694.

Galai, S., Touhami, Y., and Marzouki, M. N. (2012). Response surface methodology applied to laccases activities exhibited by *Stenotrophomonas maltophilia* AAP56 in different growth conditions. *BioResources*, 7(1), 0706-0726.

Galhaup, C., Goller, S., Peterbauer, C. K., Strauss, J., and Haltrich, D. (2002). Characterization of the major laccase isoenzyme from *Trametes pubescens* and regulation of its synthesis by metal ionsa. *Microbiology*, 148(7), 2159-2169.

Galli, C., Madzak, C., Vadalà, R., Jolivalt, C., and Gentili, P. (2013). Concerted electron/proton transfer mechanism in the oxidation of phenols by laccase. *ChemBioChem*, 14(18), 2500-2505.

Givaudan, A., Effosse, A., Faure, D., Potier, P., Bouillant, M. L., and Bally, R. (1993). Polyphenol oxidase in *Azospirillum lipoferum* isolated from rice rhizosphere: evidence for laccase activity in non-motile strains of *Azospirillum lipoferum*. *FEMS Microbiology Letters*, 108(2), 205-210.

Gorbacheva, M., Morozova, O., Shumakovich, G., Streltsov, A., Shleev, S., and Yaropolov, A. (2009). Enzymatic oxidation of manganese ions catalysed by laccase. *Bioorganic chemistry*, 37(1), 1-5.

Hatamoto, O., Sekine, H., Nakano, E., and ABE, K. (1999). Cloning and expression of a cDNA encoding the laccase from *Schizophyllum* commune. *Bioscience, biotechnology, and biochemistry*, 63(1), 58-64.

Hong, F., Meinander, N. Q., and Jönsson, L. J. (2002). Fermentation strategies for improved heterologous expression of laccase in *Pichia pastoris*. *Biotechnology and Bioengineering*, 79(4), 438-449.

Hong, Y. Z., Zhou, H. M., Tu, X. M., Li, J. F., and Xiao, Y. Z. (2007). Cloning of a laccase gene from a novel basidiomycete Trametes sp. 420 and its heterologous expression in *Pichia pastoris*. *Current Microbiology*, 54(4), 260-265.

Hou, H., Zhou, J., Wang, J., Du, C., and Yan, B. (2004). Enhancement of laccase production by *Pleurotus ostreatus* and its use for the decolorization of anthraquinone dye. *Process Biochemistry*, 39(11), 1415-1419.

Hullo, M. F., Moszer, I., Danchin, A., and Martin-Verstraete, I. (2001). CotA of *Bacillus subtilis* is a copper-dependent laccase. *Journal of bacteriology*, 183(18), 5426-5430.

Husain, M., and Husain, Q. (2007). Applications of redox mediators in the treatment of organic pollutants by using oxidoreductive enzymes: a review. *Critical Reviews in Environmental Science and Technology*, 38(1), 1-42.

Imran, M., Asad, M. J., Hadri, S. H., and Mehmood, S. (2012). Production and industrial applications of laccase enzyme. *Journal of Cell and Molecular Biology*, 10(1).

Jimenez-Juarez, N., Roman-Miranda, R., Baeza, A., Sánchez-Amat, A., Vazquez-Duhalt, R., and Valderrama, B. (2005). Alkali and halide-resistant catalysis by the multipotent oxidase from *Marinomonas mediterranea*. *Journal of biotechnology*, 117(1), 73-82.

Jones, S. M., and Solomon, E. I. (2015). Electron transfer and reaction mechanism of laccases. *Cellular and molecular life sciences*, 72(5), 869-883.

Kalme, S., Jadhav, S., Jadhav, M., and Govindwar, S. (2009). Textile dye degrading laccase from *Pseudomonas desmolyticum* NCIM 2112. *Enzyme and Microbial Technology*, 44(2), 65-71.

Kandasamy, S., Muniraj, I. K., Purushothaman, N., Sekar, A., Sharmila, D. J. S., Kumarasamy, R., and Uthandi, S. (2016). High level secretion of laccase (LccH) from a newly isolated white-rot basidiomycete, *Hexagonia hirta* MSF2. *Frontiers in microbiology*, 7.

Ko, E. M., Leem, Y. E., and Choi, H. (2001). Purification and characterization of laccase isozymes from the white-rot basidiomycete *Ganoderma lucidum*. *Applied microbiology and biotechnology*, 57(1-2), 98-102.

Koschorreck, K., Schmid, R. D., and Urlacher, V. B. (2009). Improving the functional expression of a *Bacillus licheniformis* laccase by random and site-directed mutagenesis. *BMC biotechnology*, 9(1), 1.

Kulys, J., and Vidziunaite, R. (2003). Amperometric biosensors based on recombinant laccases for phenols determination. *Biosensors and Bioelectronics*, 18(2), 319-325.

Kumar, S. V., Phale, P. S., Durani, S., and Wangikar, P. P. (2003). Combined sequence and structure analysis of the fungal laccase family. *Biotechnology and Bioengineering*, 83(4), 386-394.

Kunamneni, A., Plou, F. J., Ballesteros, A., and Alcalde, M. (2008). Laccases and their applications: a patent review. *Recent patents on biotechnology*, 2(1), 10-24.

Labat, E., Morel, M. H., and Rouau, X. (2000). Effects of Laccase and Ferulic Acid on Wheat Flour Doughs 1. *Cereal Chemistry*, 77(6), 823-828.

Biocatalysts in Biomass to Bioproducts

LaFayette, P. R., Eriksson, K. E. L., and Dean, J. F. (1999). Characterization and heterologous expression of laccase cDNAs from xylem tissues of yellow-poplar (*Liriodendron tulipifera*). *Plant molecular biology*, 40(1), 23-35.

Lee, S. K., George, S. D., Antholine, W. E., Hedman, B., Hodgson, K. O., and Solomon, E. I. (2002). Nature of the intermediate formed in the reduction of O_2 to H_2O at the trinuclear copper cluster active site in native laccase. *Journal of the American Chemical Society*, 124(21), 6180-6193.

Li, H., Webb, S. P., Ivanic, J., and Jensen, J. H. (2004). Determinants of the relative reduction potentials of type-1 copper sites in proteins. *Journal of the American Chemical Society*, 126(25), 8010-8019.

Lomascolo, A., Record, E., HerpoëlGimbert, I., Delattre, M., Robert, J. L., Georis, J., and Asther, M. (2003). Overproduction of laccase by a monokaryotic strain of *Pycnoporus cinnabarinus* using ethanol as inducer. *Journal of Applied Microbiology*, 94(4), 618-624.

Lorenzo, M., Moldes, D., Couto, S. R., and Sanromán, M. A. A. (2005). Inhibition of laccase activity from *Trametes versicolor* by heavy metals and organic compounds. *Chemosphere*, 60(8), 1124-1128.

Machczynski, M. C., Gray, H. B., and Richards, J. H. (2002). An outer-sphere hydrogen-bond network constrains copper coordination in blue proteins. *Journal of inorganic biochemistry*, 88(3), 375-380.

Martins, L.O., Durao, P., Brissos, V., and Lindley, P. F. (2015). Laccases of prokaryotic origin: enzymes at the interface of protein science and protein technology. *Cellular and molecular life sciences*, 72(5), 911-922.

Mate, D. M., and Alcalde, M. (2015). Laccase engineering: from rational design to directed evolution. *Biotechnology advances*, 33(1), 25-40.

Mate, D. M., and Alcalde, M. (2016). Laccase: a multipurpose biocatalyst at the forefront of biotechnology. *Microbial Biotechnology*.

Melo, E. P., Fernandes, A. T., Durao, P., and Martins, L. O. (2007). Insight into stability of CotA laccase from the spore coat of *Bacillus subtilis*. *Biochemical Society Transactions*, 35(6), 1579-1582.

Morozova, O. V., Shumakovich, G. P., Gorbacheva, M. A., Shleev, S. V., and Yaropolov, A. I. (2007a). "Blue" laccases. *Biochemistry (Moscow)*, 72(10), 1136-1150.

Morozova, O. V., Shumakovich, G. P., Shleev, S. V., and Yaropolov, Y. I. (2007b). Laccase-mediator systems and their applications: a review. *Applied Biochemistry and Microbiology*, 43(5), 523-535.

Mougin, C., Boyer, F. D., Caminade, E., and Rama, R. (2000). Cleavage of the diketonitrile derivative of the herbicide isoxaflutole by extracellular fungal oxidases. *Journal of agricultural and food chemistry*, 48(10), 4529-4534.

Muthukumarasamy, N. P., and Murugan, S. (2014). Production, purification and application of bacterial laccase: a review. *Biotechnology*, 13(5), 196.

Myasoedova, N. M., Chernykh, A. M., Psurtseva, N. V., Belova, N. V., and Golovleva, L. A. (2008). New efficient producers of fungal laccases. *Applied Biochemistry and Microbiology*, 44(1), 73-77.

Nyanhongo, G. S., Gomes, J., Gübitz, G., Zvauya, R., Read, J. S., and Steiner, W. (2002). Production of laccase by a newly isolated strain of *Trametes modesta*. *Bioresource Technology*, 84(3), 259-263.

Osma, J. F., Toca-Herrera, J. L., and Rodríguez-Couto, S. (2010). Uses of laccases in the food industry. *Enzyme research*, 2010.

Palmeiri, G., Giardina, P., Marzullo, L., Desiderio, B., Nittii, G., Cannio, R., and Sannia, G. (1993). Stability and activity of a phenol oxidase from the ligninolytic fungus *Pleurotus ostreatus*. *Applied Microbiology and Biotechnology*, 39(4-5), 632-636.

Piontek, K., Antorini, M., and Choinowski, T. (2002). Crystal structure of a laccase from the fungus*Trametes versicolor* at 1.90-Å resolution containing a full complement of coppers. *Journal of Biological Chemistry*, 277(40), 37663-37669.

Piscitelli, A., Pezzella, C., Lettera, V., Giardina, P., Faraco, V., and Sannia, G. (2016). Fungal laccases: structure, function and application. *Fungal Enzymes: Progress and Prospects* pp. 113-151.

Plácido, J., and Capareda, S. (2015). Ligninolytic enzymes: a biotechnological alternative for bioethanol production. *Bioresources and Bioprocessing*, 2(1), 1-12.

Pozdnyakova, N. N., Rodakiewicz-Nowak, J., Turkovskaya, O. V., and Haber, J. (2006). Oxidative degradation of polyaromatic hydrocarbons catalyzed by blue laccase from *Pleurotus ostreatus* D1 in the presence of synthetic mediators. *Enzyme and Microbial Technology*, 39(6), 1242-1249.

Ramírez, M. C., Rivera-Ríos, J. M., Téllez-Jurado, A., Gálvez, A. M., Mercado-Flores, Y., and Arana-Cuenca, A. (2012). Screening for thermotolerant ligninolytic fungi with laccase, lipase, and protease activity isolated in Mexico. *Journal of environmental management*, 95, S256-S259.

Reece, S. Y., and Nocera, D. G. (2009). Proton-coupled electron transfer in biology: Results from synergistic studies in natural and model systems. *Annual review of biochemistry*, 78, 673.

Revankar, M. S., and Lele, S. S. (2006). Enhanced production of laccase using a new isolate of white rot fungus WR-1. *Process Biochemistry*, 41(3), 581-588.

Rodríguez-Delgado, M. M., Alemán-Nava, G. S., Rodríguez-Delgado, J. M., Dieck-Assad, G., Martínez-Chapa, S. O., Barceló, D., and Parra, R. (2015). Laccase-based biosensors for detection of phenolic compounds. *TrAC Trends in Analytical Chemistry*, 74, 21-45.

Rosconi, F., Fraguas, L. F., Martínez-Drets, G., and Castro-Sowinski, S. (2005). Purification and characterization of a periplasmic laccase produced by *Sinorhizobium meliloti*. *Enzyme and Microbial technology*, 36(5), 800-807.

Sato, Y., Wuli, B., Sederoff, R., and Whetten, R. (2001). Molecular cloning and expression of eight laccase cDNAs in loblolly pine (*Pinus taeda*). *Journal of Plant Research*, 114(2), 147-155.

Selinheimo, E., Kruus, K., Buchert, J., Hopia, A., and Autio, K. (2006). Effects of laccase, xylanase and their combination on the rheological properties of wheat doughs. *Journal of Cereal Science*, 43(2), 152-159.

Senthivelan, T., Kanagaraj, J., and Panda, R. C. (2016). Recent trends in fungal laccase for various industrial applications: An eco-friendly approach-*A review*. *Biotechnology and Bioprocess Engineering*, 21(1), 19-38.

Sharma, K. K., and Kuhad, R. C. (2008). Laccase: enzyme revisited and function redefined. *Indian journal of microbiology*, 48(3), 309-316.

Sharma, P., Goel, R., and Capalash, N. (2007). Bacterial laccases. *World Journal of Microbiology and Biotechnology*, 23(6), 823-832.

Shekher, R., Sehgal, S., Kamthania, M., and Kumar, A. (2011). Laccase: microbial sources, production, purification, and potential biotechnological applications. *Enzyme research*, 2011.

Shleev, S., Tkac, J., Christenson, A., Ruzgas, T., Yaropolov, A. I., Whittaker, J. W., and Gorton, L. (2005). Direct electron transfer between copper-containing proteins and electrodes. *Biosensors and Bioelectronics*, 20(12), 2517-2554.

Singh, P., Sulaiman, O., Hashim, R., Rupani, P. F., and Peng, L. C. (2010). Biopulping of lignocellulosic material using different fungal species: a review. *Reviews in Environmental Science and Bio/Technology*, 9(2), 141-151.

Singhal, A., Jaiswal, P. K., Jha, P. K., Thapliyal, A., and Thakur, I. S. (2013). Assessment of *Cryptococcus albidus* for biopulping of eucalyptus. Preparative *Biochemistry and Biotechnology*, 43(8), 735-749.

Skálová, T., Dohnálek, J., Ostergaard, L. H., Ostergaard, P. R., Kolenko, P., Dusková, J., and Hašek, J. (2007). Crystallization and preliminary X-ray diffraction analysis of the small laccase from *Streptomyces coelicolor*. *Acta Crystallographica Section F: Structural Biology and Crystallization Communications*, 63(12), 1077-1079.

Soares, J. C., Moreira, P. R., Queiroga, A. C., Morgado, J., Malcata, F. X., and Pintado, M. E. (2011). Application of immobilized enzyme technologies for the textile industry: a review. *Biocatalysis and Biotransformation*, 29(6), 223-237.

Solomon, E. I., Augustine, A. J., and Yoon, J. (2008). O_2 reduction to H_2O by the multicopper oxidases. Dalton Transactions, (30), 3921-3932.

Strong, P. J., and Claus, H. (2011). Laccase: a review of its past and its future in bioremediation. *Critical Reviews in Environmental Science and Technology*, 41(4), 373-434.

Tadesse, M. A., D'Annibale, A., Galli, C., Gentili, P., and Sergi, F. (2008). An assessment of the relative contributions of redox and steric issues to laccase specificity towards putative substrates. *Organic and biomolecular chemistry*, 6(5), 868-878.

Teerapatsakul, C., Parra, R., Bucke, C., and Chitradon, L. (2007). Improvement of laccase production from Ganoderma sp. KU-Alk4 by medium engineering. *World Journal of Microbiology and Biotechnology*, 23(11), 1519-1527.

Tlecuitl-Beristain, S., Sánchez, C., Loera, O., Robson, G. D., and Díaz-Godínez, G. (2008). Laccases of *Pleurotus ostreatus* observed at different phases of its growth in submerged fermentation: production of a novel laccase isoform. *Mycological research*, 112(9), 1080-1084.

Tzanov, T., Basto, C., Gübitz, G. M., and CavacoPaulo, A. (2003). Laccases to improve the whiteness in a conventional bleaching of cotton. *Macromolecular materials and engineering*, 288(10), 807-810.

Upadhyay, P., Shrivastava, R., and Agrawal, P. K. (2016). Bioprospecting and biotechnological applications of fungal laccase. 3 *Biotech*, 6(1),1-12.

Uthandi, S., Saad, B., Humbard, M. A., and Maupin-Furlow, J. A. (2010). LccA, an archaeal laccase secreted as a highly stable glycoprotein into the extracellular medium by *Haloferax volcanii. Applied and environmental microbiology*, 76(3), 733-743.

Vasconcelos, A. F. D., Barbosa, A. M., Dekker, R. F., Scarminio, I. S., and Rezende, M. I. (2000). Optimization of laccase production by *Botryosphaeria* sp. in the presence of veratryl alcohol by the response-surface method. *Process Biochemistry*, 35(10), 1131-1138.

Virk, A. P., Sharma, P., and Capalash, N. (2012). Use of laccase in pulp and paperindustry. *Biotechnology progress*, 28(1), 21-32.

Viswanath, B., Rajesh, B., Janardhan, A., Kumar, A. P., and Narasimha, G. (2014). Fungal laccases and their applications in bioremediation. *Enzyme research*, 2014.

Wang, J., Feng, J., Jia, W., Chang, S., Li, S., and Li, Y. (2015). Lignin engineering through laccase modification: a promising field for energy plant improvement. *Biotechnology for biofuels*, 8(1), 1.

Witayakran, S., and Ragauskas, A. J. (2009). Synthetic applications of laccase in green chemistry. *Advanced Synthesis and Catalysis*, 351(9), 1187-1209.

Yang, Y., Ding, Y., Liao, X., and Cai, Y. (2013). Purification and characterization of a new laccase from *Shiraia* sp. SUPER-H168. *Process Biochemistry*, 48(2), 351-357.

Zumárraga, M., Camarero, S., Shleev, S., MartínezArias, A., Ballesteros, A., Plou, F. J., and Alcalde, M. (2008). Altering the laccase functionality by in vivo assembly of mutant libraries with different mutational spectra. Proteins: Structure, *Function, and Bioinformatics*, 71(1), 250-260.

4

Fungal Cellulases and Deconstruction of Lignocellulosic Biomass for Production of Biofuel and Bioproducts

Iniya Kumar Muniraj[1,3], Kalaichelvan Gurumurthy[2] and Sivakumar Uthandi[3]

[1]*Department of Crop Improvement, Kumaraguru Institute of Agriculture, Erode*
[2]*School of Bio Sciences and Technology, VIT University, Vellore – 632 014*
[3]*Biocatalysts Lab, Department of Agricultural Microbiology, Tamil Nadu Agricultural University, Coimbatore – 641 003*

1. Introduction

Lignocellulosic biomass (LCB) has been considered as one of the most sustainable and alternative renewable energy resources which can sufficiently contribute to world energy, chemical and fuels supply by combating the dependence on conventional fossil resources and reduction of atmospheric CO_2 concentration there by mitigating the climate change issues(Alcantara *et al.*, 2016). LCB primarily consists of polymers of 4 and 6 carbon sugars and phenyl proponoid units, evolved from complex systems(Bak, 2014). The polysaccharide component of plant cell wall primarily consists of cellulose, a ß-1,4 linked homopolymer of glucose, the second major polysaccharide in LCB is hemicelluloses, a branched hetero polymer of ß-1,4 linked xylan, glucoronxylan, arabinoxylan and glucomannan with linked side chains(Bhange *et al.*, 2015). Pectin is the third minor polysaccharide component of LCB which are made up of α linked polymers of galacturonic acid. LCB also has considerable portion of lignin which is made of three phenyl proponoid units linked covalently by radical coupling mechanisms during the formation of plant cell wall(Bak, 2015). Altogether the cellulose, hemicellulose and lignin consist of 20–50 per cent, 15–35 per cent, and 10–30 per cent of plant cell walls, respectively, on a dry weight basis (Kuuskeri *et al.*, 2015).

Considering its abundance in plant cell wall, cellulose represents the most abundant carbon material on the earth. Cellulose is the most complex polysaccharides when compared to other polysaccharides present in plant cell wall(Alvarez *et al.*, 2016). The sugars of cellulose represent a major feedstock for production renewable fuels and chemicals. In order to produce fuels and chemicals from the cellulose it is imperative to gain access to plant cell wall and hydrolyse the complex cellulose into fermentable sugars. To date, a variety of biochemical methods have been employed for degradation of LCB where in first, the biomass size is reduced and various chemicals would be applied to biomass to render them accessible by biocatalysts for hydrolysis to sugars and then finally the sugars are fermented to produce myriad of fuels and chemicals(Behera and Ray, 2016). In the total conversion process the pre-treatment and hydrolysis are long been considered as the costly process(Bayer *et al.*, 2007).

Although chemical hydrolytic processes are effective in monomer recovery from cellulose, the sugar loss, inhibitor generation and much expensive process conditions hinders commercial application. Enzymatic hydrolysis is being widely followed with cellulase, as they are extremely selective for glucose production from cellulose, release fewer fermentation inhibitors and operates at minimal process conditions. Cellulases are a group of inducible enzymes upon addition of cellulose. In nature, many microbes have evolved to hydrolyse the plant cell wall components. Among them, fungi play a major role in plant cell wall depolymerisation. Plant biomass was often noticed to be colonized by fungal systems and compare to bacteria their production rate is higher and exhibit a variety of mechanisms to deconstruct cellulose. Moreover, several established heterologous systems are available for fungal cellulase production(Alahakoon *et al.*, 2012; Bayer *et al.*, 1998; Bhattacharya *et al.*, 2015; Garvey *et al.*, 2013; Greene *et al.*, 2015; Hildebrand *et al.*, 2015).

In view of the above, this chapter emphasizes on fungal cellulases, historical developments in cellulase production, industrial production of cellulase on recombinant host, their mechanisms of action on deconstruction of cellulose will be discussed. The role of thermophilic cellulases, solvent tolerant cellulases and their emergence in industrial scenario will also be discussed in detail.

2. Structure of Cellulose and its Changes during Various Treatments

Cellulose is a polymer of ß-1,4 linked glucose with a reducing and non-reducing end, in reducing end the ring can open to produce an aldehyde form. The β-1,4-linked polysaccharides display an incredibly strong covalent bonds and are thus very recalcitrant when left untreated. The half-life of o glyosidic linkage is said to be 5 million years(Bayer *et al.,* 1998; Boisset *et al.,* 2001; Chen *et al.,* 2015; Cheng *et al.,* 2011a; Cheng *et al.,* 2011b). Individual cellulose chains are co-synthesized by large, membrane-bound terminal complexes, which simultaneously extrude and assemble multiple chains of cellulose into elementary micro fibrils. Cellulose micro fibrils in plants pack into tightly bound, crystalline matrixes wherein only a fraction of the chains are accessible to enzymatic attack on the micro fibril surface. The cellulose microfibers then can pack into variety of crystalline and polymorphic forms. Several studies on cellulose structure revealed that the hydrogen bond only exist in single layers and are not present in interchangeable structures, van der Walls forces are responsible for their thermodynamic nature of the plant cell wall(Ahn *et al.,* 2016).

In nature plants and other organisms produced cellulose I the detailed structure shows that, it can be further divided into 1ß and 1α polymorphs. The structural difference 1ß and 1α polymorphs lies in the hydrogen bonding and the staking of inner layer cellulose. Cellulose 1 ß forms two different layers and 1α polymorphs have single chain. Considering the above structure of cellulose, the cellulases must be incredible enzymes which can hydrolyse the strong covalent boding without using any oxygen and a metal ion for its activity.

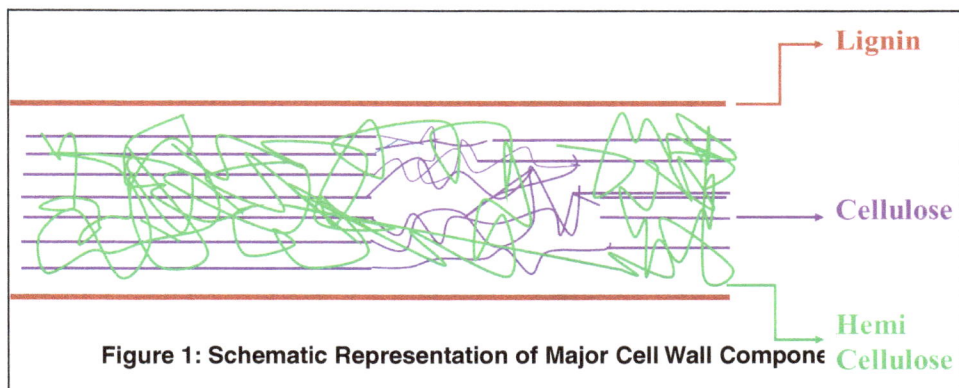

Figure 1: Schematic Representation of Major Cell Wall Compone

2.1. Changes of Cellulose Structure to Various Chemical Treatments

Since cellulose is the most complex polysaccharide of the plant cell wall, the structural changes due to chemical, physical treatments helps one to choose the right method of treatment and the knowledge on corresponding structural changes would help to achieve better saccharification of given biomass. Among the pre-treatments primarily employed, the major focus relies on removing and de-crystallizing cellulose for better saccharification. As discussed above cellulose I type is produced by the plants, the changes it undergoes due to various treatments are given below,

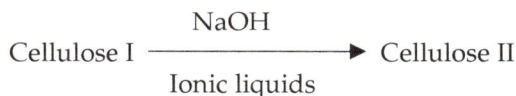

$$\text{Cellulose I} \xrightarrow[\text{Ionic liquids}]{\text{NaOH}} \text{Cellulose II}$$

Ammonia

Cellulose I or Cellulose II —————————▶ Cellulose III

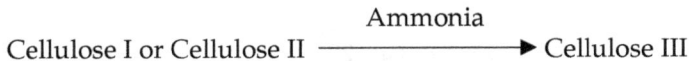

Treatment of native cellulose with NaOH or ionic liquids changes its native structure of parallel chains into antiparllel arrangement with both intra- and inter-layer hydrogen bonding. These changes lead to the formation of cellulose II. When cellulose I or cellulose II undergoes mild ammonia treatment, single layered cellulose of micro fibrils is converted to staggered layers with intra- and inter- hydrogen bonding. These resulting celluloses are known as Cellulose III. Both cellulose II and III exhibit greater digestibility to fungal cellulases(Abraham *et al.*, 2014; Agrawal *et al.*, 2011; Ali *et al.*, 2015; Alvarez *et al.*, 2016; Resch *et al.*, 2013). The right combination of pre-treatment methods is essential better saccharification of biomass.

3. Early Efforts in Fungal Cellulases

The first cellulolytic organism used for decomposition of cotton in US was *Trichoderma viride* QM6a, which was later renamed as *T. ressei* (*Hypocre jacorina*) The name change is to honour of Elwyn Reese for his extensive work on cellulases and developing the modern concept of biomass saccharification. Since then, the parent strain *T. ressei* QM6a attracted interest in industrial sector and many strain improvement strategies have been followed. The first generated mutant strain was *T. reesei* QM9414, generated from QM9123, widely used in many industries, produced two times more cellulase than parent strain. After careful selection and mutation studies RUT C30 strain was identified as potent producer from which the current strains are available. Several years of research strategies to develop a potential cellulolytic strain resulted in high cellulase producing strains and are now available with US, Finland, France, Portugal and India. The modern era of cloning and expression of genes responsible for cellulase systems show the present strain can produce up to 100mg protein/L of the fermentation medium(Amano *et al.*, 1996; Arai *et al.*, 2006; Badieyan *et al.*, 2012; Payne *et al.*, 2015).

Although, several strains are available for cellulase production in different countries, the present chapter only considers the cellulolytic mechanisms of model organisms such as *T. reesei*, and *Neurospora crassa*.

4. Structure of Fungal Cellulases

When the *T. reesi* cellulases (Trcel7A) were proteolytically cleaved with papain to yield two primary fragments of 56 kDa domain and small 10 kDa domain at C terminal end. It was noted that the 56 kDa domain lost its function on crystalline cellulose but showed catalytic function on small molecules. When the 10 kDa domain was linked manually the catalytic performance of 56 kDa domain improved on major cellulosic substrates. In addition, a small linker peptide which links the core protein to carbohydrate binding molecule (CBM) was evidenced from the proteolytic action on cellulase(Amano *et al.*, 1996; Badieyan *et al.*, 2012; Bhattacharya *et al.*, 2015; Cadena *et al.*, 2010).

Similar results were reported for the papain cleavage of TrCel6A (CBHII) where in the enzyme cleaved into two parts except that the smaller glycosylated domains were found in N terminal region. Both the enzymes lost their catalytic activity on cellulosic substrates but retaining their activity on small cellulose molecules. The Cel6A also have a carbohydrate binding molecule linker to the core enzyme by linker region consisting of small peptides.

From the structural study most of the fungal cellulases are multinodular protein with one or more catalytic domain existing with carbohydrate binding modules (Figure 4.2). Most fungal cellulases employ a family 1 CBM(Arai *et al.,* 2006; Arola and Linder, 2016; de Vries *et al.,* 2015; Hilden and Johansson, 2004).

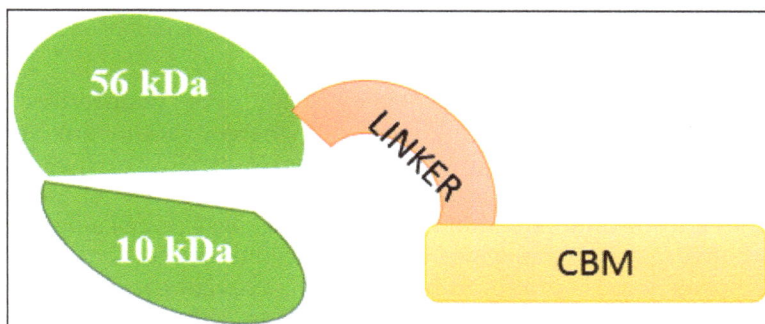

Figure 4.2: Papain Cleavage and the Structural Aspects of Cellulase (TrCel7A/CBH1).

In multi-modular cellulases, the linkers of various length with diversified sequences were observed. Characterization of fungal linkers showed that they are longer and mostly glycosylated which protects the linkers from proteolytic activity and maintains a proper distance between CD and CBM. Linkers of TrCel7A and TrCel 6A has been characterized, apart from core enzyme large tail like structure was evident which might be the cause for binding the CD with CBM. While comparing both the linkers, TrCel6A linker was longer than TrCel7A, with the exception that the length of TrCel7A can be extended in the presence of xylan in the hydrolysing environment(Malho *et al.,* 2015; Ruiz *et al.,* 2016; Textor *et al.,* 2016).

5. Mechanism of Cellulase

Hydrolysis carried out by all the cellulases belongs to two groups of catalytic behaviour namely retaining and inverting mechanisms(Ahmad *et al.,* 2016; Murashima *et al.,* 2003; Warden *et al.,* 2011; Zhou *et al.,* 2009). Under inverting mechanisms, the cellulases attack the nucleophile of anomeric carbon wherein a water molecule transfers a proton to catalytic based and a proton is transferred from catalytic acid to cleave the glyosidic linkage. The inverting mechanism involves one step catalysis where the stereochemistry of the molecule is inverted and hence molecule should be reinverted to proceed for the hydrolysis.

On the contrary, retaining mechanisms are two-step reactions, where in a proton is transferred from the catalytic acid and attack the anomeric carbon and invert form of the sugar. In the second step, water molecule enters the active site of enzyme and conducts nucleophilic attack at the anomeric carbon (Olsen *et al.,* 2016; Roberts and Davies, 2012; Zhou *et al.,* 2009).

A third type of mechanisms catalyse the oxidative cleavage of cellulose in the presence of an external electron donor. These new class of enzymes are called LPMOs (Lytic Polysaccharide Monooxygenases). They act synergistically to hydrolytic GHs and cleave the cellulose, it requires reducing agent for supply of electron donor the reducing agent is cellobiase dehydrogenases, and it is a oxygenases, requires oxygen. Finally, most of the LMPOs have copper molecule in the active site. In *N. crassa*, deletion of a single cellobiose dehydrogenase gene reduced the total cellulase activity secreted by the fungus by nearly half, indicating that a redox-active system is a significant part of the cellulolytic machinery in LPMO(Agger *et al.*, 2014; Garajova *et al.*, 2016; Hemsworth *et al.*, 2014; Jagadeeswaran *et al.*, 2016; Johansen, 2016; Kojima *et al.*, 2016; Muller *et al.*, 2015).

In contrast to multi-enzyme complex like cellulosome in anaerobic bacteria, fungi generally secrete "free enzyme" cocktails wherein various proteins diffuse independently of one another exhibiting different substrate specificities. All the enzymes work synergistically to degrade the macromulecules of biomass. The free enzymes may be unimodular or multi-modular consisting of a potent cocktail of free carbohydrate-active enzymes to degrade cellulose and hemicellulose. The majority of protein mass produced by *T.reesi* (*H. jecorina*) consists of processive (exo acting) cellulases which hydrolyse cellulose from either the reducing or non-reducing end. They are complemented by endoglucanases and oxidative enzymes that cleave cellulose at points in the interior of the chains to expose free ends for attachment and detachment of active enzymes. The detailed mechanisms of cellulose cleavage by different free enzymes are depicted in Figure 4.3 (Bhattacharya *et al.*, 2015; Borah *et al.*, 2016; Cruys-Bagger *et al.*, 2013; Fang *et al.*, 2010; Ganner *et al.*, 2012; Hiras *et al.*, 2016).

6. Thermophilic Cellulases

All the above discussed cellulase are mesophilic enzymes whose catalytic activity will be optimum at 50°C. However, many industrial operations require an enzyme with more thermotolerance and optimally active at 50°C. In order to improve the biochemical responses of bioconversions the secretomes of thermophilic fungi have been analysed. Many culturalable thermophilic fungi are present in compost where decomposition is taking place at elevated temperature (Anish *et al.*, 2007; Badieyan *et al.*, 2012; Basotra *et al.*, 2016). The GHs of the thermophilic fungi function from 60-80°C when compared to commercial mesophilic cocktails. Performance of bioconversion of cellulosic substrates at elevated temperature offers many advantages over mesophilic enzymes. Thermophilic enzymes increase the conversion rate when performed under elevated temperature, leading to shorter incubation times with reduced enzyme dosage. Further, when the bioconversion is at higher temperatures, the risk of contamination is low and downstream processes are easier to perform. With the above advantages, many thermophilic fungi have shown considerable results in bioconversion of biomass(Castiglia *et al.*, 2016; Chefetz *et al.*, 1998; de Cassia Pereira *et al.*, 2015).

Performance of two thermophilic fungal cellulases from *Thermoascus aurantiacus* and *Thielavia terrestris* on different pre-treated biomass showed that GHs activity

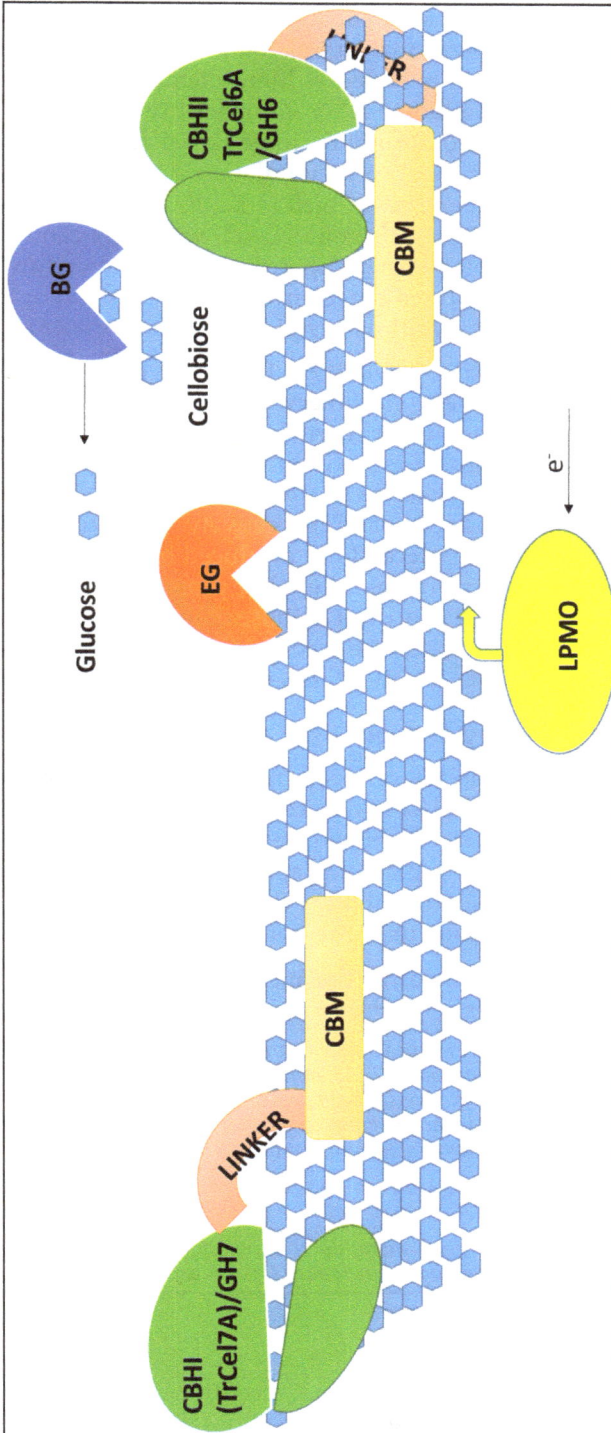

Figure 4.3: Mechanistic Action of Cellulases on Cellulose, the Sceretome of *T. reesi* has 193GHs, 93 Glycosyl Transferases, 5 Polysaccharide Lyases, 17 Carbohydrate Esterase and 41 CBM. All the three hydrolytic enzyme synergistically act on cellulose for deconstruction.

was comparable to artificial incubated substrates. When the biomass was pre-treated with ionic liquid, their performance was found to be higher with released more glucose, the enzymes while operated at higher temperature retained their activity and releasing increased glucose when compared to commercial cellulase preparation (Novozymes). In another study, ionic liquid tolerant thermophilic cellulases were screened from switch grass accumulating metagenomes. The results showed that, many of the genes expressed well in *E.coli* with good activity on IL 1-ethyl-3-methylimidazolium acetate ([C2mim][OAc]) with a concentrations of at least 10 per cent (v/v). In addition, the thermophilic enzymes showed an improved performance at elevated temperatures ranges from 50-95C with an optimum pH ranging from 5.5-7.5(Gladden *et al.*, 2014; Harnpicharnchai *et al.*, 2016).

Most archaea are extremophiles habitats environments including hot springs; over a period of time their enzymatic machinery have evolved to be thermophilic. An attempt was made to express hyper thermophilic enzymes in high cellulase producing *Talaromyces cellulolyticus* (formerly known as *Acremonium cellulolyticus*). The hyper thermophilic archaea *Pyrococcus horikoshii* and *P. furiosus* have GH family 5 and 12 hyper thermophilic endocellulase, respectively. The two kinds of hyper thermophilic endocellulases were successfully produced by *T. cellulolyticus* using expression system under the control of a glucoamylase promoter. These recombinant cellulases exhibited the same characteristics as those of the recombinant cellulases prepared in *E. coli*. The productions of the recombinant enzymes were estimated to be over 100 mg/L(van Wyk *et al.*, 2010; Vazana *et al.*, 2010; Wang *et al.*, 2016).

7. Cellulase Secretion by Fungi

Generally, cellulolytic capability of fungi has been noted in three different groups, the most predominant cellulases producers belong to group of soft rot fungi where in 25 groups of fungi know to secrete cellulases, the second major group belongs to white rot fungi these group of fungi apart from secreting cellulases they are also known for lignin degradation ability. The third but very small group of fungi are capable cellulolysis belong to brown rot fungi (Table 4.1).

Table 4.1: Cellulolytic Capability of Fungi

Soft rot Fungi	*Aspergillus niger; A. nidulans; A. oryzae; A. terreus; Fusarium solani; F. oxysporum; Humicola insolens; H.grisea; Melanocarpus albomyces; Penicillium brasilianum; P. occitanis; P. decumbans; Trichoderma reesei; T.longibrachiatum; T. harzianum; Chaetomium cellulyticum; C. thermophilum; Neurospora crassa; P. fumigosum; Thermoascus aurantiacus;Mucor circinelloides; P. janthinellum; Paecilomyces inflatus; P. echinulatum; Trichoderma atroviride*
Brown rot Fungi	*Coniophora puteana; Lanzites trabeum; Poria placenta; Tyromyces palustris; Fomitopsis* sp.
White rot fungi	*Phanerochaete chrysosporium; Sporotrichum thermophile; Trametes versicolor; Agaricus arvensis; Pleurotus ostreatus; Phlebia gigantea*

7.1. Unique Fungal Cellulases

Apart from common cellulases producers, recent research has been focused on finding out organisms capable of secreting cellulases with unique properties such as extremozymes (enzyme capable of performing under extreme conditions), anaerobic cellulases (cellulases from anaerobic fungi) and poly extremozymes (the enzymes withstand simultaneously with one are more extreme conditions). The first step in industrial conversion of lignocellulosic biomass is pre-treatment of LCB with harsh chemicals in combination with high temperature, pressure, salt and variable pH conditions. Such harsh conditions remove lignin, reduce the crystallinity of cellulose, makes pores which enables cellulases to better access the pre-treated biomass for higher glucose yield by increased saccharification. Finding out unique cellulases would have additional advantage on industrial conversion of LCB. The unique cellulases and their properties of some of fungi were given in Table 4.2.

Table 4.2: Unique Fungal Cellulases

Sl.No.	Fungi	Unique Property	Reference
1.	*Orpynomyces* sp. Y102	Anaerobic fungi isolated from cattle rumen having exo-cellulase activity (CelC7). The GHs of this fungi belong to family 8 and 48, which are otherwise generally found in aerobic fungi	(Chen *et al.*, 1998)
2.	*Neocallimastix patriciarum*	Anaerobic fungi with wide pH (5-11) and heat stability (80°C)	(Wang *et al.*, 2014)
3.	Myceliophothora thermophila	Highly thermostable cellulases producing fungi	(de Cassia Pereira *et al.*, 2015)
4.	*Cladosporium cladosporioides*	Cold adapted cellulases had 6 fold higher efficiency in converting biomass than mesophilic cellulases	(Chiranjeevi *et al.*, 2012)
5.	Wallemia, Cladosporium, Eurotium, Hortaea	Fungi can grow with very low water activity	(Zajc *et al.*, 2014)
6.	*Hortaea werneckii, Eurotium amstelodami, Wallemia ichthyophaga*	Chaophilic fungi can tolerate upto NaCl 2.1M, or Cacl$_2$ 2.0M	(Zajc *et al.*, 2014)
7.	*Pestalotiopsis* sp. NCi6	Halotolerant fungi and addition salts induces the cellulase production	(Arfi *et al.*, 2013)
8.	*Purpureocillium lilacinum*	Acidophilic cellulases producing fungi	(Aston *et al.*, 2016)
9.	*Sodiomyces alkalinus*	Alkaliphic cellulases producing fungi	(Brander *et al.*, 2014)

8. Factors Affecting Cellulase Production

8.1. Carbon Catabolite Repression

Easily metabolized carbon sources such as glucose and sucrose are used preferentially over complex sugars from plant biomass for the production of cellulase by fungal systems and this phenomenon is called carbon catabolite repression (CCR). CCR affects production of hydrolytic enzymes in *T. reesei*, *Aspergillus* spp., and *N. crassa*. Under conditions of starvation, the de-repression of genes encoding cellulase enzymes results in a low expression level. These secreted enzymes are believed to function as spies that, in the presence plant biomass, produce monomeric and/or small polymeric sugars that are subsequently transported into the cell *via* membrane-bound transporters. These sugars can function as signalling molecules that induce high expression of genes encoding enzymes required for plant biomass deconstruction. In *T. reesei* and *N. crassa*, CRE1/CRE-1 regulates both xylan and cellulose utilization However, deletion of creA homologs does not induce the expression of lignocellulolytic genes in the absence of an inducer derived from plant cell wall material(Glass *et al.*, 2013; Lindman, 2016; McNamara *et al.*, 2015; Singh *et al.*, 2015).

8.2. Light

Generally, even traces of light may affect filamentous fungi rapidly and efficiently which in turn affect the developmental processes, morphology, and metabolic pathways. In *T. reesei*, a genome-wide transcriptome study revealed a considerable number of glycoside hydrolases to be differentially expressed in light versus dark conditions (Anbar and Bayer, 2012; Andrade *et al.*, 2011).

8.3. pH

The conserved transcription factor PacC is responsible for pH regulation in filamentous fungi which functions as an activator of genes expressed in alkaline conditions and as a repressor under acidic conditions. In *N. crassa*, deletion of *pacC* causes a defect to grow on cellulose. In *T. reesei*, optima of cellulolytic enzymes vary with pH and the maximum efficiency of cellulose degradation occurs at pH 5.0. In general, different fungi have different pH optima for efficient production of cellulolytic enzymes. Gene regulation in response to pH appears to be aimed at a delicate balance between substrate degradation, uptake of the resulting sugar, metabolic activity, and enzyme activity.

8.4. Temperature

Temperature has a profound effect on cellulases. The temperature for assaying cellulase activities generally lies within 50–65 °C for a variety of microbial strains whereas growth temperature of these microbial strains was found to be 25–30 °C. Cellulose adsorption is shown to be influenced by temperature. A positive relationship between adsorption and saccharification of cellulosic substrate was observed at temperature below 60°C. The adsorption activities beyond 60°C decreased possibly because of the loss of enzyme configuration leading to denaturation of the enzyme activity. However recent thermophilic enzymes perform better than the mesophilic enzymes at industrial scale(Abe *et al.*, 2015; Abhaykumar and Dube, 1992; Fontes *et al.*, 1995).

8.5. Carbon Source

Many different substrates such as agro or industrial wastes, synthetic or naturally occurring have been evaluated as the carbon source for the production of fungal cellulase. Among the cellulosic materials, sulphite pulp, printed papers, mixed waste paper, wheat straw, paddy straw, sugarcane bagasse, jute stick, carboxy methyl cellulose, corncobs, groundnut shells, cotton, ball milled barley straw, delignified ball milled oat spelt xylan, larch wood xylan, etc. have been used as the substrates for cellulase production. The observations indicated that the production of cellulases increased with increase in substrate concentration up to 12 per cent during solid state fermentation using *Aspergillus niger*. Further increase in substrate concentration resulted in decreased production levels. This might have been due to limitation of oxygen in the central biomass of the pellets and exhaustion of nutrients other than energy sources(Dios *et al.*, 2015; Ferreira *et al.*, 2016; Gubitosi *et al.*, 2016; Hussain *et al.*, 2016; Moran-Mirabal *et al.*, 2011).

8.6. Nitrogen Source

The effect of different nitrogen sources such as ammonium sulphate, ammonium nitrate, ammonium ferrous sulphate, ammonium chloride and sodium nitrate have been studied. Among these, ammonium sulphate (0.5 g.l^{-1}) led to the maximum production of cellulases. In contrast to this, a significant reduction in enzymatic levels in the presence of ammonium salts as the nitrogen source was also observed. However, an increase in the level of ß-glucosidase was reported when corn steep liquor (0.8 per cent v/v) was added. Using corn steep liquor also led to 3-5 fold induction in endoglucanase and exoglucanase levels with synthetic cellulose (Sigma cell type-20), wheat straw and wheat bran as the substrates. Enzyme production was sensitive to corn steep liquor (0.88 g.l^{-1}), and production increased significantly when mixed nitrogen sources (corn steep liquor and ammonium nitrate) were used. However, additional incorporation of nitrogen sources into the medium escalated the cost of the process (Bilgi *et al.*, 2016; Bustamante *et al.*, 2008; Dave *et al.*, 2012).

8.7. Phosphorus Sources

Phosphorus is an essential requirement for fungal growth and metabolism. It is an important constituent of phospholipids involved in the formation of cell membranes. Besides its role in effecting the linkage between nucleotides forming the nucleic acid strands, it is also involved in the formation of numerous intermediates, enzymes and coenzymes that are essential to the metabolism of carbohydrates, as well as for many other oxidative reactions and intracellular processes. Different phosphate sources such as (Cabezas *et al.*, 2012)potassium dihydrogen phosphate, tetra-sodium pyrophosphate, sodium di-glycerophosphate and dipotassium hydrogen phosphate have been evaluated for their effect on cellulases production. It has been widely demonstrated that potassium dihydrogen phosphate is the most favourable phosphorus source for cellulase production.

8.8. Phenolic Compounds

The phenolic compounds have the ability to induce laccase that in turn stimulates the cellobiose-quinone-oxidoreductae enzyme; this enzyme possibly is involved in cellobiose (CMCase and FPase inhibitor) oxidation to cellobionic acid and thus inducing the cellulase synthesis indirectly. Among various phenolics, *e.g.* gallic acid, tannic acid, maleic acid, salicylic acid studied, salicylic acid was observed to be a better inducer of cellulases. Other phenolic compounds however, showed an inhibitory effect. Some observations indicate that the vanillin had a stimulatory effect on cellulase biosynthesis and its regulation was possibly due to cellobiono-lactone formed by the interactions of laccase, phenol, cellobiose and cellobiose-quinone-oxidoreductase. This lactone seemed to influence cellulase production(Adeboye *et al.*, 2014; Gurram and Menkhaus, 2014).

8.9. Sugars

Several investigations so far have indicated that cellulases are inducible enzymes and various carbon sources have been found to increase the synthesis. Cellobiose (2.95 mM) may act as an effective inducer of cellulases synthesis in *Nectria catalinensi*. An increased rate of endoglucanase biosynthesis in *Bacillus* sp. was reported in the presence of cellobiose or glucose (0.2 per cent) added to the culture medium. Xylanase biosynthesis was also induced by xylose or cellobiose added to the culture medium during growth. Other investigators reported that cellobiose, gentibiose at higher concentration inhibited about 80 per cent of the ß-glucosidase activity; similarly, laminaribiose and glucose also led to a 55–60 per cent inhibition in the enzymatic activity(Agrawal *et al.*, 2011; Ali *et al.*, 2015; Azevedo *et al.*, 2002).

8.10. Cellulases for Biorefineries

The deconstruction of LCB can be made efficient by a combination of chemical and biological agents. Many options of pretreatment are available for LCB (Kudakasseril Kurian *et al.*, 2013). Chief among them are the ionic liquids (IL) which provides hydrolysable cellulose (Brandt *et al.*, 2013). IL based pretreatments are preferred over acid or alkali based treatments since they are soft on the downstream operations involving organisms or enzymes (Qiu *et al.*, 2012). But all cellulases are tolerant to the traces of ionic liquids that remain after pretreatment of LCB (Elgharbawy *et al.*, 2016). Bacteria (Sriariyanun *et al.*, 2016) and Fungi (Xu *et al.*, 2015; Xu *et al.*, 2014) are known to synthesize cellulases are tolerant to IL at various levels. It has been observed that the level of negative charge exhibited on the surface of these enzymes play a major role in determining their tolerance to IL (Jaeger *et al.*, 2015). An added advantage of these IL-tolerant cellulases is that they also thermo-tolerant (Xu *et al.*, 2016). Employing such IL tolerant cellulases is a further step in making the industrial adoption easier and greener. Besides identifying natively tolerant cellulases, other physical and chemical processes such as immobilization on polyethylene glycol/glutaraldehyde, covalent linking to succinylation, coating with hydrophobic IL are some simple strategies to stabilize the cellulases used in biorefineries.

9. Conclusions

Given that many fungal systems evolved as efficient degraders of lignocellulose biomaterials by producing myriad of cellulase enzyme cocktails, the demand for production of fuels and chemicals from biomass is growing rapidly and many countries around the world is realising the alternative renewable, sustainable fuels production from lignocellulosic biomass. Undoubtedly fungal cellulase would be the corner stone in the industrial biotechnology for diversified applications in various bio catalytic processes involving not only cellulose but also associated hemicelluloses sugars. However, in order to improve the performance, their catalytic activity at higher temperatures need to be studied. In addition, considering the serious efforts on biochemistry, molecular biology coupled with systems biology and proteomics studies have given detailed understanding at least existing fungal cellulase systems. However, over the years more recent new family of enzymes both in GHs and LMPOs are emerging and their mechanisms of action in detail need detailed studies., For industrial adoption, protein engineering studies are needed to formulate working enzymatic cocktails which tolerate organic solvents for production of newer chemicals and fuels.

References

Abe, M., Kuroda, K., Sato, D., Kunimura, H., Ohno, H. 2015. Effects of polarity, hydrophobicity, and density of ionic liquids on cellulose solubility. *Phys Chem Chem Phys*, **17**(48), 32276-82.

Abhaykumar, V.K., Dube, H.C. 1992. Cellulases of Vibrio agar-liquefaciens isolated from sea mud. *World J Microbiol Biotechnol*, **8**(3), 313-5.

Abraham, R.E., Verma, M.L., Barrow, C.J., Puri, M. 2014. Suitability of magnetic nanoparticle immobilised cellulases in enhancing enzymatic saccharification of pretreated hemp biomass. *Biotechnol Biofuels*, **7**, 90.

Adeboye, P.T., Bettiga, M., Olsson, L. 2014. The chemical nature of phenolic compounds determines their toxicity and induces distinct physiological responses in Saccharomyces cerevisiae in lignocellulose hydrolysates. *AMB Express*, **4**, 46.

Agger, J.W., Isaksen, T., Varnai, A., Vidal-Melgosa, S., Willats, W.G., Ludwig, R., Horn, S.J., Eijsink, V.G., Westereng, B. 2014. Discovery of LPMO activity on hemicelluloses shows the importance of oxidative processes in plant cell wall degradation. *Proc Natl Acad Sci U S A*, **111**(17), 6287-92.

Agrawal, P., Verma, D., Daniell, H. 2011. Expression of Trichoderma reesei beta-mannanase in tobacco chloroplasts and its utilization in lignocellulosic woody biomass hydrolysis. *PLoS One*, **6**(12), e29302.

Ahmad, I., Rouf, S.F., Sun, L., Cimdins, A., Shafeeq, S., Le Guyon, S., Schottkowski, M., Rhen, M., Romling, U. 2016. BcsZ inhibits biofilm phenotypes and promotes virulence by blocking cellulose production in Salmonella enterica serovar Typhimurium. *Microb Cell Fact*, **15**(1), 177.

Ahn, Y., Kwak, S.Y., Song, Y., Kim, H. 2016. Physical state of cellulose in BmimCl: dependence of molar mass on viscoelasticity and sol-gel transition. *Phys Chem Chem Phys*, **18**(3), 1460-9.

Alahakoon, T., Koh, J.W., Chong, X.W., Lim, W.T. 2012. Immobilization of cellulases on amine and aldehyde functionalized Fe2O3 magnetic nanoparticles. *Prep Biochem Biotechnol*, **42**(3), 234-48.

Alcantara, M.A., Dobruchowska, J., Azadi, P., Garcia, B.D., Molina-Heredia, F.P., Reyes-Sosa, F.M. 2016. Recalcitrant carbohydrates after enzymatic hydrolysis of pretreated lignocellulosic biomass. *Biotechnol Biofuels*, **9**, 207.

Ali, N., Ting, Z., Li, H., Xue, Y., Gan, L., Liu, J., Long, M. 2015. Heterogeneous Expression and Functional Characterization of Cellulose-Degrading Enzymes from Aspergillus niger for Enzymatic Hydrolysis of Alkali Pretreated Bamboo Biomass. *Mol Biotechnol*, **57**(9), 859-67.

Alvarez, C., Reyes-Sosa, F.M., Diez, B. 2016. Enzymatic hydrolysis of biomass from wood. *Microb Biotechnol*, **9**(2), 149-56.

Amano, Y., Shiroishi, M., Nisizawa, K., Hoshino, E., Kanda, T. 1996. Fine substrate specificities of four exo-type cellulases produced by Aspergillus niger, Trichoderma reesei, and Irpex lacteus on (1—>3), (1—>4)-β-D-glucans and xyloglucan. *J Biochem*, **120**(6), 1123-9.

Anbar, M., Bayer, E.A. 2012. Approaches for improving thermostability characteristics in cellulases. *Methods Enzymol*, **510**, 261-71.

Andrade, J.P., Bispo, A.S., Marbach, P.A., do Nascimento, R.P. 2011. Production and Partial Characterization of Cellulases from Trichoderma sp. IS-05 Isolated from Sandy Coastal Plains of Northeast Brazil. *Enzyme Res*, **2011**, 167248.

Anish, R., Rahman, M.S., Rao, M. 2007. Application of cellulases from an alkalothermophilic Thermomonospora sp. in biopolishing of denims. *Biotechnol Bioeng*, **96**(1), 48-56.

Arai, T., Kosugi, A., Chan, H., Koukiekolo, R., Yukawa, H., Inui, M., Doi, R.H. 2006. Properties of cellulosomal family 9 cellulases from *Clostridium cellulovorans*. *Appl Microbiol Biotechnol*, **71**(5), 654-60.

Arfi, Y., Chevret, D., Henrissat, B., Berrin, J.-G., Levasseur, A., Record, E. 2013. Characterization of salt-adapted secreted lignocellulolytic enzymes from the mangrove fungus *Pestalotiopsis* sp. *Nature Communications*, **4**, 1810.

Arola, S., Linder, M.B. 2016. Binding of cellulose binding modules reveal differences between cellulose substrates. *Sci Rep*, **6**, 35358.

Aston, J.E., Apel, W.A., Lee, B.D., Thompson, D.N., Lacey, J.A., Newby, D.T., Reed, D.W., Thompson, V.S. 2016. Degradation of phenolic compounds by the lignocellulose deconstructing thermoacidophilic bacterium Alicyclobacillus Acidocaldarius. *J Ind Microbiol Biotechnol*, **43**(1), 13-23.

Azevedo, H., Bishop, D., Cavaco-Paulo, A. 2002. Possibilities for recycling cellulases after use in cotton processing: part II: Separation of cellulases from reaction products and released dyestuffs by ultrafiltration. *Appl Biochem Biotechnol*, **101**(1), 77-91.

Badieyan, S., Bevan, D.R., Zhang, C. 2012. Study and design of stability in GH5 cellulases. *Biotechnol Bioeng*, **109**(1), 31-44.

Bak, J.S. 2015. Lignocellulose depolymerization occurs via an environmentally adapted metabolic cascades in the wood-rotting basidiomycete Phanerochaete chrysosporium. *Microbiologyopen*, **4**(1), 151-66.

Bak, J.S. 2014. Process evaluation of electron beam irradiation-based biodegradation relevant to lignocellulose bioconversion. *Springerplus*, **3**, 487.

Basotra, N., Kaur, B., Di Falco, M., Tsang, A., Chadha, B.S. 2016. *Mycothermus thermophilus* (Syn. *Scytalidium thermophilum*): Repertoire of a diverse array of efficient cellulases and hemicellulases in the secretome revealed. *Bioresour Technol*, **222**, 413-421.

Bayer, E.A., Chanzy, H., Lamed, R., Shoham, Y. 1998. Cellulose, cellulases and cellulosomes. *Curr Opin Struct Biol*, **8**(5), 548-57.

Bayer, E.A., Lamed, R., Himmel, M.E. 2007. The potential of cellulases and cellulosomes for cellulosic waste management. *Curr Opin Biotechnol*, **18**(3), 237-45.

Behera, S.S., Ray, R.C. 2016. Solid state fermentation for production of microbial cellulases: Recent advances and improvement strategies. *Int J Biol Macromol*, **86**, 656-69.

Bhange, V.P., William, S.P., Sharma, A., Gabhane, J., Vaidya, A.N., Wate, S.R. 2015. Pretreatment of garden biomass using Fenton's reagent: influence of Fe(2+) and H2O2 concentrations on lignocellulose degradation. *J Environ Health Sci Eng*, **13**, 12.

Bhattacharya, A.S., Bhattacharya, A., Pletschke, B.I. 2015. Synergism of fungal and bacterial cellulases and hemicellulases: a novel perspective for enhanced bioethanol production. *Biotechnol Lett*, **37**(6), 1117-29.

Bilgi, E., Bayir, E., Sendemir-Urkmez, A., Hames, E.E. 2016. Optimization of bacterial cellulose production by *Gluconacetobacter* xylinus using carob and haricot bean. *Int J Biol Macromol*, **90**, 2-10.

Boisset, C., Petrequin, C., Chanzy, H., Henrissat, B., Schulein, M. 2001. Optimized mixtures of recombinant Humicola insolens cellulases for the biodegradation of crystalline cellulose. *Biotechnol Bioeng*, **72**(3), 339-45.

Borah, A.J., Agarwal, M., Poudyal, M., Goyal, A., Moholkar, V.S. 2016. Mechanistic investigation in ultrasound induced enhancement of enzymatic hydrolysis of invasive biomass species. *Bioresour Technol*, **213**, 342-9.

Brander, S., Mikkelsen, J.D., Kepp, K.P. 2014. Characterization of an alkali- and halide-resistant laccase expressed in E. coli: CotA from *Bacillus clausii*. *PLoS One*, **9**(6), e99402.

Brandt, A. *et al.*, 2013. Deconstruction of lignocellulosic biomass with ionic liquids. *Green Chemistry*, 15(3), pp.550–583.

Bustamante, M.A., Moral, R., Paredes, C., Perez-Espinosa, A., Moreno-Caselles, J., Perez-Murcia, M.D. 2008. Agrochemical characterisation of the solid by-products and residues from the winery and distillery industry. *Waste Manag*, **28**(2), 372-80.

Cabezas, L., Calderon, C., Medina, L.M., Bahamon, I., Cardenas, M., Bernal, A.J., Gonzalez, A., Restrepo, S. 2012. Characterization of cellulases of fungal endophytes isolated from *Espeletia* spp. *J Microbiol*, **50**(6), 1009-13.

Cadena, E.M., Chriac, A.I., Pastor, F.I., Diaz, P., Vidal, T., Torres, A.L. 2010. Use of cellulases and recombinant cellulose binding domains for refining TCF kraft pulp. *Biotechnol Prog*, **26**(4), 960-7.

Castiglia, D., Sannino, L., Marcolongo, L., Ionata, E., Tamburino, R., De Stradis, A., Cobucci-Ponzano, B., Moracci, M., La Cara, F., Scotti, N. 2016. High-level expression of thermostable cellulolytic enzymes in tobacco transplastomic plants and their use in hydrolysis of an industrially pretreated Arundo donax L. biomass. *Biotechnol Biofuels*, **9**, 154.

Chefetz, B., Chen, Y., Hadar, Y. 1998. Purification and characterization of laccase from Chaetomium thermophilium and its role in humification. *Appl Environ Microbiol*, **64**(9), 3175-9.

Chen, H., Li, X.L., Blum, D.L., Ljungdahl, L.G. 1998. Two genes of the anaerobic fungus Orpinomyces sp. strain PC-2 encoding cellulases with endoglucanase activities may have arisen by gene duplication. *FEMS Microbiol Lett*, **159**(1), 63-8.

Chen, L., Li, A., He, X., Han, L. 2015. A multi-scale biomechanical model based on the physiological structure and lignocellulose components of wheat straw. *Carbohydr Polym*, **133**, 135-43.

Cheng, G., Liu, Z., Murton, J.K., Jablin, M., Dubey, M., Majewski, J., Halbert, C., Browning, J., Ankner, J., Akgun, B., Wang, C., Esker, A.R., Sale, K.L., Simmons, B.A., Kent, M.S. 2011a. Neutron reflectometry and QCM-D study of the interaction of cellulases with films of amorphous cellulose. *Biomacromolecules*, **12**(6), 2216-24.

Cheng, G., Varanasi, P., Li, C., Liu, H., Melnichenko, Y.B., Simmons, B.A., Kent, M.S., Singh, S. 2011b. Transition of cellulose crystalline structure and surface morphology of biomass as a function of ionic liquid pretreatment and its relation to enzymatic hydrolysis. *Biomacromolecules*, **12**(4), 933-41.

Chiranjeevi, T., Rani, G., Chandel, A.K., Sekhar, P.V.S., Prakasham, R.S., Addepally, U. 2012. Optimization of holocellulolytic enzymes production by Cladosporium cladosporioidesusing taguchi-L'16 orthogonal array. *J Biobased Mater Bioenergy*, **6**.

Cruys-Bagger, N., Elmerdahl, J., Praestgaard, E., Borch, K., Westh, P. 2013. A steady-state theory for processive cellulases. *FEBS J*, **280**(16), 3952-61.

Dave, B.R., Sudhir, A.P., Pansuriya, M., Raykundaliya, D.P., Subramanian, R.B. 2012. Utilization of Jatropha deoiled seed cake for production of cellulases under solid-state fermentation. *Bioprocess Biosyst Eng*, **35**(8), 1343-53.

de Cassia Pereira, J., Paganini Marques, N., Rodrigues, A., Brito de Oliveira, T., Boscolo, M., da Silva, R., Gomes, E., Bocchini Martins, D.A. 2015. Thermophilic fungi as new sources for production of cellulases and xylanases with potential use in sugarcane bagasse saccharification. *J Appl Microbiol*, **118**(4), 928-39.

de Vries, M., Scholer, A., Ertl, J., Xu, Z., Schloter, M. 2015. Metagenomic analyses reveal no differences in genes involved in cellulose degradation under different tillage treatments. *FEMS Microbiol Ecol*, **91**(7).

Dios, P., Pernecker, T., Nagy, S., Pal, S., Devay, A. 2015. Influence of different types of low substituted hydroxypropyl cellulose on tableting, disintegration, and floating behaviour of floating drug delivery systems. *Saudi Pharm J*, **23**(6), 658-66.

Elgharbawy, A.A. *et al.*, 2016. Ionic liquid pretreatment as emerging approaches for enhanced enzymatic hydrolysis of lignocellulosic biomass. Biochemical Engineering Journal, 109, pp.252–267.

Fang, X., Qin, Y., Li, X., Wang, L., Wang, T., Zhu, M., Qu, Y. 2010. [Progress on cellulase and enzymatic hydrolysis of lignocellulosic biomass]. *Sheng Wu Gong Cheng Xue Bao*, **26**(7), 864-9.

Ferreira, D.C., Bastos, G.S., Pfeifer, A., Heinze, T., El Seoud, O.A. 2016. Cellulose carboxylate/tosylate mixed esters: Synthesis, properties and shaping into microspheres. *Carbohydr Polym*, **152**, 79-86.

Fontes, C.M., Hall, J., Hirst, B.H., Hazlewood, G.P., Gilbert, H.J. 1995. The resistance of cellulases and xylanases to proteolytic inactivation. *Appl Microbiol Biotechnol*, **43**(1), 52-7.

Ganner, T., Bubner, P., Eibinger, M., Mayrhofer, C., Plank, H., Nidetzky, B. 2012. Dissecting and reconstructing synergism: in situ visualization of cooperativity among cellulases. *J Biol Chem*, **287**(52), 43215-22.

Garajova, S., Mathieu, Y., Beccia, M.R., Bennati-Granier, C., Biaso, F., Fanuel, M., Ropartz, D., Guigliarelli, B., Record, E., Rogniaux, H., Henrissat, B., Berrin, J.G. 2016. Single-domain flavoenzymes trigger lytic polysaccharide monooxygenases for oxidative degradation of cellulose. *Sci Rep*, **6**, 28276.

Garvey, M., Klose, H., Fischer, R., Lambertz, C., Commandeur, U. 2013. Cellulases for biomass degradation: comparing recombinant cellulase expression platforms. *Trends Biotechnol*, **31**(10), 581-93.

Gladden, J.M., Park, J.I., Bergmann, J., Reyes-Ortiz, V., D'Haeseleer, P., Quirino, B.F., Sale, K.L., Simmons, B.A., Singer, S.W. 2014. Discovery and characterization of ionic liquid-tolerant thermophilic cellulases from a switchgrass-adapted microbial community. *Biotechnol Biofuels*, **7**(1), 15.

Glass, N.L., Schmoll, M., Cate, J.H., Coradetti, S. 2013. Plant cell wall deconstruction by ascomycete fungi. *Annu Rev Microbiol*, **67**, 477-98.

Greene, E.R., Himmel, M.E., Beckham, G.T., Tan, Z. 2015. Glycosylation of Cellulases: Engineering Better Enzymes for Biofuels. *Adv Carbohydr Chem Biochem*, **72**, 63-112.

Gubitosi, M., Duarte, H., Gentile, L., Olsson, U., Medronho, B. 2016. On cellulose dissolution and aggregation in aqueous tetrabutylammonium hydroxide. *Biomacromolecules*, **17**(9), 2873-81.

Gurram, R.N., Menkhaus, T.J. 2014. Continuous enzymatic hydrolysis of lignocellulosic biomass with simultaneous detoxification and enzyme recovery. *Appl Biochem Biotechnol*, **173**(6), 1319-35.

Harnpicharnchai, P., Pinngoen, W., Teanngam, W., Sornlake, W., Sae-Tang, K., Manitchotpisit, P., Tanapongpipat, S. 2016. Production of high activity Aspergillus niger BCC4525 beta-mannanase in Pichia pastoris and its application for mannooligosaccharides production from biomass hydrolysis. *Biosci Biotechnol Biochem*, **80**(12), 2298-2305.

Hemsworth, G.R., Henrissat, B., Davies, G.J., Walton, P.H. 2014. Discovery and characterization of a new family of lytic polysaccharide monooxygenases. *Nat Chem Biol*, **10**(2), 122-126.

Hildebrand, A., Kasuga, T., Fan, Z. 2015. Production of cellobionate from cellulose using an engineered Neurospora crassa strain with laccase and redox mediator addition. *PLoS One*, **10**(4), e0123006.

Hilden, L., Johansson, G. 2004. Recent developments on cellulases and carbohydrate-binding modules with cellulose affinity. *Biotechnol Lett*, **26**(22), 1683-93.

Hiras, J., Wu, Y.W., Deng, K., Nicora, C.D., Aldrich, J.T., Frey, D., Kolinko, S., Robinson, E.W., Jacobs, J.M., Adams, P.D., Northen, T.R., Simmons, B.A., Singer, S.W. 2016. Comparative Community Proteomics Demonstrates the Unexpected Importance of Actinobacterial Glycoside Hydrolase Family 12 Protein for Crystalline Cellulose Hydrolysis. *MBio*, **7**(4).

Hussain, P.R., Rather, S.A., Suradkar, P., Parveen, S., Mir, M.A., Shafi, F. 2016. Potential of carboxymethyl cellulose coating and low dose gamma irradiation to maintain storage quality, inhibit fungal growth and extend shelf-life of cherry fruit. *J Food Sci Technol*, **53**(7), 2966-2986.

Jagadeeswaran, G., Gainey, L., Prade, R., Mort, A.J. 2016. A family of AA9 lytic polysaccharide monooxygenases in Aspergillus nidulans is differentially regulated by multiple substrates and at least one is active on cellulose and xyloglucan. *Appl Microbiol Biotechnol*, **100**(10), 4535-47.

Jaeger, V., Burney, P. and Pfaendtner, J., 2015. Comparison of Three Ionic Liquid-Tolerant Cellulases by Molecular Dynamics. *Biophysical Journal*, 108(4), pp.880–892.

Johansen, K.S. 2016. Lytic Polysaccharide Monooxygenases: The Microbial Power Tool for Lignocellulose Degradation. *Trends Plant Sci*.

Kojima, Y., Varnai, A., Ishida, T., Sunagawa, N., Petrovic, D.M., Igarashi, K., Jellison, J., Goodell, B., Alfredsen, G., Westereng, B., Eijsink, V.G., Yoshida, M. 2016. A Lytic Polysaccharide Monooxygenase with Broad Xyloglucan Specificity from the Brown-Rot Fungus Gloeophyllum trabeum and Its Action on Cellulose-Xyloglucan Complexes. *Appl Environ Microbiol*, **82**(22), 6557-6572.

Kuuskeri, J., Makela, M.R., Isotalo, J., Oksanen, I., Lundell, T. 2015. Lignocellulose-converting enzyme activity profiles correlate with molecular systematics and phylogeny grouping in the incoherent genus Phlebia (Polyporales, Basidiomycota). *BMC Microbiol*, **15**, 217.

Kudakasseril Kurian, J. *et al.*, 2013. Feedstocks, logistics and pre-treatment processes for sustainable lignocellulosic biorefineries: A comprehensive review. *Renewable and Sustainable Energy Reviews*, 25, pp.205–219.

Lindman, B. 2016. From surfactant to cellulose and DNA self-assembly. A 50-year journey. *Colloid Polym Sci*, **294**(11), 1687-1703.

Malho, J.M., Arola, S., Laaksonen, P., Szilvay, G.R., Ikkala, O., Linder, M.B. 2015. Modular architecture of protein binding units for designing properties of cellulose nanomaterials. *Angew Chem Int Ed Engl*, **54**(41), 12025-8.

McNamara, J.T., Morgan, J.L., Zimmer, J. 2015. A molecular description of cellulose biosynthesis. *Annu Rev Biochem*, **84**, 895-921.

Moran-Mirabal, J.M., Bolewski, J.C., Walker, L.P. 2011. Reversibility and binding kinetics of Thermobifida fusca cellulases studied through fluorescence recovery after photobleaching microscopy. *Biophys Chem*, **155**(1), 20-8.

Muller, G., Varnai, A., Johansen, K.S., Eijsink, V.G., Horn, S.J. 2015. Harnessing the potential of LPMO-containing cellulase cocktails poses new demands on processing conditions. *Biotechnol Biofuels*, **8**, 187.

Murashima, K., Kosugi, A., Doi, R.H. 2003. Solubilization of cellulosomal cellulases by fusion with cellulose-binding domain of noncellulosomal cellulase engd from Clostridium cellulovorans. *Proteins*, **50**(4), 620-8.

Olsen, J.P., Alasepp, K., Kari, J., Cruys-Bagger, N., Borch, K., Westh, P. 2016. Mechanism of product inhibition for cellobiohydrolase Cel7A during hydrolysis of insoluble cellulose. *Biotechnol Bioeng*, **113**(6), 1178-86.

Payne, C.M., Knott, B.C., Mayes, H.B., Hansson, H., Himmel, M.E., Sandgren, M., Stahlberg, J., Beckham, G.T. 2015. Fungal cellulases. *Chem Rev*, **115**(3), 1308-448.

Qiu, Z., Aita, G.M. and Walker, M.S., 2012. Effect of ionic liquid pretreatment on the chemical composition, structure and enzymatic hydrolysis of energy cane bagasse. *Bioresource Technology*, 117, pp.251–256.

Resch, M.G., Donohoe, B.S., Baker, J.O., Decker, S.R., Bayer, E.A., Beckham, G.T., Himmel, M.E. 2013. Fungal cellulases and complexed cellulosomal enzymes exhibit synergistic mechanisms in cellulose deconstruction. *Energy and Environmental Science*, **6**(6), 1858-1867.

Roberts, S.M., Davies, G.J. 2012. The crystallization and structural analysis of cellulases (and other glycoside hydrolases): strategies and tactics. *Methods Enzymol*, **510**, 141-68.

Ruiz, D.M., Turowski, V.R., Murakami, M.T. 2016. Effects of the linker region on the structure and function of modular GH5 cellulases. *Sci Rep*, **6**, 28504.

Singh, G., Kaur, K., Puri, S., Sharma, P. 2015. Critical factors affecting laccase-mediated biobleaching of pulp in paper industry. *Appl Microbiol Biotechnol*, **99**(1), 155-64.

Sriariyanun, M. *et al.*, 2016. Production, purification and characterization of an ionic liquid tolerant cellulase from Bacillus sp. isolated from rice paddy field soil. Electronic Journal of Biotechnology, 19, pp.23–28.

Textor, T., Derksen, L., Gutmann, J.S. 2016. Employing ionic liquids to deposit cellulose on PET fibers. *Carbohydr Polym*, **146**, 139-47.

van Wyk, N., den Haan, R., van Zyl, W.H. 2010. Heterologous co-production of Thermobifida fusca Cel9A with other cellulases in *Saccharomyces cerevisiae*. *Appl Microbiol Biotechnol*, **87**(5), 1813-20.

Vazana, Y., Morais, S., Barak, Y., Lamed, R., Bayer, E.A. 2010. Interplay between Clostridium thermocellum family 48 and family 9 cellulases in cellulosomal versus noncellulosomal states. *Appl Environ Microbiol*, **76**(10), 3236-43.

Wang, C., Dong, D., Wang, H., Muller, K., Qin, Y., Wang, H., Wu, W. 2016. Metagenomic analysis of microbial consortia enriched from compost: new insights into the role of Actinobacteria in lignocellulose decomposition. *Biotechnol Biofuels*, **9**, 22.

Wang, H.-C., Chen, Y.-C., Hseu, R.-S. 2014. Purification and characterization of a cellulolytic multienzyme complex produced by *Neocallimastix patriciarum* J11. *Biochemical and Biophysical Research Communications*, **451**(2), 190-195.

Warden, A.C., Little, B.A., Haritos, V.S. 2011. A cellular automaton model of crystalline cellulose hydrolysis by cellulases. *Biotechnol Biofuels*, **4**(1), 39.

Xu, J. *et al.*, 2015. A novel ionic liquid-tolerant *Fusarium oxysporum* BN secreting ionic liquid-stable cellulase: Consolidated bioprocessing of pretreated lignocellulose containing residual ionic liquid. *Bioresource Technology*, 181, pp.18–25.

Xu, J. *et al.*, 2014. An ionic liquid tolerant cellulase derived from chemically polluted microhabitats and its application in *in situ* saccharification of rice straw. *Bioresource Technology*, 157, pp.166–173.

Xu, J., Xiong, P. and He, B., 2016. Advances in improving the performance of cellulase in ionic liquids for lignocellulose biorefinery. *Bioresource Technology*, 200, pp.961–970.

Zajc, J., Džeroski, S., Kocev, D., Oren, A., Sonjak, S., Tkavc, R., Gunde-Cimerman, N. 2014. Chaophilic or chaotolerant fungi: a new category of extremophiles? *Frontiers in Microbiology*, 5, 708.

Zhou, F., Olman, V., Xu, Y. 2009. Large-scale analyses of glycosylation in cellulases. *Genomics Proteomics Bioinformatics*, 7(4), 194-9.

5

Cellulosome: A Multi-domain Complex for Effective Biomass Hydrolysis

Salom Gnana Thanga[1], Balagurusamy Nagamani[2] and Kumarasamy Ramasamy[3]

[1]Department of Environmental Sciences, University of Kerala, Thiruvanathapuram, [2]Universidad Autonoma de Coahuila, Mexico [3]Department of Agricultural Microbiology, Tamil Nadu Agricultural University, Coimbatore

1. Introduction

Biological conversion of cellulosic biomass to fuels and chemicals offers the high yields to products vital to economic success and the potential for very low costs. Nevertheless, many microorganisms in nature, mostly bacteria and fungi are capable of producing biomass-degrading enzymes (Yang *et al.*, 2011). Enzymatic hydrolysis is influenced by both structural features of cellulose and the mode of enzyme action. Two types of systems occur in regard to plant cell wall degradation by microorganisms. In one type, the organisms produce a set of free enzymes that act synergistically to degrade plant cell walls. In the second type, the degradative enzymes are organized into an enzyme complex located in cellular surface called the cellulosome. This complex is very effective in degrading plant cell walls (Doi, 2008). The occurrence of a cellulosome was first observed in the thermophilic bacterium

Clostridium thermocellum and has been described in a number of mesophilic anaerobic bacteria and with some anaerobic fungi, particularly *Piromyces* sp. (Doi 2008; Lamed *et al.*, 1987). This paper focuses on the review of the current understanding of key features of cellulosome and its role in biomass hydrolysis.

2. Biomass Hydrolysis

Biomass is primarily the product of CO_2 and light and is produced by the photosynthetic apparatus in bacteria, algae and plants. 2000 Gt (1Gt = 10^9 t) of biomass, dead or living has been calculated to be around the surface of the continents. Plant biomass, the greater part of whole biomass is composed of cellulose, hemicelluloses, lignin, proteins and some other substances and is called lignocellulosic biomass (LCB).Cellulose is the far most abundant carbohydrate and polysaccharide on earth. About 40 Gt of cellulose are produced per year on land. This means that 40 Gt of cellulose are naturally degraded in the biosphere by enzymatic process.

The process of converting the biomass polymers to fermentable sugars is called hydrolysis. There are two major categories of methods employed. The first and older method uses acids as catalysts, while the second uses enzymes called cellulases. Biomass conversion is being expanded beyond fuel production; the concept of biorefinery is being actively pursued so that a wide range of useful materials (for chemicals, energy, food, healthcare and other industry) can be derived from biomass (Sweeney and Xu 2012). It is a major challenge to convert or degrade at industrial scale highly complex and heterogenous lignocellulosic biomass into simple carbohydrate, phenolics, aromatics and other more transformable substances.

Table 5.1: Composition of Lignocellulosic Materials (Kumar *et al.*, 2009)

Lignocellulosic Materials	Cellulose (Per cent)	Hemicellulose (Per cent)	Lignin (Per cent)
Coastal Bermuda grass	25	35.7	6.4
Corn cobs	45	35	15
Cotton seed hairs	80-95	5-20	0
Grasses	25-40	35-50	10-30
Hardwoods stem	40-55	24-40	18-25
Leaves	15-20	80-85	0
Newspaper	40-55	25-40	18-30
Nutshells	25-30	25-30	30-40
Paper	85-99	0	0-15
Softwoods stems	45-50	25-35	25-35
Wheat straw	30	50	15

3. Degradation of Cellulose

The main commercial purpose of enzymatic hydrolysis of cellulosics is to deconstruct cellulose and other carbohydrate polymers into fermentable sugars, including glucose and/or oligomers that can be further converted into valuable products through biological or chemical approaches. Cellulose particles consist of crystalline, paracrystalline (disordered) and amorphous structures. Amorphous cellulose has been reported to be rapidly degraded to cellobiose by cellulases, while the hydrolysis of crystalline cellulose is much slower. Hence, it is proposed that hydrolysis rate depends on cellulose crystallinity (Yang *et al.*, 2011). The definitive enzymatic degradation of cellulose to glucose is generally accomplished by the synergistic action of three distinct classes of enzymes (Himmel *et al.*, 2010):

The endo-β-(1-4)-glucanases or β-(1-4)-D-glucan-4-glucanohdrolases (EC 3.2.1.4), which act randomly on soluble and insoluble β-(1-4)-glucan substrates and are commonly measured by detecting the reducing groups released from carboxymethylcellulose.

The exo-β-(1-4)-D-glucanases, including both the β-(1-4)-D-glucanglucohydrolases (EC 3.2.1.74), which liberate D-glucose from β-(1-4)-D-glucans and hydrolyze D-cellobiose slowly, and β-(1-4)-D-glucancellobiohydrolase (EC 3.2.1.91), which liberates D-cellobiose in a 'processive' manner (successive cleavage of product) from β-(1-4)-glucans;

The β-D-glucosidases or β-D-glucosideglucohydrolases (EC 3.2.1.21), which act to release D-glucose units from cellobiose and soluble cellodextrins, as well as an array of glycosides.

4. Microbiology of Cellulose Degradation

Lignocellulosic biomass consists of morphologically different cellulose, structurally and compositionally complex hemicellulose, recalcitrant lignin, diverse proteins, different lipids and other substances that interact with each other. Since cellulose is very difficult to degrade as a component of plant cell walls, only a few microorganisms specialized for plant cell wall degradation can hydrolyze cellulose. Among these, anaerobic and aerobic genera of Domain bacteria and fungi of Domain Eukarya are included.Anaerobic and aerobic bacteria have different strategies to degrade cellulolytic substrate; whereas anaerobic bacteria degrade cellulose, using cellulosome. Aerobic bacteria secrete enzymes capable of degrading cellulose that freely diffuse to reach the substrate.

Anaerobic bacteria of the order Clostridiales (phylum Firmicutes) are generally found in soils, decaying plant waste, the rumen of ruminant animals, composts, waste water and wood processing plants; these bacteria have also been found in insects like termites (Isopteran) bookworm (Lepidopteran) and so, in a symbiotic relationship in their guts responsible for cellulosic feed digestion. Anaerobic hydrolysis represents 5 to 10 per cent of global cellulose degradation (Ransom-Jones *et al.*, 2012; Shwarz 2001).Some anaerobic bacteria with cellulolytic activity are *Butyrivibriofibrisolvens, Fibrobactersuccinogens, Ruminococcusflavifaciens, Clostridium cellulovorans, C.cellulolyticum* and *C.thermocellum* (Lin and Thomson 1991; Murty and Chandra 1992). Adhesion represents a major step in the colonization of cellulose by anaerobic bacterial It is considered to be a prerequisite for the digestion of complex substrates such as plant cell wall or cellulose. Ramasamy *et al.* (1992) characterized anaerobic cellulolytic bacteria associated with adsorption to biogas slurry solids. Nagamani and Ramasamy (1999) opined that binding of bacteria to cellulose has direct relation to biogas production. The adherence studies of three cellulolytic anaerobic bacteria (*Acetivibrio* sp., *Bacteroides*sp. and *Clostridium* sp.) revealed that the extend to which these bacteria adhere to the substrate depends on the type of cellulosic substrate (Thanga Vincent and Ramasamy 1996; 1998; 2001).

Due to significant diversity in the physiology of cellulolytic bacteria, sometimes is difficult to classify bacteria as mentioned above, therefore, on this basis, they can be placed into three diverse physiological groups (1) fermentative anaerobes, typically gram-positive (Clostridium and Ruminococcus), but with a few gram negative species (Butyvibrio and Acetivibrio) that are phylogenetically related to the Clostridium assemblage of Fibrobacter; (2) aerobic gram-positive bacteria (Cellulomonas and Thermobifida) and (3) aerobic gliding bacteria (Cytophaga and Sporocytophaga) (Lynd *et al.*, 2002; Ransem-Jones *et al.*, 2012). Moreover, the capability of anaerobic fungi to utlize a broad range of plant biomass including lignocellulosic biomass substrates renders anaerobic fungi appealing candidates for utilization in biomass conversion (Couger *et al.*, 2015).

5. Cellulose Degrading Strategy by Cellulolytic Organisms: Cellulases and Cellulosomes

Different strategies for the cellulose degraders are used by the cellulose producing microorganisms : aerobic bacteria and fungi secrete soluble extracellular enzymes known as non-complexedcellulase system; anaerobic cellulolytic microorganisms produce complexed cellulose systems, called cellulosomes.

5.1. Non Complexed Cellulase System

One of the fully investigated non-complexedcellulase system in the *Trichodermareesei* model.*T. reesei* is a saprobic fungus, known as an efficient producer of extracellular enzymes (Bayer *et al.*, 1988). Its non-complexedcellulase system includes two cellobiohydrolases, at least seven endglucanases and several β-glucosidases. However, in *T.reesei* cellulases, the amount of β-glucosidases is lower than that needed for the efficient hydrolysis of cellulose into glucose. As a result, the major product of hydrolysis is cellobiose. This is a dimer of glucose with strong inhibition toward endo- and exo-glucanases so that the accumulation of cellobiose significantly slows down the hydrolytic process (Gilkes *et al.*, 1991).

5.2. Complexed Cellulase System: The Cellulosomes

When bacteria adhere to the substrate, a complex of enzymes located on the bacterial cell wall surface can be presented to the surface in optimal configuration. In a cellulolytic thermophile, such a complex, consisting of both enzymes and specific adhesion proteins is known as the cellulosome (Lamed *et al.*, 1983). Cellulosome are large extracellular enzyme complexes capable of degrading cellulose, hemicelluloses and pectin; they may be the largest extracellular enzyme complexes found in nature, although the individual cellulosome size range from 0.65 MDa to 25 MDa, some polycellulosome have been reported to be as large as 100 MDa (Doi and Kosugi 2004). Cellulosomes are produced mainly by anaerobic bacteria, but their presence have also been described in few anaerobic fungi from species such as *Neocallimastix*,

Piromyces and *Orpinomyces* (Tatsumi *et al.*, 2006; Watanabe and Tokuda 2010). In the domain bacteria, organisms possessing cellulosomes are only found in this phylum Firmicutes, class clostridia, order Clostridiales and in the Lachnospiraceae and Clostridium families.

Cellulosomes are protrubances produced on the cell wall of the cellulolytic bacteria grown on cellulosic materials. These protrubences are stable enzyme complexes tightly bound to the bacterial cell wall, but flexible enough to bind strongly to cellulose (Lenting and Warmoeskerken 2001). A cellulosome consists of two types of subunits : non-catalytic subunits, called scaffoldins and enzymatic subunits. The scaffoldin is a functional unit of cellulosome, which contain multiple copies of cohesions that interact selectively with domains of the enzymatic subunits, CBD (Cellulose binding domains) and CBM (cellulose binding molecules). These have complementary cohesions called dockerins, which are specific for each bacterial species (Gillian and Reese 1954; Lynd *et al.*, 2002; Arai *et al.*, 2006).

The concept of cellulosome was firstly discovered in the thermophilic cellulolytic and anaerobic bacterium *Clostridium thermocellum* (Wyman 1996). It consists of a large number of proteins including several cellulases and hemicellulases. Other enzymes that can be included in the cellulosomes are lichenases. For the bacterial cells, the biosynthesis of cellulosome enables a specific adhesion to the substrate of interest without competition with other microorganisms. The cellulosome allows several advantages 1) synergism of the cellulose 2) absence of unspecific adsorption (McCarter and Whiters 1994; Zhang and Lynd 2004). Thanks to its intrinsic lego-like architecture, cellulosome may provide great potential in the biofuel industry. Free enzymes are more active on pretreated biomass; in contrast, cellulosomes are much more active on purified cellulose. Although free cellulases and cellulosomes employ very different physical mechanisms to break down recalcitrant polysaccharides, when combined, these systems display dramatic synergistic enzyme activity on cellulose (Resch *et al.*, 2013).

6. Structure of Cellulosome

The cellusosome structure is characterized by two components (a) the non-enzymatic scaffolding proteins with enzyme binding sites called cohesions (b) enzymes with dockerins proteins interacting with cohesions in the scaffolding protein (Figure 5.1). Depending on the bacterial species, the scaffolding protein varies in the number of cohesions and cellulose binding modules (CBM) that binds the cellulosome tightly to the substrate and concentrates the enzymes to a particular site of the substrate. A more complex cellulosome structure with multiple scaffolding proteins that allows the binding of as much as enzymes has been revealed (Doi 2008). The underlying mechanism for cellulosome's high catalytic efficiency and overall stability are identified to be spatial proximity of the enzymes and the enzyme-substrate targeting (Dou *et al.*, 2015).

The cohesion-dockerin interconnect the different scaffolding component, whereby the specificities among the individual cohesion-dockerin complexes dictate the overall supramolecular architecture of the participating components (Bayer *et al.*, 2008).In short, the enzymatic cellulosome system may exceed the potential of

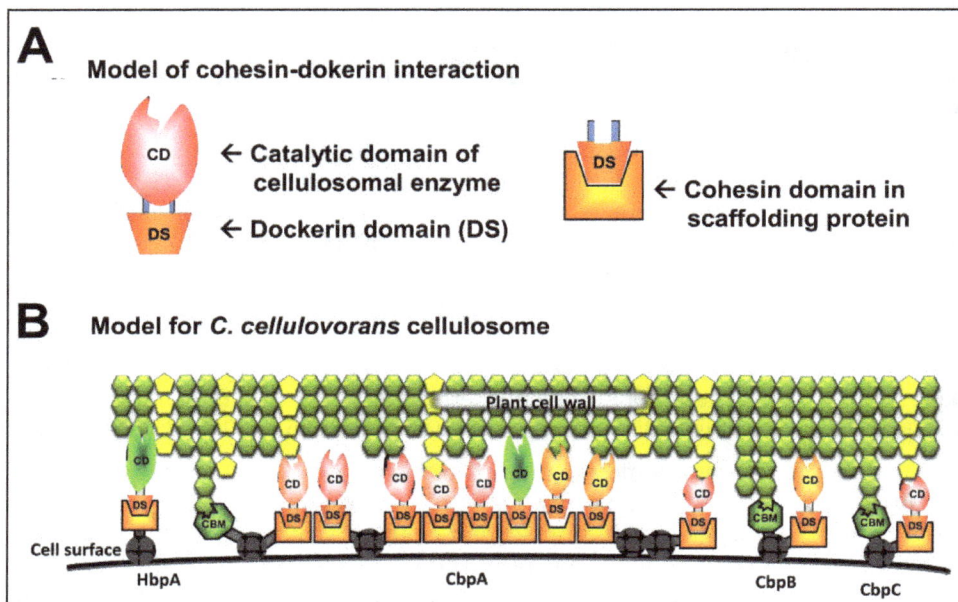

Figure 5.1: Model for *C. cellulovorans* Cellulosome (A) Model of Cohesion-Dockerin Interactions (B) Recent Model of Cellulosomes Attached to its Substrate and Cell Surface (Tamaru and Lopez-Contreras 2013).

non-cellulosomal degradative system due to its structural organization, efficient binding to the substrate, variety of hydrolytic enzymes act synergistically (Doi, 2008). Cellulosomes are protein networks designed by nature to degrade lignocellulosic biomass. These networks comprise intricate assemblies of conserved subunits including catalytic domains, scaffold proteins, carbohydrate binding modules (CBMs), cohesions (Cohs), dockerins (Docs) and X-modules (XMods) of unkown function (Schoeler 2014).

7. Carbohydrate Binding Modules (CBMs)

The cellulosome contains a noncatalytic subunit called scaffoldin that binds the insoluble substrate via a cellulose-specific carbohydrate-binding module (CBM). Cellulases cannot work on crystalline cellulose without the aid of binding modules. A wide range of CBMs have been identified : 28 families, from 75 to 150 amino acid residues. They usually bind the substrate with an arrangement of hydrophobic tryptophane residues. They are either in the loose vicinity of the catalytic module, connected by a flexible arm or closely attached to it. The CBMs may bind to flat crystal surfaces or bind single cellulose molecules (Carrard *et al.*, 2000). Some CBMs are tightly attached to the catalytic module and seem to have lost their binding capacity-they rather perform a stabilizing function. Another CBM is connected to the scaffoldin of the cellulosome and obviously holds the huge complex on the surface of the substrate (Irwin *et al.*, 1998). The *C. thermocellum* scaffoldin contains a set of nine subunit-binding modules coined cohesions that mediate the specific incorporation and organization of the catalytic subunits through a complementary binding module (dockerin) that is carried by each enzymatic subunit. The scaffoldin contains another type of dockerin (type II) at its C terminus that mediates the attachment of the cellulosome to the cell wall through a selective binding interaction with a set of cell-anchoring proteins (Gefen *et al.*, 2012). The major role of the CBMs might be to hold the catalytic module in the close vicinity of the substrate, thus increasing the local substrate concentration.

There is much interest in exploiting the properties of cellulosomes for practical purposes. The specific cohesion-dockerin interaction, the strong cellulose-binding property of the CBD domain, the potential for transforming non-cellulose degraders to cellulose degraders and the construction and use of 'designer' cellulosomes for specific for specific degradation activities are important properties of the cellulosome that can be used in biotechnology (Doi and Kosugi 2004). The detailed understanding of cellulosomal network components may help in the development of biocatalysts for production of fuels and chemicals from renewable plant-derived biomass.

8. Summary and Conclusion

Lignocellulosic biomass such as agricultural, industrial and forestry residues constitute renewable and abundant resources with greatpotential for a low-cost and uniquely sustainable bioconversion to value-added products. Thus, many organic cuels and chemicals that can be obtained from lignocellulosic biomass can reduce greenhouse gas emissions, enhance energy security, improve the economy, dispose of problematic solid wastes, and improve air quality. In nature, a wide varietyof microorganisms bacteria and fungi have the ability to degrade lignocellulosic biomass to C-5 or C-6 sugars. Several clostridial species produce an extracellular enzyme complex called the cellulosomes. The cellulosomes are particularly designed for efficient degradation of plant cell wall polysaccharides such as cellulose, hemicelluloses and pectins. Cellulosomes with specific dockerin-binding capacities can be created by using cohesions from various scaffoldings. By creating chimeric scaffolding proteins with tandem dockerin-specific cohesions, it should be possible to construct designer cellulosomes and mini-cellulosomes with enzymatic pathways in which efficient substrate channeling occurs. This could lead to the development of complex and sophisticated bioreactors and biosensors.

References

Arai,T., Kosugi,A., Chan,H., Koukiekolo,R., Yukawa,H., Inui,M. and Doi,R.H. 2006. Properties of cellulosomal family 9 cellulases from *Clostridium cellulovorans*. *Applied Microbiology and Biotechnology*, 71 (5), 654-600.

Bayer,E.A., Belaich,J.P., Shoham,Y. and Lamed,R. 2004. The Cellulosomes: multienzyme machine for degradation of plant cell wall polysaccharides. *Annual Review of Microbiology*,58, 521-54.

Bayer,E.A., Chanzy,H., Lamed,R., Shoham,Y. 1998. Cellulose, cellulases and cellulosomes.*Current Opinion in Structural Biology*, 8(5), 548-557.

Carrard,G., Koivula,A., Soderlund,H. and Beguin,P. 2000. Cellulose-binding domains promote hydrolysis of different sites on crystalline cellulose. *Proceedings of National Academy of Sciences*, USA, 97, 10342-10347.

Couger,M.B., Youssef,N.H., Struchtemeyer,C.G., Liggenstoffer,A.S. and Elshahed,M.S. 2015. Transcriptomicanalysis of lignocellulosic biomass degradation by the anaerobic fungal isolate *Orpinomyces* sp. strain C1A. Biotechnology for Biofuels, 8, 208.

Doi,R. 2008. Cellulases of mesophilic microorganisms: cellulosome and non-cellulosome producers. Annals of the New York Academy of Sciences, 1125, 269-79.

Doi,R. and Kosugi,A. 2004. Cellulosomes: plant cell wall degrading enzyme complexes. *Nature Reviews Microbiology*.2, 541-51.

Dou,T., Luan,H., Ge,G., Dong,M., Zou,H., He,Y., Cui,P., Wang,J., Hao,D., Yang,S and Yang,L. 2015. Functional and structural properties of a novel cellulosome like multienzyme complex: efficient glycoside hydrolysis of water-insoluble 7-xylosyl-10-deacetylpaclitaxel. *Science Reports*,5, 13768.

Gefen,G., Anbar,M., Morag,E., Lamed,R. and Bayer,E.A. 2012. Enhanced cellulose degradation by targeted integration of a cohesion-fused B-glucosidase into the Clostridium thermocellumcellulosome. *Proceedings of National Academy of Sciences*, 26, 10298-10303.

Gilkes,N.R., Henrissat,B., Kilburn,DG., Miller,R.C.Jr and Warren,R.A.J. 1991. Domains in microbial 1,4-glucanases: sequence conservation, function and enzyme families. *Microbiological Reviews*, 55(2), 303-315.

Gilligan,W. and Reese,R.T. 1954.Evidence for multiple components in microbial cells. *Canadian Journal of Microbiology*, 1(2), 90-107.

Himmel,M.E., Xu,Q., Luo,Y., Ding,S., Lamed,R. and E.A.Bayer. 2010. Microbial enzyme systems for biomass conversion: emerging paradigms, *Biofuels*, 1(2), 323 – 341.

Irwin,D., Shin,D,H., Zhang,S., Barr,B.K., Sakon,H., Karplus,P.A. and Wilson,D.B. 1998. Roles of the catalytic domain and two cellulose binding domains of *Thermomonosporafusca* E4 in cellulose hydrolysis. *Journal of Bacteriology*, 180, 1709-1714.

Kumar,P., Barrett,D.M., Delwiche,M.J. and Stroeve,P. 2009. Methods for pretreatment of lignocellulosic biomass for efficient hydrolysis and biofuel production. *Industrial and Engineering Chemistry Research*, 48 (8) 3713-3729.

Lamed,R. Naimark,J., Morgenstern,E., and Bayer,E.A. 1987. Specialized cell surface structures in cellulolytic bacteria. Journal of Bacteriology, 169, 3792-800.

Lamed,R., Setter,E., Bayer,E.A. 1983. Characterization of a cellulose-binding cellulose-containing complex in *Clostridium thermocellum. Journal of Bacteriology*, 156, 828-836.

Lenting,H.B.M. and Warmoeskerken,M.M.C.G. 2001. Mechanism of interaction between cellulase action and applied shear force, an hypothesis. *Journal of Biotechnology*, 89 (2-3), 217-226.

Lin,L. and Thomson,J. 1991. An analysis of the extracellular xylanases and cellulases of *Butyrivibriofibrisolvens* H17c.*FEMS Microbiology Letters*.84:197-204.

Lynd,L.R., Weimer,P.J., van Zyl,W.H. and Pretorius,I.S. 2002. Microbial cellulose utilization: Fundamentals and Biotechnology. *Microbiology and Molecular Biology Reviews*.66(3), 506-577.

McCarter,J.D. and Whiters,S.G. 1994. Mechanisms of enzymatic glycoside hydrolysis. *Current Opinion in Structural Biology*, 4(6); 885-890.

Murty,M.V.S. and Chandra,T.S. 1991. Purification and properties of an extra cellular xylanase enzyme of Clostridium strain SAIV. Antonie van Leewenhoek.1, 35-41.

Nagamani,B. and Ramasamy.K. 1999. Biogas Production Technology, An Indian Perspective, *Current Science*, 77.

Ramasamy,K., Kalaichelvan,G. and Nagamani,B. 1992. Working with anaerobes: Methanogens – A laboratory manual. Fermentation laboratory, Tamil Nadu Agricultural University, Coimbatore, India, 87p.

Ransom-Jones,E., Jones,D., McCarthy,A. and McDonald,J. 2012. The Fibrobacteres: an important phylum of cellulose-degrading bacteria. *Microbial Ecology*, 63:267-81.

Resch,M.G., Donohoe,B.S., Baker,J.O., Decker,S.R., Bayer,E.A>, Beckham,G.T., Himmer,M.E. 2013. Fungal cellulases and complexedcellulosomal enzymes exhibit synergistic mechanism in cellulose deconstruction. *Energy and Environmental Science*, 6, 1858-1867.

Schoeler,C., Malinowska,K.H., Bernardi,R.C., Milles,L.F., Jobst,M.A., Durner,E., Ott,W., Fried,D.B., Bayer,E.A., Schulten,K., Gaub,H.E., and Nash,M.A. 2014. Ultrastablecellulosome-adhesion complex tightens under load. *Nature Communications*, 5, 5635.

Sweeney,M.D. and Xu,F. 2012. Biomass converting enzymes as industrial biocatalysts for fuels and chemicals : Recent Developments. *Catalysts*, 2, 244-263.

Tamaru,Y. and Lopez-Conterars. 2013. Lignocellulosic biomass utilization toward biorefinery using mesophilic *Clostridial* species. In: Cellulose – Biomass Conversion. Imtech publishers.

Tatsumi,H., Katano,H. and Ikeda,T. 2006. Kinetic analysis of enzymatic hydrolysis of crystalline cellulose by cellobioydrolase using an amperometric biosensor. *Analytical Biochemistry*, 357 (2), 257-261.

Thanga Vincent,S.G. and Ramasamy, K. (2001). Cellulose binding proteins of cellulolytic anaerobic bacteria. Asian Journal of Microbiology, *Biotechnology and Environmental Sciences* 4, 25-29.

Thanga Vincent,S.G. and Ramasamy, K. (1998). Bacterial adherence to cellulose : Can it be capitalized? *Madras Agricultural Journal* 85, 1-16.

Thanga Vincent,S.G. and Ramasamy, K. (1996). Cellulolysis in relation to adherence of anaerobic cellulolytic bacteria. *Madras Agricultural Journal* 83, 375-380.

Watanabe,H. and Tokuda,G. 2010. Cellulolytic system in insects. *Annual Review of Entomoloty*, 55, 609-632.

Wyman,C.E. 2007. What is (and is not) vital to advancing cellulosic ethanol.*Trends in Biotechnology*, 25(4), 153-157.

Yang,B., Dai,Z., Ding,S., Wyman,C.E. 2011. Enzymatic hydrolysis of cellulosic biomass.*Biofuels*. 2, 421-450.

Zhang,Y.P and Lynd,L.R. 2004. Toward an aggregated understanding of enzymatic hydrolysism of cellulose: noncomplexed cellulose systems. *Biotechnology and Bioengineering*, 88(7), 797-823.

6

Thermophilic Cellulases: Novel Agents for Enhanced Saccharification in Biorefinery Application

R. Priyadharshini[1,2], P.Gunasekaran[2,3],
K. Ramasamy[1] and U. Sivakumar[1]

[1]Department of Agricultural Microbiology,
Tamil Nadu Agricultural University, Coimbatore – 641 002
[2]Department of Microbiology, School of Biological Science,
Madurai Kamaraj University, Madurai – 625 021
[3]Vellore Institute of Technology, Vellore – 632 017

Introduction

Biomass derived from trees, agro-forest residues, grasses, plants, aquatic plants and crops are versatile and important renewable feed stock for chemical industry. Through photosynthetic process, plants convert carbon dioxide and water into primary and secondary metabolites. Primary metabolites are carbohydrate (simple sugar, cellulose, hemicellulose, starch, etc.) and lignin called lignocellulose present in high volume in biomass (Naik *et al.*, 2010). Lignocellulosic biomass (LCB) is one of earth's most abundant resources on earth.LCB is primarily composed of cellulose (40-50 per cent), hemicellulose (25-35 per cent), lignin (15-20 per cent) (Alonso *et al.*, 2010, Kumar *et al.*, 2009),with small quantities of pectin, protein, extractives (nonstructural sugars, nitrogenous material, chlorophyll and waxes) and ash. The compositions of LCB vary significantly depending on types and geographical origin. The degradation of plant biomass is an expensive process, which currently requires 3 steps: i) physicochemical pretreatments, ii) enzymatic hydrolysis and iii) fermentation. Among these processes, cellulases have a potential central role in the bioconversion of renewable LCB involving the hydrolysis in order to produce the simple monomeric glucose from cellulose. Such enzymatic hydrolysis or saccharification is one of the most critical in lignocellulosic utilization and represents one of the main technology development areas.

LCB processing includes pretreatment followed by saccharification which is conventionally accomplished through physical, chemical, and biological methods. However, saccharification by cellulases is always the main bottleneck for commercialization. Although physical and chemical methods can effectively hydrolyze lignocellulose, most efficient method of LCB hydrolysis is through enzymatic saccharification using cellulases. Current industrial processes are performed under harsh conditions, including extremely high and low temperatures, acidic or basic pHs, and elevated salinity where the standard enzymes are easily denatured. In many industries, traditional chemical solutions are still the only viable option and hence there is a clear need for more sustainable and environmentally friendly methods to replace the current potentially harmful chemical processes. The identification of more novel enzymes with properties that can cope with industrial processing conditions is the key to the future of biocatalysis (Sarmiento *et al.*, 2015). The unique properties of microbial enzymes, *e.g.*, diversity, consistency, reproducibility, high yields, and economic feasibility have elevated their biotechnological interest and application to different industrial areas (Gurung *et al.*, 2013). Significant information has also been gained about the physiology of thermophilic microbes producing cellulases and process development for enzyme production and biomass saccharification. The main focus of this chapter is on thermophilic enzymes from microbes and their advantage, some highlights on improvement of thermostability of cellulases is also discussed.

Cellulases: Types and Action

Enzymes that catalyze the depolymerization of cellulose are broadly classified as cellulases. For the conversion of LCB into value-added products, cellulose needs to be depolymerized, which can be attained by the combined application of three types of cellulose-degrading enzymes (cellulases): Endocellulases or endoglucanases (EC 3.2.1.4) randomly cleave internal β-1,4 linked glycosidic bonds, primarily at the amorphous regions of the polymer. The generated chain ends are processed by exocellulases or exoglucanases including cellobiohydrolases (CBHs) (EC 3.2.1.91), yielding mainly disaccharide units. Finally, β-glucosidases (BG) (EC 3.2.1.21) hydrolyze the remaining glycosidic bonds to form glucose, which can then simultaneously or in a separate step converted to the desired product by a fermenting microorganism or enzyme (Lynd *et al.*, 2002).

Enzymes that mediate the release of monosaccharides from cellulose and hemicellulose belongs to Glycoside hydrolases(GHs). GHs are a large class of enzymes that exhibit both broad and stringent substrate specificities. GHs selectively catalyze reactions that produce smaller carbohydrate units from polysaccharides.

These enzymes are exquisite catalysts that accelerate the rate of hydrolysis of glycosidic linkages by up to 17 orders of magnitude over the uncatalyzed hydrolysis (Wolfenden *et al.*, 1998). Polypeptides associated with plant cell wall hydrolysis commonly harbor a catalytic GH domain and a carbohydrate-binding module (CBM). The catalytic modules of cellulases have been classied into numerous families based on their amino acid sequences and crystal structures (Henrissat, 1991).The endoglucanases are found in GH families 5-9, 12, 44, 45, 48, 51, 61 and 74. The exoglucanases belongs to GH families 5, 6, 7, 9 and 48. The β-glucosidases have been classified into GH families 1, 3, 5, 9, 30 and 116.There are multifunctional GHs that possess cellulases, hemicellulases, pectinase activities and many more specificities have been identified. Cellulases are produced by fungi, bacteria, protozoans, plants, and animals. Microbial enzymes have certain advantages like ease of microbe culturing to produce more enzymes within a short time and the microbial proteins are stable when compared to enzymes derived from plant or animal sources (Headon *et al.*, 1994).Hence most of the commercial enzymes are derived from the microorganisms (Walsh *et al.*, 1994).

Of the cellulases, the endoglucanases also termed as primary cellulases, are most crucial in enabling their host organism to efciently utilize crystalline cellulose. Primary cellulolytic enzymes are dened as those that contain a catalytic domain and carbohydrate-binding domains and can efciently hydrolyze crystalline cellulose. Such enzymes often contain a second catalytic domain of different activity, and/ or specicity and domains of unknown function may also be present. While many 'cellulases' have been reported from thermophilic organisms, very few are of the primary type. Two different types of primary cellulolytic enzymes are currently known and both are found in thermophilic anaerobes: those contained in the cellulosome and the free-acting cellulases (Sara *et al.*, 2008, 2014).Cellulases contain noncatalytic CBMs and/or other functionally known or unknown modules, which may be located at the N- or C-terminus of a catalytic module. To hydrolyze and metabolize insoluble cellulose, the microorganisms must secrete the cellulases (possibly except BG) that are either free or cell-surface-bound (Bayer *et al.*, 2004).

Cellulosome

Many anaerobic bacteria and fungi secrete a high molecular mass multi-protein complex called the cellulosome, initially discovered in *Clostridium thermocellum* (Doi, 2008, Gilbert, 2007). More than 30 years ago, observation of the adherence of *C. thermocellum* to insoluble cellulose, even when grown under conditions of constant agitation, was noted, and subsequently the mechanism was characterized by Bayer *et al.* (1983). The complex responsible was isolated (Lamed *et al.*, 1983a) and described as a 'cellulosome' (Lamed *et al.*, 1983b). The *C. thermocellum* cellulosome (Ct-cellulosome) has been shown to be capable of efficiently degrading plant cell wall polysaccharides, including crystalline cellulose. Most cellulosomes display a range of plant-biomass deconstruction catalytic activities, including xylanase, mannanase, arabino furanosidase, lichenase, and pectate lyase, in addition to endoglucanase and exoglucanase.

Many members of the order Clostridiales possess the machinery to produce cellulosomes, although *C. thermocellum* and *C. clariflavum* are the only two described cellulosomal species isolated capable of efficient crystalline cellulose hydrolysis at higher temperatures ($T_{opt} \geq 55$–60 °C). The key protein is an enzymatically inactive scaffoldin (CipA) composed of nine highly similar type I cohesin domains (cohI), a C-terminal type II cohesin domain (cohII) that interacts with cell surface proteins, and an internal family CBM3 that binds the cellulosome to cellulose. Assembly of the complex occurs by a specic high-afnity interaction between CohI domains and special dockerin domains (DD), present as part of the catalytic subunits. The cellulosome is simultaneously bound to cellulose and to the cell wall providing efcient consumption of hydrolysis products. The mechanism of incorporation of catalytic subunits into the cellulosome is not clear and different types of cellulosomes are produced. For example, the genome of *C. thermocellum* encodes over 70 DD-containing components with several overlapping activities (Zverlov *et al.*, 2005), in addition to 20 non-cellulosomal glycosyl hydrolase-type enzymes (Berger *et al.*, 2007). Nevertheless, the cellulosome is very efficient in hydrolysis of crystalline cellulose. Its components display signicant synergism since the individual proteins have low activity against cellulose, presumably because they lack the CBMs necessary to bring catalytic sites into close proximity to the insoluble substrate. Cellulosomes are produced by anaerobic bacteria and anaerobic fungi, but are also in some aerobic microorganisms.

Thermophilic Cellulases: Advantages and Challenges

A great deal of attention has been focused on proteins from hyper-thermophiles as thermostable enzymes have gained importance in biotechnological processes. From an industrial viewpoint, hyper-thermophilic enzymes possess certain advantages over their mesophilic counterparts. These enzymes are active and efficient under high temperatures, extreme pH values, high substrate concentrations and high pressure. They are also highly resistant to denaturing agents and organic solvents (Unsworth *et al.*, 2007). In addition, hot extremozymes perform faster reactions and are easier to separate from other heat-labile proteins during purification steps. Due to their overall activity and stability at high temperatures, hyper-thermophilic enzymes are attractive for several important industrial activities. Higher process temperatures minimize the potential for biological contamination, reduce viscosity of substrate and product streams, and increase solubility or bioavailability of substrates that can increase reaction rates (Egorova and Antranikian, 2005).

Temperature might also be used to accelerate lignocellulose deconstruction in a thermophilic bioreactor. Thermophilic bioprocessing will act as a mild hot water pretreatment, and in one reported case, it acts in concert with the organism to solubilize all components of plant biomass, including lignin. Another potential use of thermophilic processes is *in planta* expression of thermostable biomass-degrading enzymes by the plant feed stocks. Here, the biomass feedstock would be brought up to the optimal temperature of the transgenic enzyme(s) to facilitate saccharication prior to or during fermentation.

Biological production of fuels and chemicals from plant biomass involves one of three described processing strategies: separate hydrolysis and (co-)fermentation (SH[c]F), simultaneous saccharication and (co-)fermentation (SS[c]F), or consolidated bioprocessing (CBP). Multiple vessels are often required for saccharication and fermentation in the SHF scheme, thereby incurring capital costs above that for SSF, which combines saccharication and fermentation into one bioreactor. Overall, the economic costs of procuring or producing separate enzymes for SHF and SSF cannot be ignored (Lynd *et al.*, 2008) and are one of the main drivers behind developing commercial technology that combines enzyme production, hydrolysis, and fermentation into one reactor vessel, that is, CBP. Other advantages of using thermophiles in SSF or CBP are that the optimal temperatures of the enzymes and fermentative organism are closely matched boosting the efciency of saccharication, in addition to concurrent saccharication and fermentation relieving product inhibition by mono- and disaccharides. Enhanced thermal stability of liberated oligosaccharides is also a benet of concurrent saccharication and fermentation because liberated oligosaccharides have been shown to better promote growth at very high temperatures (Driskill *et al.*, 1999).

Drawbacks of Thermophilic Cellulases

Some drawbacks of thermophilc enzyme include the difficulty to obtain pure enzymes from source microorganisms and their yields are typically low when cultivated on a large scale for industrial applications. Efforts to overcome these problems have focused on cloning and expressing hyper-thermophilic enzymes in mesophilic hosts without losing their activity and thermostability. Thermostable enzymes expressed in mesophilic hosts can typically be purified easily and the degree of purity obtained is suitable for industrial applications. Working with extremophiles and/or extremozymes requires the adaptation and creation of new methodologies, assays, and techniques that operate under non-standard conditions. Many of the tools that are currently used in classical microbiology and biochemistry experimentation cannot be applied toward extremophilic research because they do not possess the chemical and/or mechanical properties to withstand extreme conditions (Turner *et al.*, 2007). Similarly, techniques for researching common microorganisms need to be further adjusted to fit the requirements of extremophiles. A classic example is the plating of hyper-thermophiles on a solid surface. Conventional streaking on agar-based media is impracticable because agar melts and water evaporates quickly at such high temperatures. Alternative solidifying agents are used to grow thermophiles and hyperthermophiles, such as silicagel, starch, and Gelrite, a low-acetyl gel langum made from *Pseudomonas*. Additionally, the large technical gap between producing an enzyme under laboratory conditions and obtaining a final product to commercial industrial scale is still a problem for the development of novel biocatalysts. Several scientific challenges need to be solved before fully realizing the potential of extremozymes. Therefore, improved cultivation approaches have to be applied, such as providing the most environmentally close conditions for cultivation (Kublanov *et al.*, 2009), utilization of novel substrates and/or electron acceptors, presence or absence of growth factors, as well as the inhibition of cultured fast-growing microorganisms.

Thermophiles and Thermophilic Cellulases from Bacteria, Archaea and Fungi

To produce a wide variety of cellulases and hemicellulases, both fungi and bacteria have been heavily exploited. However, the focus has been more towards the fungi because of their capacity to excrete abundant amount of non-complex cellulases and hemicellulases. Recently, this trend is shifting towards bacteria, due to their higher growth rates, presence of more complex multi-enzymes and their presence in wide variety of environmental niches. Not only can these bacteria survive the harsh conditions, but they often produce stable enzymes which may increase rates of bioconversion processes (Acharya and Chaudhary,2012).

Thermophilic Cellulases of Bacterial Origin

Since lignocellulosic biomasses are highly variable from site to site and even season to season, the most attractive biomass conversion technologies will be those that are insensitive to uctuations in feedstock and robust in the face of biologically challenging process-operating conditions. Given these specications, it makes sense to consider microorganisms that grow at extreme temperatures, and enzymes derived from them, for key roles in biomass conversion processes. Several thermophilic bacteria contain 'free-acting' cellulases that are not part of a cellulosome complex. These include *Ca. saccharolyticus* (T_{opt} 70°C) (Rainey *et al.*, 1994), the genome of which was recently sequenced, and the most thermophilic cellulose-degrading organism known to date, *Anaerocellum thermophilum* (T_{opt} 75°C). The cellulases of these two extreme thermophiles are multi-domain and multi-functional. They contain CBMs of different families often duplicated and in some cases two catalytic domains of different function and/or activity. For example, the *Ca. saccharolyticus* genome encodes a putative bifunctional cellulase CelB, which is composed of an N-terminal endoglucanase catalytic domain of family GH10, a triplet of CBM3, and a C-terminal exocellulase catalytic domain of family GH5. This molecular complex combines all necessary enzymatic components and CBMs needed to hydrolyze crystalline cellulose and a family CBM3 needed to bind to cellulose. CelA of this organism has a similar domain arrangement: a GH9 endocellulase domain, a triplet of CBM3s, and a GH48 exocellulase; the recombinant version displayed endoglucanase activity (Teo *et al.*, 1995). The presence of CelA and CelB in *Caldicellulosiruptor saccharolyticus* is thought to be responsible for its growth on crystalline cellulose.

One cellulolytic enzyme (CelA) isolated from the *A. thermophilum* is very similar in domain structure to CelA of *Ca. saccharolyticus*. This cellulase contains an N-terminal GH9 domain, a triplet of CBMs and a C-terminal GH48 domain, highly thermostable and able to bind to and efciently hydrolyze crystalline cellulose. A truncated version composed of only the GH9 domain and a CBM was also isolated, but it was much less active on crystalline cellulose, although it hydrolyzed soluble carboxymethyl cellulose (CMC) at the same rate as the holoenzyme. This suggests that multiple CBMs and both endoacting and exoacting domains are required for efcient hydrolysis of the crystalline substrate (Zverlov *et al.*, 1998).

Many CBHs have been characterized from bacteria of the genus *Clostridium*. In fact, most thermostable CBHs derive from cellulosomal complexes produced by members of this genus. *C. thermocellum*, in particular, appears to be a significant resource for CBHs with up to four GH family 5, 15 GH family 9, and four GH family 48 CBHs having been described. *Clostridial* CBHs typically have optimal activity at temperatures ranging from 60 to 75°C and pH 5.0 to 6.5 (Table 6.1). To date, the most thermostable CBH has been isolated from the culture supernatant of the thermophilic bacterium *Thermotoga* sp. strain FjSS3-B1. The enzyme has maximal activity at 105°C and maintains a half-life of 70 min at 108°C (Ruttersmith and Daniel, 1991). CBH from *Thermotoga*sp. is active on amorphous cellulose and carboxymethyl cellulose (CMC) with cellobiose as the sole product. However, the enzyme exhibits limited activity against filter paper or Sigmacell 20, highlighting the need for this enzyme to work synergistically with an endoglucanase enzyme exhibiting activity on the crystalline substrate. GH family 5 CBH, CelO, from *C. thermocellum* shows a marked preference for crystalline substrate still maintains activity on cellodextrins, barley β-glucan, and CMC (Zverlov *et al.*, 2002).

Table 6.1: Thermostable Exoglucanases from Microbes

Microorganism	Optimum Temperature (ºC)	Substrate Specificity	References
Clostridium stercorarium	75	Cdex, AVI, PASC, OSX	Bronnenmeier *et al.* (1991)
C. thermocellum (Ct)	65	PNPC, CMC, Cdex	Kataeva *et al.* (1999)
Ct	65	CMC	Cornet *et al.* (1983)
Ct	65	LICH, CMC, Xylan	Tuka *et al.* (1990)
Ct CelSCt CbhA	70	Amorphous and crystalline cellulose, CMC	Kruus *et al.* (1995) Schubot *et al.* (2004)
Talaromyces emersonii	78	AVI	Tuohy *et al.* (2002)
Thermoascus aurantiacus	65	PASC, AVI	Hong *et al.* (2003)
Thermotoga sp.	100-105	CMC, AC, WFP, MCC	Ruttersmith and Daniel (1991)
Thermococcus sp. *strain 2319*	60	CMC, Amorphouscellulose, CE-cellulose, β-glucan, lichenan, xylo glucan, avicel, xylan (beech, birch), locust bean gum	Gavrilov *et al.* (2016)
Thermomonospora. fuska YX	55	CMC	Zhang *et al.* (1995)

AC: Amorphous cellulose; AVI: Avicel; Cdex: Cellodextrins; CMC: Carboxymethylcellulose; CNPG: -2-chloro-4-nitrophenyl-L-cellobioside; LICH: Lichenan; MCC: Microcrystalline cellulose; MLC: 4-methylumbelliferyl-L-cellooligosaccharides; OSX: Oat spelt xylan; PNPG: p-nitrophenyl-b-D-glucoside; PASC: Phosphoric acid swollen cellulose.

The order Clostridiales, which is much less themophilic, includes numerous species that utilize crystalline cellulose, as well as hemicellulose, as growth substrates, and these organisms are of great importance to biomass deconstruction. The most thermophilic species are *Caldicellulosiruptor kristjanssonii* (T_{opt} 78°C) and *A. thermophilum* (T_{opt} 75°C), while the most extensively studied is *C. thermocellum*

(T_{opt} 60°C). These organisms grow on crystalline cellulose, yielding lactate, ethanol, acetate, H_2, and CO_2 (Bredholt *et al.*, 1999, Svetlichnyi *et al.*, 1990, Freier *et al.*, 1988, Bolshakova *et al.*, 1994). *C. thermocellum* was originally isolated from a cotton bale and is able to grow on a wide array of puried cellulose substrates and will slowly degrade natural untreated cellulosic material. A higher efciency of cellulose degradation was demonstrated in *C. thermocellum* cultures as compared with cellfree cellulase mixtures [36]. In contrast to *C. thermocellum*, *A. thermophilum* utilizes a broader spectrum of substrates, such as glucose, galactose, and arabinose, allowing for synergy with other plant biomass degrading microorganisms (Lu *et al.*, 2006).

Thermostable enzymes from a number of moderately thermophilic actinobacteria, most notably *Acidothermus cellulolyticus* and *Thermobifidafusca* (formerly '*Thermomonospora fusca*'), have been characterized (Tucker *et al.*, 1989; Baker *et al.*, 1994; Wilson, 2004). Currently, the only enzymes from *Thermobispora bispora* or *Thermomonospora curvata* that have been biochemically characterized are a β-glucosidase (Wright *et al.*, 1992) and an endo-glucanase (Lin and Stutzenberger, 1995), respectively.

The most studied, *Pyrococcus furiosus* (T_{opt} 100°C), metabolizes α-linked glucosides, such as starch and pullulan, and β-linked glucosides, such as laminarin, barley glucan and chitin generating hydrogen, carbon dioxide, and acetate as primary fermentative products (Driskill *et al.*, 1999, Fiala *et al.*, 1986). The closely related species, *P. horikoshii*, reportedly does not grow α- or β- linked glycosides, yet produces several glycoside hydrolases (Kaper *et al.*, 2002). The genome sequences of *P. furiosus*, *P. horikoshii*, and two other members of the Thermococcales, *P. abysii* and *Thermococcus kodakaraensis* KOD1 encode a variety of glucosidases and glucanases, some of which have been biochemically characterized (Ando *et al.*, 2002, Bauer *et al.*, 1999, Gueguen *et al.*, 1997, Kashima *et al.*, 2005).

Thermophilic Cellulases of Fungal Origin

Thermostable CBHs also occur widely in fungi, such as *Thermoascus aurantiacus*, *Talaromyces emersonii*, and *Cladosporium* spp. Biological saccharification of lignocellulosic biomass by wild-type fungi, such as *T. reesei*, *Fusarium oxysporum*, *Piptoporus betulinus*, *Penicillium echinulatum*, *Penicillium purpurogenum*, *Aspergillus niger*, and *Aspergillus fumigatus*, have been widely researched for saccharification.

The benefits of thermophilic cellulases from thermophilic bacterial origin super cede the thermophilic cellulases of fungal origin. This can be understood from the following examples. The specific cellulolytic activity, the cellulose system of the thermophilic, anaerobic bacterium *C. thermocellum* has been reported to degrade cellulose more effectively than fungal enzyme systems. Additionally, several anaerobic bacteria, such as *Caldicellulosiruptor saccharolyticus*, *Caldicellulosiruptor lactoaceticus*, *Ruminococcus albus*, and *Clostridium cellulofermentans*, which have evolved distinct enzyme systems, were reported to efficiently and directly degrade lignocellulose. Using multi-enzyme complexes, these bacteria can saccharify lignocellulose during cultivation, which would greatly reduce operating costs (Sheng *et al.*, 2016).

Thermostable Endoglucanases

Thermostable endoglucanases have been isolated from a number of thermophilic bacteria, and archaea as well as mesophilic and moderately thermophilic filamentous fungi (Table 6.2). Recombinant versions of many of these endoglucanases have been heterologously expressed and characterized. To date several enzymes have been reported that display optimal enzymatic activity at, or above, 100°C, with the hyperthermophilic archaeon, *P. furiosus*, possessing a GH family 12 endoglucanase that maintains enzymatic integrity up to 112°C. This endoglucanase has a signal peptide indicating that it may be exported from the cell and act extracellularly. The enzyme, however, lacks a CBM, a feature apparently common to many thermostable cellulases including those described for GH family 7 endo- and exo-glucanases from the ascomycete fungus *Melanocarpus albomyces*, and the GH family 8 endoglucanase of *Aquifex aeolicus*.

Thermostable β-Glucosidases

β-Glucosidases are ubiquitous, occurring in organisms representing all domains of life ranging from bacteria to highly evolved mammals. *Thermotoga* species represent an important source of hyper-thermophilic GHs, and many thermophilic β-glucosidases have been obtained from these organisms. Examples include two GH 1 family β-glucosidases from *T. maritime* and *T. neapolitana* that were both active at 100°C (Park *et al.*, 2005). However, the most thermostable β-glucosidase reported to date comes from the hyper-thermophilic archaeon *P. furiosus*. This β-glucosidase shows optimum activity at 102–105°C with half-lives of 85 h at 100°C and 13 h at 110°C (Kengen *et al.*, 1993). One β-glucosidase with particular biotechnological applicability has been isolated from *Thermus* sp. Z1. This β-glucosidase displays optimal activity at 85°C,has a broad optimal pH range (4.5–7.0), and exhibits a half-life of 5 days at 75°C.

Improvement of Thermostability of Cellulases

Although cellulases from thermophilic fungi are thermostable, the potential to increase their thermostability further would be beneficial for industrial applications. Cellulase engineering have been carried out using directed evolution, rational design, and the reconstitution designer cellulosome or cellulase mixtures. Directed evolution is a robust protein-engineering tool independent of knowledge of the protein structure andof the interaction between enzymes and the substrate, success of a directed-evolution experiment greatly depends on the method chosen to screen the best mutant enzyme. Rational design on the other hand requires detailed knowledge of protein structure, of the structural causes of biological catalysis or structural-based molecular modeling, and of the ideal structure-function relationship. The modification of amino acid sequence can be achieved through site-directed mutagenesis. A designer cellulosome combines multiple enzymes and forms a single macromolecular complex, which is useful for understanding cellulosome action and for biotechnological applications. Designer cellulosomes include the construction of chimeric scaffoldins that contain divergent cohesins and matching dockerin-bearing enzymes.

Table 6.2: Thermostable Endoglucanases from Microbes

Microorganism	Optimum Temperature (°C)	Substrate Specificity	References
A. cellulolyticus	83	AVI, CMC	Tucker et al. (1989)
Anaerocellum thermophilum	95-100	AVI, PASC, CMC,BBG, OSX	Zverlov et al.(1998a)
Archaeon EBI-244	109	CMC, Avicel, WFP	Graham et al. (2011)
Bacillus sp. KSM-S237	45	LICH, CMC	Hakamada et al.(1997)
Bacillus sphaericus	60	LICH, CMC, LAM,AVI, MCC, WFP	Singh et al. (2004)
Chaetomium thermophilum	60	CMC, PASC, WFP,MCC	Li et al. (2003)
Clostridium stercorarium 481	90	BBG, CMC, PASC,AVI	Bronnenmeier and Staudenbauer (1990)
Clostridium thermocellum	70	C5, C4, CMC	Fauth et al. (1991)
Clostridium thermocellum	70	PASC, AVI, MCC	Reverbel-Leroyetal.(1997)
Clostridium thermocellum	83	C5, C4, CMC	Fauth et al. (1991)
Enrichment culture (Ignisphaera)	109	LICH,ANI,CMC, β-glucan	Graham et al. (2011)
Environmental (Thermococcus)	92	CMC, β-glucan, LICH, AMC	Leis et al. (2015)
Fervidobacterium nodosum	80	BBG, CMC	Zheng et al. (2009)
Mytilus edulis	30-50	CMC, PASC, C5, C6	Xu et al. (2000)
Pyrococcus furiosus	100	BBG, LICH, CMC,WFP, Cdex	Bauer et al. (1999)
Pyrococcus horikoshii	97	CMC, AVI, LICH	Ando et al. (2002)
Pyrococcus horikoshii	95	CMC, PASC	Kang et al. (2007)
Rhodothermus marinus	95	CMC, Cdex	Hreggvidsson et al. (1996)
Rhodothermus marinus	100	GSM,CMC, LICH	Halldorsdottir et al. (1998)
Sporotrichum sp.	70	CMC, AVI	Ishihara et al.(1999)
Sulfolobus solfataricus	80	CMC	Huang et al. (2005)
Talaromyces emersonii	80	BBG, LICH	Murray et al. (2001)
T. fusca YX	77	CMC	Posta et al. (2004)
Thermotoga maritima	95	BBG, CMC, AVI	Bronnenmeier et al. (1995)
T. maritima	80	β-glucan, CMC GSM	Chhabra et al. (2002)
Thermotoga neapolitana	95	CMC, PASC, WFP	Bok et al. (1998)

AVI, Avicel; BBG, barley β-glucan; C4, cellotetraose, C5, cellopentaose, C6, cellohexaose; Cdex, cellodextrins; CMC, carboxymethylcellulose; GSM, glucosomannan; LAM, laminarin; LICH, lichenan; MCC, microcrystalline cellulose; OSX, oat spelt xylan; PASC, phosphoric acid swollen cellulose; WFP, Whatmann filter paper.

Some of the engineering efforts in cellulases are discussed. Improvement of *M. albomyces* Cel7B has been pursued by error-prone PCR (Voutilainen *et al.*,

2007). Two positive thermostable mutants, Ala30Thr and Ser290Thr, showed improvements in unfolding temperatures (T_m) by 1.5 and 3.5°C, respectively. Recently, Cel7A cellobiohydrolase from the thermophilic fungus *T. emersonii* was engineered using rational mutagenesis to improve its thermostability and activity (Voutilainen *et al.*, 2010). Structural analysis of *H. grisea* Cel12A, a thermostable endoglucanase stability was improved by site-directed mutagenesis (Sandgren *et al.*, 2005). Chimeric CBHs were produced through recombination of CBH genes from the fungi *Chaetomium thermophilum*, *Humicola insolens*, and *Hypocrea jecorina* (Heinzelman *et al.*, 2009). Mutant with 7–15°C increase in temperature optimum over the parental enzymes was obtained. The exchange of identifiable modules led to increases in thermostability and hydrolytic activity. The addition of CBMs to enzymes lacking this module has been shown to improve performance, particularly against crystalline substrate. Chung *et al.*, 2015 engineered *Caldicellulosiruptor bescii*, a bacterium able to degrade lignocellulosic biomass by itself by homologous expression of a multimodular cellulase, called CelA, which contains GH9 and GH48 domains. Remarkably, this particular work shows a successful way to improve and enhance the cellulolytic activity in this important thermophilic bacterium with a potential for biotechnological applications in industry.

Several thermostable mutants have been identified from a random mutant library of the *Paenibacillus polymyxa* BG. The most thermostable mutant A17S has 11-fold increase in the half-life of thermo-inactivation at 50°C (Liu *et al.*, 2009). An open reading frame Cphy1163 encoding a family 5 GH (Cel5A) from an anaerobic cellulolytic bacterium *Clostridium phytofermentans* ISDg was cloned and expressed in *E. coli*. Cel5A has been displayed on the surface of *E. coli* by using *Pseudomonas syringae* INP as an anchoring motif for identification of Cel5A mutants with improved thermostability. The half-life times of thermo-inactivation of the wild-type and mutant Cel5A fused to a CBM3 have been significantly prolonged when the fusion proteins are pre bound before thermo inactivation to the substrates, especially to the insoluble substrates. The Cel5 mutant had increase in half-life time of thermo-inactivation on CMC, RAC, and Avicel (1.92, 1.36, and 1.46-fold, respectively). By the integration of computer modeling and site-directed mutagenesis, Escovar-Kousen *et al.*, 2004 have improved the activity of the *T. fusca* endocellulase/exocellulase Cel9A on soluble and amorphous cellulose by 40 per cent.

Summary

There are large numbers of thermostable enzymes available for the utilization of LCB. However more insight on understanding the mechanism of action and its binding ability with the substrate for enhanced conversion is much needed so as to use the thermophilic enzyme over mesophiles. It is important to understand more about the synergistic contributions of extremely thermophilic enzymes to biomass deconstruction, since nature has optimized multi-enzyme processes for growth substrate acquisition. The lack of a better understanding of the mechanisms of individual cellulases and their synergistic actions is yet to be overcome for large-scale commercial applications of cellulases. Knowledge on the inhibitors generated during pretreatment steps and lignin breakdown is another challenge to be addressed towards bio based product development to address global bioeconomy.

A considerable number of enzymes applicable to lignocellulose depolymerization have been investigated. More research is needed to realize the complete diversity of these enzymes in nature. This review has considered thermostable enzymes required for the depolymerization of lignocellulosic substrates to longer chain sugars. It is also important to know other enzymes involved in the LCB utilization including hemicellulases and lignolytic enzymes.

There has been tremendous progress in the area of biomass conversion technology and below are few examples of the same. Progress has been made in genetic strategies to enhance plant biomass yield and quality-related traits to develop dedicated and highly specialized plant varieties that meet targeted applications (Lauria *et al.*, 2015). Metatranscriptomic and metagenomic studies are attempted to disclose the expression activity of genes that are involved in thermophilic cellulose decomposition (Xia *et al.*, 2014). Engineered organism to produce high ethanol titers in the presence of inhibitors (Herring *et al.*, 2016). Surface display technology and whole cell catalysts have been developed (Tozakidis *et al.*, 2016). Metabolically engineering solventogenic thermophiles with additional biomass-deconstruction enzymes by incorporation of heterologous pathways from microorganisms into the metabolic engineering host (Zhang *et al.*, 2011). With the increasing resource and novel technologies that are being developed the conversion of LCB to valuable products will be accelerated in the years to come. The future processing technology will also focus to produce valuable products from plant biomass at lower cost and throughput.

References

Acharya S, Chaudhary A (2012). Bioprospecting thermophiles for cellulase production: a review. *Brazilian Journal of Microbiology*. 43(3):844-856.

Alonso DM, Bond JQ, Dumesic JA (2010). Catalytic conversion of biomass to biofuels. *Green Chem.*12:1493–513.

Ando S, Ishida H, Kosugi Y, Ishikawa K(2002). Hyperthermostable endoglucanase from *Pyrococcus horikoshii*. *Appl Environ Microbiol* 68:430-433.

Baker J, Adney W, Nleves R, Thomas S, Wilson D and Himmel M (1994). A new thermostable endoglucanase, *Acidothermus cellulolyticus* E1. *Appl Biochem Biotechnol* 45–46: 245–256.

Bauer M. W, Driskill L. E, Callen W, Snead M. A, Mathur, E. J and Kelly, R. M (1999). An endoglucanase, EglA, from the hyperthermophilic archaeon *Pyrococcus furiosus* hydrolyzes β-1, 4 bonds in mixed-linkage (1-3),(1-4)-β-D-glucans and cellulose. *J. Bacteriol.* 181, 284–290.

Bauer M. W, Driskill L. E, Callen W, Snead M. A, Mathur E. J and Kelly R. M (1999). An endoglucanase, EglA, from the hyperthermophilic archaeon *Pyrococcus furiosus* hydrolyzes β-1, 4 bonds in mixed-linkage (1-3),(1-4)-β-D-glucans and cellulose. *J. Bacteriol.*, 181, 284–290.

Bayer EA, Kenig R and Lamed R (1983). Adherence of *Clostridium thermocellum* to cellulose. *J. Bacteriol.*156: 818–827.

Berger E, Zhang D, Zverlov VV, Schwarz WH (2007). Two noncellulosomal cellulases of *Clostridium thermocellum*, Cel9I and Cel48Y, hydrolyse crystalline cellulose synergistically. *FEMS Microbiol. Lett.*268:194-201.

Bok J. D, Yernool D. A and Eveleigh D. E (1998). Purification, characterization, and molecular analysis of thermostable cellulases CelA and CelB from *Thermotoga neapolitana. Appl. Environ. Microbiol.* 64, 4774–4781.

Bolshakova EV, Ponomariev AA, Novikov AA, Svetlichnyi VA, Velikodvorskaya GA (1994). Cloning and expression of genes coding for carbohydrate degrading enzymes of *Anaerocellum thermophilum* in Escherichia coli. *Biochem. Biophys. Res. Commun.*202:1076-1080.

Bredholt S, Sonne-Hansen J, Nielsen P, Mathrani IM, Ahring BK (1999). *Caldicellulosiruptor kristjanssonii* sp. *nov.*, a cellulolytic, extremely thermophilic, anaerobic bacterium. *Int. J. Syst. Bacteriol.*49:991-996.

Brian H. Davison, Lee R. Lynd and David A. Hogsett (2016). Strain and bioprocess improvement of a thermophilic anaerobe for the production of ethanol from wood. *Biotechnol Biofuels* 9:125.

Bronnenmeier K and Staudenbauer W. L (1990). Cellulose hydrolysis by a highly thermostable endo-1, 4-b-glucanase (Avicelase I) from *Clostridium stercorarium. Enzyme Microbiol. Technol.* 12, 431–436.

Bronnenmeier K, Rucknagel K P and Staudenbauer W L (1991). Purification and properties of a novel type of exo-1, 4-β-glucanase (Avicelase II) from the cellulolytic thermophile *Clostridium stercorarium. Eur. J. Biochem.* 200, 379–385.

Bronnenmeier K, Kern A, Liebl W and Staudenbauer, W. L (1995). Purification of *Thermotoga maritima* enzymes for the degradation of cellulosic materials. *Appl. Environ. Microbiol.* 61, 1399–1407.

Chhabra SR and Kelly RM (2002). Biochemical characterization of *Thermotoga maritima* endoglucanase Cel74 with and without a carbohydrate binding module (CBM).*FEBS Lett.* 531: 375–380.

Christopher D. Herring, William R. Kenealy, A. Joe Shaw1, Sean F. Covalla, Daniel G. Olson, Jiayi Zhang, W. Ryan Sillers, Vasiliki Tsakraklides, John S. Bardsley, Stephen R. Rogers, Philip G. Thorne, Jessica P. Johnson, Abigail Foster, Indraneel D. Shikhare, Dawn M. Klingeman, Steven D. Brown, Brian H. Davison, Lee R. Lynd and David A. Hogsett (2016). Strain and bioprocess improvement of a thermophilic anaerobe for the production of ethanol from wood. Biotechnol. Biofuels. 9:125.

Chung D, Cha M, Guss AM, *et al.* (2015). Direct conversion of plant biomass to ethanol by engineered *Caldicellulosiruptor bescii. P. Natl. Acad. Sci.*U S A 111: 8931–8936.

Cornet, P., Tronik, D., Millet, J., and Aubert, J. P. (1983).Cloning and expression in Escherichia coli of *Clostridium thermocellum* genes coding for amino acid synthesis and cellulose hydrolysis.*FEMS Microbiol.Lett.* 16, 137–141.

Doi RH (2008). Cellulases of mesophilic microorganisms: cellulosome and non-cellulosome producers. Ann NY Acad Sci. doi: 10.1038/ncomms1373, 1125:267-279.

Driskill LE, Bauer MW, Kelly RM (1999). Synergistic interactions among β-laminarinase, β-1,4-glucanase, and b-glucosidase from the hyperthermophilic archaeon *Pyrococcus furiosus* during hydrolysis of β-1,4, β-1,3-, and mixed-linked polysaccharides. *Biotechnol Bioeng*, 66:51-60.

Egorova K and Antranikian G (2005). Industrial relevance of thermophilic Archaea. *Curr Opin Microbiol*. 8: 649–655.

Fauth U, Romaniec M. P, Kobayashi T, and Demain A. L (1991). Purification and characterization of endoglucanase Ss from *Clostridium thermocellum.Biochem. J*. 279, 67–73.

Fiala G, Stetter KO (1986). *Pyrococcus furiosus* sp. nov. represents a novel genus of marine heterotrophic archaebacteria growing optimally at 100 degrees C. *Arch Microbiol*, 145:56-61.

Freier D, Mothershed CP, Wiegel J (1988). Characterization of *Clostridium thermocellum* JW20. *Appl Environ Microbiol*, 54:204-211.

Gilbert HJ (2007). Cellulosomes: microbial nanomachines that display plasticity in quaternary structure. *Mol Microbiol*, 63:1568- 1576.

Graham J.E, Clark M.E, Nadler,D.C, Huffer S, Chokhawala H.A, Rowland S.E. *et al.*, (2011). Identification and characterization of a multidomain hyperthermophilic cellulase from an archaeal enrichment. *Nat.Commun*. 2:375.

Gueguen Y, Voorhorst WG, van der Oost J, de Vos WM (1997). Molecular and biochemical characterization of an endo-b-1,3-glucanase of the hyperthermophilic archaeon *Pyrococcus furiosus. J Biol Chem*, 272:31258-31264.

Gurung N, Ray S, Bose S, Rai V (2013). A broader view: microbial enzymes and their relevance in industries, medicine, and beyond. *Biomed Res. Int*. 2013, 329121.10.1155/2013/329121

Hakamada Y, Koike, K, Yoshimatsu T, Mori H, Kobayashi T and Ito S (1997). Thermostable alkaline cellulase from an alkaliphilic isolate, *Bacillus* sp. KSM-S237. *Extremophiles* 1, 151–156.

Halldorsdottir S, Thorolfsdottir E. T, Spilliaert R, Johansson M, Thorbjarnardottir S. H, Palsdottir A, Hreggvidsson G., Kristja´nsson J. K, Holst O and Eggertsson G. (1998). Cloning, sequencing and overexpression of a *Rhodothermus marinus* gene encoding a thermostable cellulase of glycosyl hydrolase family 12. *Appl. Microbiol. Biotechnol*. 49, 277–284.

Headon D.R., Walsh G (1994). The industrial production of enzymes. *Biotechnol. Adv*. 12(4):635–646.

Heinzelman P, Snow CD, Smith MA, Yu X, Kannan A, Boulware K, Villalobos A, Govindarajan S, Minshull J, Arnold FH (2009a). SCHEMA recombination of a fungal cellulase uncovers a single mutation that contributes markedly to stability. *J. Biol. Chem*. 284: 26229 – 26233.

Heinzelman P, Snow CD, Wu I, Nguyen C, Villalobos A, Govindarajan S, Minshull J, Arnold FH (2009b). A family of thermostable fungal cellulases created by structure-guided recombination. *Proc. Natl. Acad. Sci.* U.S.A. 106: 5610-5615.

Henrissat B.A (1991). classification of glycosyl hydrolases based on amino acid sequence similarities. *Biochem J.* 280:309–316.

Hong J, Tamaki H, Yamamoto K and Kumagai H. (2003).Cloning of a gene encoding thermostable cellobiohydrolase from *Thermoascus aurantiacus* and its expression in yeast. *Appl. Microbiol. Biotechnol.* 63, 42–50.

Hreggvidsson G. O, Kaiste E, Holst O, Eggertsson G, Palsdottir A and Kristjansson J. K. (1996). An extremely thermostable cellulase from the thermophilic eubacterium *Rhodothermus marinus. Appl. Environ. Microbiol.* 62, 3047–3049.

Huang Y, Krauss G, Cottaz S, Driguez H and Lipps G. (2005).A highly acid-stable and thermostable endo-β-glucanase from the thermoacidophilic archaeon *Sulfolobus solfataricus.Biochem. J.* 385, 581–588.

Iasson E. P. Tozakidis, Tatjana Brossette, Florian Lenz, Ruth M. Maas and Joachim Jose (2016). Proof of concept for the simplified breakdown of cellulose by combining Pseudomonas putida strains with surface displayed thermophilic endocellulase, exocellulase and β-glucosidase. *Microb Cell Fact.* 15:103

Ishihara M, Tawata S and Toyama S. (1999). Disintegration of uncooked rice by carboxymethyl cellulase from *Sporotrichum* sp. HG-I. *J. Biosci. Bioeng.* 87, 249–251.

Jessica P. Johnson, Abigail Foster, Indraneel D. Shikhare, *et al.* (2016). Strain and bioprocess improvement of a thermophilic anaerobe for the production of ethanol from wood. *Biotechnol Biofuels* 9:125.

José M. Escovar-Kousen and David WilsonDiana Irwin (2004). Integration of computer modeling and initial studies of site-directed mutagenesis to improve cellulase activity on Cel9A from *Thermobifida fusca.Applied Biochemistry and Biotechnology.*113:287.

Kang H. J, Uegaki K, Fukada, H and Ishikawa K (2007). Improvement of the enzymatic activity of the hyperthermophilic cellulase from *Pyrococcus horikoshii. Extremophiles* 11, 251–256.

Kaper T, van Heusden HH, van Loo B, Vasella A, van der Oost J, de Vos WM (2002). Substrate specificity engineering of β-mannosidase and β-glucosidase from *Pyrococcus* by exchange of unique active site residues. *Biochemistry* 41:4147-4155.

Kashima Y, Mori K, Fukada H, Ishikawa K (2005). Analysis of the function of a hyperthermophilic endoglucanase from *Pyrococcus horikoshii* that hydrolyzes crystalline cellulose. *Extremophiles*, 9:37-43.

Kataeva I, LiX. L, Chen H, Choi S. K and Ljungdahl L. G (1999). Cloning and sequence analysis of a new cellulase gene encoding CelK, a major cellulosome component of *Clostridium thermocellum*, evidence for gene duplication and recombination. *J. Bacteriol.* 181, 5288–5295.

Kengen SWM, Luesink EJ, Stams AJM, Zehnder AJB (1993). Purification and characterization of an extremely thermostable β-glucosidase from the hyperthermophilic archaeon *Pyrococcus furiosus*. *Eur. J. Biochem.* 213:305–312.

Kruus K, Andreacchi A, Wang W. K, and Wu J. H. (1995). Product inhibition of the recombinant CelS, an exoglucanase component of the *Clostridium thermocellum* cellulosome. *Appl. Microbiol. Biotechnol.* 44, 399–404.

Kublanov IV, Perevalova AA, Slobodkina GB *et al.* (2009). Biodiversity of thermophilic prokaryotes with hydrolytic activities in hot springs of Uzon Caldera, Kamchatka (Russia). *Appl Environ Microbiol* 75: 286–291.

Kumar P, Barrett DM, Delwiche MJ, Stroeve P (2009). Methods for pretreatment of lignocellulosic biomass for efficient hydrolysis and biofuel production. *Ind Eng Chem Res.* 48:3713–29.

Lamed R, Setter E and Bayer EA (1983a). Characterization of a cellulose-binding, cellulase-containing complex in *Clostridium thermocellum*. *J Bacteriol* 156: 828–836.

Lamed R, Setter E, Kenig R and Bayer EA (1983b). The cellulosome - a discrete cell surface organelle of *Clostridium thermocellum* which exhibits separate antigenic, cellulose-binding and various cellulolytic activities. *Biotechnol Bioeng Symp* 13: 163–181.

Li D C, Lu M, Li Y. L and Lu J. (2003). Purification and characterization of an endocellulase from the thermophilic fungus *Chaetomium thermophilum* CT2. *Enzyme Microbiol.*

Lin S-B and Stutzenberger FJ (1995). Purification and characterization of the major β-1,4-endoglucanase from *Thermomonospora curvata*. *J Appl Microbiol* 79: 447–453.

Liu W, Hong J, Bevan D R, and Zhang Y H (2009). Fast identification of thermostable β-glucosidase mutants on cellobiose by a novel combinatorial selection/screening approach. *Biotechnol.Bioeng.* 103, 1087–1094.

Lu Y, Zhang YH, Lynd LR (2006). Enzyme-microbe synergy during cellulose hydrolysis by *Clostridium thermocellum*. *Proc Natl Acad Sci U S A*, 103:16165-16169.

Lynd LR, Weimer PJ, van Zyl WH, Pretorius IS (2002). Microbial cellulose utilization: fundamentals and biotechnology. *Microbiol Mol Biol Rev.* 66:506–77.

Lynd LR, Laser MS, Bransby D, Dale BE, Davison B, Hamilton R, Himmel M, Keller M, McMillan JD, Sheehan J *et al.* (2008). How biotech can transform biofuels. *Nat Biotechnol* 26:169-172.

Massimiliano Lauria, Francesco Molinari and Mario Motto (2015). Genetic Strategies to Enhance Plant Biomass Yield and Quality-Related Traits for *BioRenewable Fuel and Chemical Productions*. http://dx.doi.org/10.5772/61005. Chap 5.

Murray P G, Grassick A, Laffey CD, Cuffe M M, Higgins T, SavageA V, Planas A and Tuohy M. G (2001). Isolation and characterization of a thermostable endo-β-glucanase active on 1, 3–1, 4-β-D-glucans from the aerobic fungus *Talaromyces emersonii* CBS. *Enzyme Microbiol. Technol.* 29, 90–98.

Naik SN, Vaibhav V. Goud, Prasant K. Rout, Ajay K. Dalai (2010). Production of first and second generation biofuels: A comprehensive review. *Renewable and Sustainable Energy Reviews* 14 578–597.

Park TH, Choi KW, Park CS, Lee SB, Kang HY, Shon KJ, Park JS, Cha J. Substrate specificity and transglycosylation catalyzed by a thermostable β-glucosidase from marine hyperthermophile *Thermotoga neapolitana*. *Appl. Microbiol. Biotechnol*.2005; 69:411–422.

Posta K, Beki E, Wilson DB, Kukolya J and Hornok L (2004). Cloning, characterization and phylogenetic relationships of cel5B, a new endoglucanase encoding gene from *Thermobifida fusca*. *J Basic Microbiol* 44: 383–399.

Rainey FA, Donnison AM, Janssen PH, Saul D, Rodrigo A, Bergquist PL, Daniel RM, Stackebrandt E, Morgan HW (1994). Description of *Caldicellulosiruptor saccharolyticus* gen. nov. sp. nov: an obligately anaerobic, extremely thermophilic, cellulolytic bacterium. *FEMS Microbiol Lett* 120: 263-266.

Reverbel-Leroy C, Pages S, Belaich A, Belaich J P and Tardif C (1997). The processive endocellulase CelF, a major component of the *Clostridium cellulolyticum* cellulosome, purification and characterization of the recombinant form. *J. Bacteriol.* 179, 46–52.

Ruttersmith L. D, and Daniel R. M (1991). Thermostable cellobiohydrolase from the thermophilic eubacterium *Thermotoga* sp. strain FjSS3-B1. *Biochem. J.* 277, 887–890.

Sandgren M, Stahlberg J, Mitchinson C (2005). Structural and biochemical studies of GH family 12 cellulases: improved thermal stability, and ligand complexes. *Progress in Biophysics and Molecular Biology.* 89(3):246–291.

Sara E Blumer-Schuette, Irina Kataeva, Janet Westpheling, Michael WW Adams and Robert M Kelly (2008). Extremely thermophilic microorganisms for biomass conversion: status and prospects. *Current Opinion in Biotechnology*, Volume 19, Issue 3, Pages 210-217.

Sara E. Blumer-Schuette, Steven D. Brown, Kyle B. Sander, Edward A. Bayer, Irina Kataeva, Jeffrey V. Zurawski, Jonathan M. Conway, Michael W. W. Adams, Robert M. Kelly (2014). Thermophilic lignocellulose deconstruction. 38, Pages 393–448.

Sarmiento F, Peralta R, Blamey JM (2015). Cold and Hot Extremozymes: Industrial Relevance and Current Trends. Frontiers in Bioengineering and Biotechnology, 3:148. doi:10.3389/fbioe.2015.00148.

Schubot FD, Kataeva IA, Chang J, Shah AK, Ljungdahl LG,Rose JP and Wang B-C (2004). Structural basis for the exocellulase activity of the cellobiohydrolase CbhA from *Clostridium thermocellum*. *Biochemistry* 43: 1163–1170.

Sergey N. Gavrilov, Christina Stracke, Kenneth Jensen, Peter Menzel, Verena Kallnik, Alexei Slesarev, Tatyana Sokolova, Kseniya Zayulina, Christopher Bräsen, Elizaveta A. Bonch-Osmolovskaya, Xu Peng, Ilya V. Kublanov and

Bettina Siebers (2016). Isolation and Characterization of the First Xylanolytic Hyperthermophilic Euryarchaeon *Thermococcus* sp.Strain 2319x1 and Its Unusual Multidomain Glycosidase. *Frontiers in Microbiology*, 7,552.

Sheng T, Zhao L, Gao L-F, *et al.* (2016). Lignocellulosic saccharification by a newly isolated bacterium, *Ruminiclostridium thermocellum* M3 and cellular cellulase activities for high ratio of glucose to cellobiose. *Biotechnology for Biofuels*, 9:172. doi:10.1186/s13068-016-0585-z.

Singh J, Batra N and Sobti R. C. (2004). Purification and characterization of alkaline cellulase produced by a novel isolate,*Bacillus sphaericus* JS1. *J. Ind. Microbiol. Biotechnol.* 31, 51–56.

Svetlichnyi VA, Svetlichnaya TP, Chernykh NA, Zavarzin GA (1990). *Anaerocellum thermophilum* gen. nov sp. an extremely thermophilic cellulolytic eubacterium isolated from hot springs in the Valley of Geysers. *Microbiology*, 59:598-604.

Te'o VS, Saul DJ, Bergquist PL(1995). celA, another gene coding for a multidomain cellulase from the extreme thermophile *Caldocellum saccharolyticum. Appl Microbiol Biotechnol,* 43:291-296.

Tucker MP, Mohagheghi A, Grohmann K and Himmel ME (1989). Ultra-thermostable cellulases from *Acidothermus cellulolyticus*: comparison of temperature optima with previously reported cellulases. *Nat Biotechnol* 7: 817–820.

Tuka K, ZverlovV V, Bumazkin, B. K, Velikodvorskaya G. A and Strongin A. Y (1990). Cloning and expression of *Clostridium thermocellum* genes coding for thermostable exoglucanases (cellobiohydrolases) in *Escherichia coli* cells.*Biochem. Biophys. Res. Comm.* 169, 1055–1060.

Tuohy M. G, Walsh D. J, Murray P. G, Claeyssens M, Cuffe M. M, Savage A. V and Coughlan, M. P. (2002). Kinetic parameters and mode of action of the cellobiohydrolases produced by *Talaromyces emersonii. Protein Struct. Mol. Enzyme* 1596, 366–380.

Turner P, Mamo G, Karlsson EN (2007). Potential and utilization of thermophiles and thermostable enzymes in biorefining. *Microbial Cell Factories.* 6:9. doi: 10.1186/1475-2859-6-9.

Unsworth LD, van der Oost J and Koutsopoulos S (2007). Hyperthermophilic enzymes – stability, activity and implementation strategies for high temperature applications. *FEBS J* 274: 4044–4056.

Voutilainen SP, Boer H, Linder MB, *et al.* (2007). Heterologous expression of *Melanocarpus albomyces* cellobiohydrolase Cel7B, and random mutagenesis to improve its thermostability. *Enzyme and Microbial Technology.* 41 (3):234–243.

Voutilainen SP, Murray PG, Tuohy MG, Koivula A (2010). Expression of *Talaromyces emersonii* cellobiohydrolase Cel7A in *Saccharomyces cerevisiae* and rational mutagenesis to improve its thermostability and activity.*Protein Engineering, Design and Selection.* 23(2):69–79.

Walsh G and Headon D. Wiley, UK: Chicester (1994). *Protein Biotechnology.*

Wilson DB (2004). Studies of *Thermobifida fusca* plant cell wall degrading enzymes. *Chem Rec* 4: 72–82.

Wolfenden R, Lu X, Young G (1998), Spontaneous hydrolysis of glycosides. *J. Am. Chem. Soc.* 120:6814–6815.

Wright RM, Yablonsky MD, Shalita ZP, Goyal AK and Eveleigh DE (1992). Cloning, characterization, and nucleotide sequence of a gene encoding *Microbispora bispora* BglB, a thermostable β-glucosidase expressed in *Escherichia coli*. *Appl Environ Microbiol* 58: 3455–3465.

Xu, B., Hellman, U., Ersson, B., and Janson, J. C. (2000). Purification, characterization and amino-acid sequence analysis of a thermostable, low molecular mass endo-ß-1,4-glucanase from blue mussel, *Mytilus edulis*. *Eur. J. Biochem.* 267, 4970–4977.

Yu Xia, Yubo Wang, Herbert H. P. Fang, Tao Jin, Huanzi Zhong and Tong Zhang (2014). Thermophilic microbial cellulose decomposition and methanogenesis pathways recharacterized by metatranscriptomic and metagenomic analysis. *Nature scientific reports,*4 : 6708, pg 1-9.

Zhang S, Lao G and Wilson DB (1995). Characterization of a *Thermomonospora fusca* exocellulase. *Biochemistry* 34: 3386-3395.

Zhang F, Rodriguez S and Keasling JD (2011). Metabolic engineering of microbial pathways for advanced biofuels Zhang F, Rodriguez S and Keasling JD. Metabolic engineering of microbial pathways for advanced biofuels production. *Curr Opin Biotechnol.* 22: 775–783.

Zheng B, Yang W, Wang Y, Feng Y and Lou Z (2009). Crystallization and preliminary crystallographic analysis of thermophilic cellulase from *Fervidobacterium nodosum* Rt17-B1. Acta Crystallogr. Sect. F. Struct. Biol. *Cryst. Commun.* 65, 219–222.

Zverlov V, Mahr, S, Riedel K, and Bronnenmeier K (1998). Properties and gene structure of a bifunctional cellulolytic enzyme (CelA) from the extreme thermophile *Anaerocellum thermophilum* with separate glycosyl hydrolase family 9 and 48 catalytic domains. *Microbiology* 144, 457–465.

Zverlov VV, Kellermann J, Schwarz WH (2005). Functional subgenomics of *Clostridium thermocellum* cellulosomal genes: identification of the major catalytic components in the extracellular complex and detection of three new enzymes. *Proteomics*, 5:3646-3653.

7

Development and Application of Nanocatalyst for Algal Biofuel Production

N. Thajuddin[1,2,], G. Vinitha[2], D. Mubarak Ali[1] and F. Lewis Oscar[1,2]*

[1]National Repository for Microalgae and Cyanobacteria-Freshwater (DBT Sponsored), [2]Department of Microbiology, Bharathidasan University, Tiruchirappalli – 620 024, Tamil Nadu

1. INTRODUCTION

An increased global depletion of fossils based fuel resources has rapidly raised the exploration rate of algal based biofuels, particularly biodiesel. The availability, cost effectiveness and ease in cultivation of algae have made them a superior contender. The major hurdles in the production of algal based biodiesel are transesterifcation process. The transesterification is the conversion of triglycerides to esters and glycerol by using various catalysts like acids, bases, enzymes and heterogeneous catalyst. Apart from the catalyst, nanoparticles are emerging as a potential catalyst today. Nanocatalysts are nano-sized particles which provides high rate of catalytic activity compared to other catalyst. The superiority of nanocatalyst lies in easy separation, reusability and less pollutant producing property. Several metal oxides, bimetallic oxides, hydrocalcites and $Ca/Al/Fe_3O_4$ magnetic composites were used as solid base catalysts. Compared to heterogeneous catalysts, nanocatalysts have versatile advantages such as easy separation, high reaction temperature short reaction time and high catalytic stability. Therefore, the present review clearly deciphers the potentiality of nanocatalysts in the transesterifcation process for sustainable production of biofuel most economic and eco-friendly ways.

An emerging global requirement of biofuel is at the peak; the replacement of fossil based fuel derivatives can be only done using microalgal biofuels. Microalgae are one of the most primitive phototrophic organisms considered as future fuel substrate. The products from microalgae are of higher molecular weight making higher energy density fluids like diesel; therefore they are target for fossil fuel derived diesel (Tran *et al.*, 2010).Microalgae are promising feedstocks for the production of renewable and alternative diesel fuels in view of energy security. They also possess an environmental protection with great potential of carbon dioxide (CO_2) reduction from the entire cycle of biodiesel production (Unpaprom *et al.*, 2015). Numerous modified techniques have been reported in order to produce biodiesel such as microemulsions, pyrolysis and transesterification (Farobie *et al.*, 2015; Shuit *et al.*, 2012; Avhad and Marchetti 2015). Among the reported processes transesterification is the most simple and common method of biodiesel production, it involves alcoholysis of the oils to form esters and glycerol by the influence of catalyst (Singh *et al.*, 2008).Catalyst are important agents involved in the production of biodiesel as it enhances the reaction rate of transesterification process and also aids in producing high yields. For this, conventional homogenous and or heterogenous catalyst has been used, which leads to saponification reaction and low yield of biodiesel (Vincente *et al.*, 2003). Hence, it requires additional downstream process which reflects the high production cost (Sidra *et al.*, 2016).

Nanomaterials either natural or man-made particles, with at least one dimension of 100 nm or less, have been widely used in many fields, mostly in industry, such as in cosmetics, chemical catalysts, drug delivery, medical devices (used to remove tumor cells), antimicrobial agents, optoelectronics, electronics, and magnetic (MubarakAli *et al.*, 2011, 2012, 2013, 2015, Kango *et al.*, 2013; Ferreira *et al.*, 2013; Cai *et al.*, 2013; Gopinath *et al.*, 2012, 2014; Saravanakumar *et al.*, 2015; LewisOscar *et al.*, 2015, 2016). Recently, nanocatalysts have gained special attention for biodiesel production. Catalyst activity strongly depends on the size thus the smaller particles possess higher catalytic activity, they can easily spread evenly within the liquid (Selvan *et al.*, 2009). Hence, the nanosized materials were used as the catalyst for biofuel production. Nanocatalyst play an important role in improving product quality and achieving optimal operating conditions. Nanocatalyst with high specific surface area (Singh *et al.*, 2008) and high catalytic activity and good rigidity may solve the most common problems of heterogenous catalysts such as mass transfer resistance time consumption, fast deactivation and inefficiency (Akia *et al.*, 2014).

This study is specifically intended to provide a basic platform for application of nanoparticle as a catalyst for transesterification process of biofuel production from various feedstocks.

1.1. Microalgae

Microalgae are a diverse group of eukaryotic or prokaryotic photosynthetic microorganisms that grow rapidly due to their simple structure which colonizes both marine and freshwater environments. Their photosynthetic mechanism is similar to land based plants, have efficient access to water, CO_2 and other nutrients. They are generally more efficient in converting solar energy into biomass. Each species of microalga produces different ratio of lipids, carbohydrates and proteins in the account of the biomass. These cellular components were used for human welfare (Ilavarasi *et al.*, 2011).

Many of them are exceedingly rich in oil and they are also fast growing hence making it suitable for biodiesel production. The advantage of the algae based feedstock is that the lipid productivity can be 15-300 times (respect to the dry weight of biomass) larger than that derived from plant (Lam *et al.*, 2010; Lee *et al.*, 2010). Algal oil is more attractive because of the algae have the capacity to yield more oil without requiring large area of arable lands when compared to terrestrial plants and it have better scope for better strain improvement and the capacity to enhance the value through co-products (Ramaraj *et al.*, 2016).

1.2. Advantages of Microalgae

The advantages of microalgae over higher plants as a source of transportation biofuels are numerous (Rodalfi *et al.*, 2009).

1. Oil yield per acre of microalgae cultures could greatly exceed they yield of the best oilseed crops

2. Microalgae grow in an aquatic medium, but need less water than terrestrial crops

3. Microalgae can be cultivated in seawater or brackish water on non-arable, and do not compete for resources with conventional agriculture.

4. Microalgae biomass production may be combined with direct biofixation of waste CO_2 (1 kg of dry algal biomass sequestrate about 1.8 kg of CO_2).

5. Algae cultivation does not need herbicides or pesticides.

6. The residual algal biomass after oil extraction maybe used as feed or fertilizer, or fermented to produce ethanol or methane.

7. The biochemical composition of the algal biomass can be modulated by varying growth conditions and the oil content can be highly enhance.

1.3. Lipids in Microalgae

Microalgal oil is enriched with high proportions of long chain fatty acids (*i.e.*, C20, and C22) with a high degree of un-saturation (20:5), thus making it feasible for food making industry (Packer 2009).Microalgae survive heterotrophically, because the exogenous carbon sources offer prefabricated chemical energy and it is stored as lipid droplets in the cells (Ratledge, 2004). The schematic representation of lipid biosynthesis pathway in microalgae were shown in Figure 7.1.

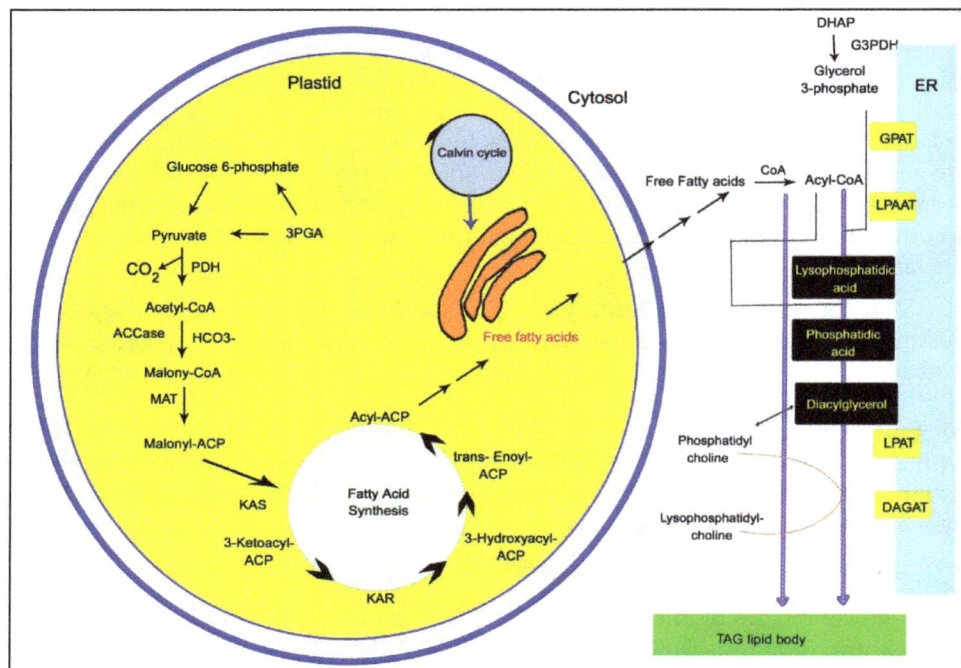

Figure 7.1: Schematic Representation of Lipid Biosynthesis Pathway in Microalgae (Re-designed from Radakovits *et al.*, 2010).

The natural or artificially created condition by the man for the excess accumulation of lipids in microalgae is under stress. Under nutrient deprived condition (preferably nitrogen) the algal cells stops almost all cellular mechanisms and the production of cellular components. In such condition, the rate of lipid synthesis remains higher, which leads to the excess accumulation of oil in starved cells (Sheehan *et al.*, 1998). During daily light-dark cycle, many microalgae initiate TAG storage during the day and deplete those stores at night to support cellular ATP demands and/or cell division. Consequently inhibit β-oxidation (fatty acid catabolism) would prevent the loss of TAG during the night, but most likely at the cost of reduced growth. This strategy therefore may not be beneficial for microalgae grown in outdoor open ponds, but it may be a valid strategy to increase lipid production in microalgae grown in photo bioreactors with exogenous carbon sources and/or continuous light.

2. Biofuel and Biodiesel

Fatty Acids Methyl Ester (FAME) is mainly produced from edible vegetable oils, animal fats and microbial biomass. It can be defined as a biofuel comprising of mono-alkyl esters of long chain fatty acids, produced from renewable lipids via transesterification process (Singh *et al.,* 2016). In general, biodiesel could be produced from any type of oil such as plant oils, animal fats or algal oil (especially microalgal oil) (Ramaraj *et al.,* 2016). Transesterification is the routine procedure for converting algal oil to biodiesel (Miao and Wu 2006).Feed stocks like *Jatropha*, Karanja, Mahua and castor oils are the most often used in biodiesel (Ivana *et al.,* 2012) synthesis. Various oils extracted from seeds or kernels of non-edible crops are potential feed stocks for biodiesel production (Ivana *et al.,* 2012). Biodiesel can be produced from edible oils *e.g.* soybean, sunflower, rapeseed, palm etc. as well as non-edible oils *e.g. Jatropha*, Jojoba, Tallow, *Pongamia* etc. (Sujatha 2006). From socioeconomic reasons, edible oils should be substituted by reliable and low cost inedible sources (algal oil) for biodiesel production (Raja *et al.,* 2011).

Based on development and the production of biofuels using biomass as sustainable biological resources which is categorized as first generation and second generation is to be fuel (Ilavarasi *et al.,* 2011).

2.1. Transesterification Processes

Transesterification process is the reaction of a triglyceride with an alcohol to form esters and glycerol with the presence of catalyst. It involves the displacement of alcohol from an ester where the preferred alcohol is methanol due to cheaper cost and polar nature, understanding the reaction mechanism tend to design the reaction conditions for maximum biodiesel production (Mohammed *et al.,* 2014; Ren *et al.,* 2014; Vincenzo *et al.,* 2015; Baskar and Aishwarya 2015). This process reduces the viscosity of oils which is higher than petro-diesel, selecting a suitable alcohol and catalyst is important. Various alcohols such as methanol, butanol, and amyl alcohol can be used for transesterification (Auwal *et al.,* 2012).

Transesterification consists of three consecutive reactions. The first step is conversion of triglycerides into diglycerides followed by conversion of diglycerides convert into monoglycerides and finally monoglycerides into glycerol (Enciner *et al.,* 2007; Fan *et al.,* 2009).

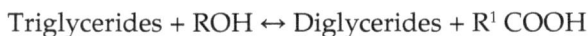

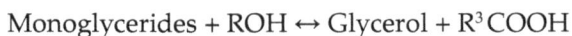

$$Triglycerides + ROH \leftrightarrow Diglycerides + R^1 COOH$$

$$Diglycerides + ROH \leftrightarrow Monoglycerides + R^2 COOH$$

$$Monoglycerides + ROH \leftrightarrow Glycerol + R^3 COOH$$

3. Catalyst and Types

Catalysts are chemical or biological substances, which enhance or stimulate or speed up the transesterification process in biofuel production from various feedstocks. Currently biodiesel is mainly synthesized through the catalytic method. Zhang *et al.* (2013) and Sharma *et al.* (2011) reported that the transesterification process of the catalyst mainly belongs to the categories of homogenous (acid, base and enzyme) and heterogenous were shown in Figure 7.2.

3.1. Acidic Catalyst

An acidic catalyst such as H_2SO_4 and HCl are usually used as catalyst in small scale studies. Some studies have pointed out that the acid catalyzed process

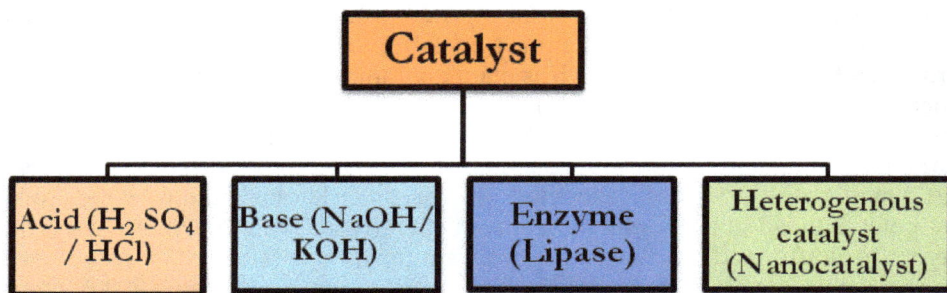

Figure 7.2: Types of Catalyst.

needs extra care in reactor due to the aggressive characteristic of employed acid. Additionally, it usually requires excess methanol (molar ration of methanol to oil is around 60:1) (Xu *et al.,* 2006; Jon 2005).

3.2. Basic Catalyst

Basic catalyst such as NaOH and KOH are usually employed in industrial biodiesel production. While base catalyzed biodiesel production consumes base due to the soap formation (Ullah *et al.,* 2009).

3.3. Enzymatic Catalyst

The enzymatic catalyst such as lipases can be used as a catalyst for the production of biofuel. Compared to the base and acid is ecofriendly, but their efficiency is very less and it is highly cost effective. However, the use of costly raw material for lipase production inhibited the enzymatic biodiesel production (Pizzaro and Park 2003). Enzymatic catalyst slows down the reaction rate and is deactivated when alcohol is used as acyl acceptor (Lam *et al.,* 2010). It may also exhibit unstable behavior at various conditions those are the main drawbacks to use enzymes as an effective catalyst (Degirmenbasi *et al.,* 2016).

3.4. Heterogeneous Catalyst

A new trend in the preparation of biodiesel is to use "green" method based on heterogenous catalyst (Martyanov and Sayari 2008; Granados *et al.,* 2007). Heterogenous catalyst tends to overcome the problem with homogenous and enzymes as catalyst, it has great advantages such as it requires mild conditions, easy to separate, reuse and regenerate.Thus the production cost can be reduced to a great extent (Lam *et al.,* 2010). Heterogenous catalyst such as alkali earth oxides, hydrocalcites, alkali-doped metal oxides, bimetallic oxides and ion resins were reported to be used for the transesterification process. The catalytic activity of heterogenous catalyst was improved by doping those elements (Yu *et al.,* 2011).

A major disadvantage of using heterogeneous catalysts is a low reaction rate caused by diffusion limitations in the three-phases (oil–alcohol–catalyst) of reaction mixture, as well as the complex catalyst preparation followed by a significant contribution to the environmental impact in some cases. Recent researches have been focused towards low cost and eco-friendly heterogeneous catalysts with a high catalytic activity. Generally, the preparation of these types of catalyst involves washing, drying, crushing/powdering and calcinating at high temperatures (Ivana *et al.*, 2012).

Table 7.1: Nanoparticles Used as Catalyst in Biofuel Production from Various Feedstocks

Sl.No.	Catalyst (Nanoparticle)	Used Plant/Algae	References
1.	KF/CaO	Chinese Tallow Seed Oil	Wen *et al.*, 2010
2.	TiO_2-MgO	Waste Cooking Oil	Wen *et al.*, 2010
3.	$ZrO_2/C_4H_4O_6HK$	Soybean Oil	Qiu *et al.*, 2011
4.	CaO/KF	Cinder (Solid Waste from Coal)	Hui *et al.*, 2012
5.	Fe_3O_4	*Nannochloropsis maritime*	Hu *et al.*, 2013
6.	TiO_2-ZnO	Palm Oil	Madhuvilakku and Piraman, 2013
7.	Fe_3O_4 @ Silica core Shell nanoparticles	Microalgae	Dong *et al.*, 2014
8.	Copper doped Zinc Oxide (CZO)	Waste cooking Oil	Baskar and Aiswarya, 2015
9.	Ca Fe Al (Layered double oxides)	Soybean Oil	Lu *et al.*, 2015
10.	Copper doped Zinc Oxide (CZO)	Neem Oil	Baskar and Aiswarya 2015
11.	CaO	Jatropha Oil	Reddy *et al.*, 2015
12.	Mg Fe_2O_4	Soybean Oil	Liu *et al.*, 2016
13.	Cao based/Au nanoparticles	Sunflower Oil	Elsie *et al.*, 2016
14.	Fe_3O_4	*Chlorella salina*	Surendiran *et al.*, 2016
15.	SrO @ SiO_2	Cooking Oil	Alex *et al.*, 2016
16.	Zr supported CdO	Soybean Oil	Patil and Pratap2016
17.	CaO-Al_2O_3	*Jatropha curcus* oil	Sidra *et al.*, 2016
18.	Cu-Co	Biomass	Chen *et al.*, 2016
19.	Ca (OCH_3)	*Nannochloropsis* sp.	Siowet al.2016
20.	Iron doped Zinc Oxide	Castor Oil	Baskar and Soumiya 2016

4. Nanocatalyst

Presently many of the nanoparticles were used as a catalyst for biofuel production. They are sub-microscopic or nano-sized particles with size ranges from 1 to 100 nm in diameter. The use of nanocatalyst provides higher catalytic activity, easier to separate and reusable, simple operational procedures, and regenerates of less pollution (Hu *et al.*, 2012). Metal oxides CaO, SrO, MgO, mixed metal oxides Al_2O_3, LI/CaO, Al_2O_3, K_2CO_3/Al_2O_3, KF/C-Al_2O_3, hydrotalcites and Ca/Al/Fe_3O_4 magnetic composites were used as solid base catalysts. Although these heterogeneous catalysts have some advantages such as easier catalyst separation most of them have limitations. They require high reaction temperature (>170 / C), long reaction time (>24 h) and low catalytic stability with significant leaching of catalyst components to the deactivation of the catalyst (Dehkordi and Ghasemi, 2012). The detailed evidence on the nanoparticle used as catalyst in biofuel production from various feedstock was listed in Table 7.1.

4.1. Transesterification of Algal Oil using Nanocatalyst

Transesterification is reversible reaction; there are certain parameters which shifts the equilibrium to the forward direction. The different parameters name water content, temperature, time, catalyst (nanocatalyst) and molar ratio of methanol to oils (Patil and Pratap, 2016).

4.2. Advantages of Nanocatalyst

Nanocatalyst possesses long term durability and stability than any other catalyst which were used in the Transesterification process. From the economic point of view the cost of the catalyst accounts for large. Therefore, the stability and sustainable activity of catalyst are of great importance in industrial application. The nanocatalyst KF/CaO exhibits high catalytic activity when compared to ZrO_2 in biofuel production. They were identitified by after completion of reaction; the reaction mixtures were centrifuged; separated the catalyst and again it can be reused. The KF/CaOnanocatalyst results 91 per cent biofuel conversion upto16 cycles. The ZrO_2 manifest minor decrease in conversion due to deposition of glycerol intermediates and blocks the active sites of the nanocatalyst (Wen *et al.*, 2010; Patil and Pratap 2016). The schematic process for biofuel production using nanocatalyst was shown Figure 7.3.

4.3. Reusability of Nanocatalyst

Hu *et al.* (2011) examined the reusability of KF/CaO- Fe_3O_4 by carrying out reaction cycles. The catalyst, after 3 h transesterification, they were separated by a permanent magnet. Then it washed with anhydrous methanol and dried at 105 /C for calculating the catalyst recovery (catalyst recovery is defined as the per cent of the mass of collected catalyst relative to that of fresh catalyst). Then, the catalysts were for the transesterification process in a same reaction condition of 1st cycle. This process is repeated upto14 time's usage was observed 10 wt. per cent of the catalyst. After 16 times of reuse, however, the catalyst loses their activity considerably. Although the catalyst is reduced again from 90.2 wt. per cent to 84.4 wt. per cent, catalyst deterioration probably due to its failure to maintain the formation of CaO (Hu *et al.*, 2011).

5. Research Plan and Future Prospects

In early, current studies by the researchers is mainly focused on the mixed oxide based catalytic (heterogenous) system for transesterification of oils from various feedstock. Among the current scenario of nanocatalyst mainly focused on selectable catalyst because of their ease availability and high FAME yield when compared to other particles and they were used in many forms for the production

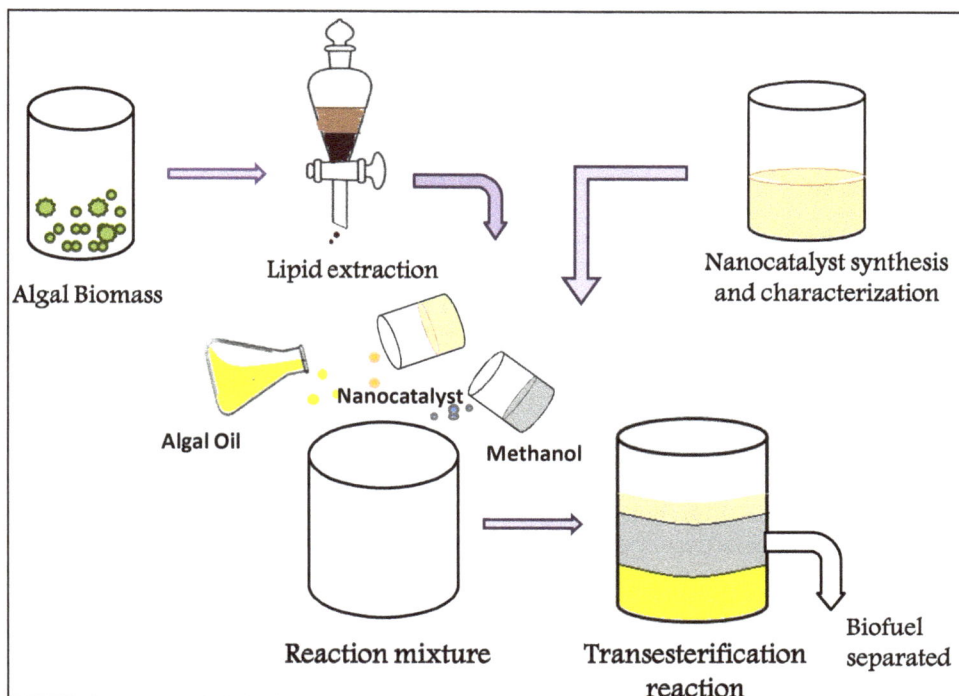

Figure 7.3: Schematic Process for Biofuel Production using Nanocatalyst.

of biodiesel. The more attention will be needed for the all other particles and the urge to understand the capabilities of metal oxides such as ZrO_2, ZnO, CuO, MgO. Among all the nanocatalyst, it could losses their effectiveness in further recycle process. In future, there must be further research required to reduce the problem encountered. There has been reported that the calcination temperature could reduce the binding sites blocked by reactants.

6. Conclusion

For a successful commercial catalyst, it must have the eco-friendly, high stability under various conditions, reusability and also cost effective. According to this, nanocatalyst is very effective catalyst when compared to other chemical or biological catalyst for biofuel production. In this review, microalgae considered as a feedstock for the biofuel production. A handful of studies have been reported on the direction. The nanocatalyst can be synthesized by various techniques and they were used as a catalyst in transesterification process. The catalyst activity mainly depends on the temperature, catalyst amount, methanol to oil molar ratio and reaction time. By altering these parameters there could be an effective result will be attained in a future.

Acknowledgement

All the authors are gratefully acknowledged to Department of Biotechnology (Govt. of India) by sanctioning a major project, NRMC-F (BT/PR7005/PBD/26/357/2012).

References

Akia, M., Yazdani, F., Motaee, E., Han, D and H. Arandiayan (2014). A review on conversion of biomass to biofuel by nanocatalysts. *Biofuel Res J.* I: 16-25.

Arul MozhiSelvan, V., Anand, RB and M. Udayakumar (2009). Effects of cerium oxide nanoparticle addition in diesel and diesel biodiesel-ethanol blends on the performance and emission characteristics of a CI engine. *ARPN J. Eng. Appl. Sci.* 4:1–7.

Auwal, A., Elisha, L and H. Abdulhamid (2012). Production of Biodiesel via NaOH Catalysed Transesterification of Mahogany Seed Oil. *Adv Appl Sci Res.* 3: 615-618.

Avhad, MR and JM. Marchetti (2015). A review on recent advancement in catalytic materials for biodiesel production. *Renew Sustain Ener Rev.* 50: 696–718.

Baskar, G and R. Aishwarya (2015). Biodiesel production from waste cooking oil using copper doped zinc oxide nanocomposite as heterogenous catalyst. *Bioresour Technol.* 188: 124-127.

Baskar, G and S. Soumiya (2016). Production of biodiesel from castor oil using iron (II) doped zinc oxide *Nanocatalyst. Renew Energy.* 98:101-107.

Cai, S., Wang, D., Niu, Z and Y. Li (2013). Progress in organic reactions catalyzed by bimetallic nanomaterials. *Chin J Catal.* 34:1964–74.

Chen, B., Li, F., Huang, Z and G. Yuan (2016). Carbon-coated Cu-Co bimetallic nanoparticles as selective and recyclable catalysts for production of biofuel 2,5-dimethylfuran. *Appl. Catal. B: Environmental.* 200:192–199.

Davoodbasha, M.A., Lee, S.Y., Kim, S.C. and J.W. Kim (2015). One-step synthesis of cellulose/silver nanobiocomposites using a solution plasma process and characterization of their broad spectrum antimicrobial efficacy. *RSC Adv.* 5: 35052-35060

Degirmenbasi, N., Coskun, S., Boz, N and DM. Kalyon (2015). Biodiesel synthesis from canola oil via heterogeneous catalysis using functionalized CaO nanoparticles. Fuel. 153:620–7.

Dehkordi, A.M and M. Ghasemi (2012). Transesterification of waste cooking oil to biodiesel using Ca and Zr mixed oxide as heterogenous base catalysts. *Fuel Process Technol.* 97: 45–51.

Elsie, B.M., Khalil, F., Yaghoub, M., Ali Mohammed, N., Rahim, M and F. Mehrdad (2016). Application of CaO-based/Au nanoparticles as heterogenous nanocatalysts in biodiesel production. *Fuel.* 164: 119-127.

Encinar, J.M., González, J.F. and A. R. Rodríguez (2007). Ethanolysis of Used Frying Oil. Biodiesel Preparation and Characterization, *Fuel Process Tech.* 88: 513–522.

Fan, X., Rachel, B and A. Greg (2009). Preparation and Characterization of Biodiesel Produced from Recycled Canola Oil, *The Open Fuels Ener Sci J.* 02: 113-118.

Farobie O and Y.A. Matsumura (2015). Comparative study of biodiesel production using methanol, ethanol, and tert-butyl methyl ether (MTBE) under supercritical conditions. *Bioresour Technol.* 191:306–11.

Ferreira, AJ., Cemlyn-Jones, J and CR. Cordeiro (2013). Nanoparticles, nanotechnology and pulmonary nanotoxicology. *Rev. Port Pneumol.* 19:28–37.

Gopinath, V., Priyadarshini, S., Loke, M.F., Arunkumar, J. and E. Marsili (2014). Biogenic synthesis, characterization of antibacterial silver nanoparticles and its cell cytotoxicity. *Arabian J Chem.* 1-11.dx.doi.org/10.1016/j.arabjc.2015.11.011

Gopinath, V., MubarakAli, D., Priyadarshini, S. and N.M. Priyadharsshini (2012) .Biosynthesis of silver nanoparticles from *Tribulusterrestris* and its antimicrobial activity: a novel biological approach. *Coll surf B.* 96: 69-74.

Granados, ML., Poves, MDZ., Alonoso, DM., Mariscal, R., Galisteo, FC and RM. Tost (2007). Biodiesel from sunflower oil by using activated calcium oxide. *Appl Catal B.* 73:317-26.

Hu, S., Guan, Y., Wang, Y and H. Han (2011). Nano-magnetic catalyst KF/CaO–Fe3O4 for biodiesel production. *Appl Energ.* 88:2685–2690.

Hui, L., Lingyan, S., Yong, S and Z. Lubin (2012). Biodiesel production catalyzed by cinder supported CaO/KF particle catalyst. *Fuel.* 97:651-657.

Ilavarasi, A., Mubarakali, D., Praveenkumar, R., Baldev, E and N. Thajuddin (2011). Optimization of various growth media to freshwater microalgae for biomass production. *Biotechnol.* 10: 540-545.

Ivana, B., Bankovic-llic, Olivera, S., Stamenkovic, Vlada B. Veljkovic, (2012). Biodiesel production from non-edible plant oils, *Renew. Sustain. Ener Rev.* 16:3621-3647.

Jahirul L. Mohammad I., Wenyong, K., Richard, B.J., Wijitha, S., Ian, H and M. Lalehvash (2014). Biodiesel Production from Non-Edible Beauty Leaf (*Calophyllum inophyllum*) Oil: Process Optimization Using Response Surface Methodology (RSM), *Energies.*07: 5317-5331.

Jon, VG (2005). Biodiesel processing and production. *Fuel Process Technol.* 86:1097–107.

Kango, S., Kalia, S., Celli, A., Njuguna, J., Habibi, Yand R. Kumar (2013). Surface modification of inorganic nanoparticles for development of organic–inorganic nanocomposites—a review. *Prog Polym Sci.* 38: 1232–61.

Lam, M.K., Lee, K.T and A.R. Mohamed, (2010). Homogenous, heterogenous and enzymatic catalysis for transesterification of high free fatty acid oil (waste cooking oil) to biodiesel: a review. *Biotechnol Adv.* 28:500-518.

Lee, J.Y., Yoo, C., Jun, S.Y., Ahn, C.Y and H.M. Oh (2010). Comparison of several methods for effective lipid extraction from microalgae. *Bioresour Technol.* 101: 75–77.

LewisOscar, F., MubarakAli, D., Nithya, C., Priyanka, R. and V. Gopinath, N. Thajuddin (2015). One pot synthesis and anti-biofilm potential of copper nanoparticles (CuNPs) against clinical strains of *Pseudomonas aeruginosa*. *Biofouling*. 31: 379-391.

LewisOscar, F., Vismaya, S., Arunkumar, M., Thajuddin, N., Dhanasekaran, D. and C. Nithya (2016). Algal Nanoparticles: Synthesis and biotechnological potentials. InTech - open science. 7: 157-152.

Liu, Y., Zhang, P., Fan, M and P. Jiang (2016). Biodiesel production from soybean oil catalyzed by magnetic nanoparticle $MgFe_2O_4@CaO$. *Fuel*. 164:314–321.

Lu, Y., Zhang, Z., Xu, Y., Liu, Q and G. Qian (2015). CaFeAl mixed oxide derived heterogeneous catalysts fortransesterification of soybean oil to biodiesel. *Bioresour Technol*. 190:438–441.

Madhuvilakku, R and S. Piraman (2013). Biodiesel synthesis by TiO_2–ZnO mixed oxide nanocatalyst catalyzedpalm oil transesterification process. *Bioresour Technol*. 150:55–59.

Martyanov, IN and A. Sayari (2008). Comparative study of triglyceride transesterification in the presence of catalytic amounts of sodium, magnesium, and calcium methoxides. *Appl. Catal A*. 339:45-52.

Miao, X. L and Q. Y. Wu (2006). Biodiesel production from heterotrophic microalgal oil. *Bioresour Technol*. 97:841-846.

MubarakAli, D., Sasikala, M., Gunasekaran, M and N. Thajuddin (2011). Synthesis and characterization of silver nanoparticles using marine cyanobacterium, *Phormidium willei* NTDM01, Digest J. Nanomater. *Biostruc*. (6): 385-390.

MubarakAli, D., Arunkumar, J., HarishNag, K, SheikSyedIshack, K.A. and E. Baldev, N. Thajuddin (2013). Gold nanoparticles from Pro and eukaryotic photosynthetic microorganisms-Comparative studies on synthesis and its application on biolabelling. *Coll Surf B*. 103: 166-173.

MubarakAli, D., Gopinath, V., Rameshbabu, N and N. Thajuddin (2012). Synthesis and characterization of CdS nanoparticles using C-phycoerythrin from the marine cyanobacteria. *Mater Lett*. 74: 8-11

MubarakAli, D., Thajuddin, N., Jeganathan, K and M. Gunasekaran (2011). Plant extract mediated synthesis of silver and gold nanoparticles and its antibacterial activity against clinically isolated pathogens. *Col Surf B*. 85: 360-365.

Packer, M (2009). Algal capture of carbon dioxide; biomass generation as a tool forgreenhouse gas mitigation with reference to New Zealand energy strategy andpolicy. *Energy Policy*. 37:3428–3437.

Patil, P and A. Pratap (2016). Preparation of Zirconia Supported Basic Nanocatalyst: A Physicochemical and Kinetic Study of Biodiesel Production from Soybean Oil. *J Oleo Sci*. 1-4.

Pizarro, AV and EY. Park (2003). Lipase –catalyzed production of biodiesel fuel from vegetable oils contained in waste activated bleaching earth. *Process Biochem*. 38:1077–82.

Qiu, F., Li, Y., Yang, D., Li, X and P. Sun (2011). Heterogeneous solid base nanocatalyst: Preparation, characterization and application in biodiesel production. *Bioresour Technol.* 102:4150–4156.

Raja, A.S., Robinson, D.S and R.L.C. Lee (2011). Biodiesel production from *Jatropha* oil and its characterization. *Res.J. Chem. Sci.*1: 81-87.

Ramaraj, K., Kawaree, R and Y. Unpaprom (2016). Direct transesterification of Microalga *Botryococcus braunii* biomass for biodiesel production. *Emer Life Sci Res.* 2: 1-7.

Ratledge, C (2004). Fatty acid biosynthesis in microorganisms being used for single cell oil production. *Biochimie.* 86: 807–815.

Reddy, ANR., Saleh, A.A., Saiful Islam, Md., Hamdan, S and Md. Abdul Maleque (2016). Biodiesel production from crude *Jatropha* oil using a highly active heterogeneous nanocatalyst by optimizing transesterification reaction parameters. *Energy Fuels.* 30: 334–343.

Radakovits, R., Jinkerson, R.E., Darzins, A. and M.C. Posewitz (2010). Genetic engineering of algae for enhamced biofuel production. *Euk Cell.* 9: 486-501.

Ren, Q., Zuo, T., Pan, J., Chen. C and L. Weimin (2014). Preparation of biodiesel from soybean catalyzed by basic ionic liquids OH, Materials. 07: 8012-8023.

Rodalfi, L., Zittelli, G., Bassi, N., Padovani, G., Biondi, N., Bonini, G and M.R. Tredici (2009). Microalgae for oil: Strain selection, induction of lipid synthesis and outdoor mass cultivation in low cost photobioreactor. *Biotech and Bioeng.* 102:100-112.

Saravanakumar, K., MubarakAli, D., Kathiresan, K., Thajuddin, N. and N.S. Alharbi (2015) Biogenic metallic nanoparticles as catalyst for bioelectricity production: A novel approach in microbial fuel cells. *Mat Sci Eng: B.* 203: 27-34.

Sharma, Y.C., Singh, B and J. Korstad (2011). Latest developments on application of heterogenous basic catalysts for an efficient and eco-friendly synthesis of biodiesel: *A review. Fuel.* 90: 1309-1324.

Sheehan, J., Dunahay, T., Benemann, J and P. Roessler (1998). A look back at the US Department of energy's aquatic species program: Biodiesel from Algae Golden. National Renewable Energy Laboratory, Colorado, TP-580-24190.

Shuit, S., Ong, Y., Lee, K., Subhash, B and S. Tan (2012).Membrane technology as a promising alternative in biodiesel production: a review. *Biotechnol Adv.* 30:1364–80.

Sidra, H., Sumbal, G., Tariq, M., Umar, N and F. Hadayatullah (2016). Biodiesel Production by using CaO-Al2O3 Nano Catalyst. *IJOER.* 2: 2395-6992.

Singh, Ak and SD. Fernando (2008). Transesterification of soybean oil using heterogenous catalyst. *Energy Fuels.* 22:2067-9.

Singh, R., Raj, J., Saxena, G and U.S. Sharma (2016).Synthesis of biodiesel by transesterification using homogeneous and heterogeneous catalysts: *A Review. Internatl Journl of Advan Resrch in Sci, Eng and Tech (IJARSET).* 6: 2238-2249.

Siow, HT., Aminul, I and HYT. Yun (2016). Algae derived biodiesel using nano catalytic transesterification process.*Chem Engineer Resear Design.* 111:362–370.

Sujatha, M (2006). Genetic improvement of *Jatropha curcas* L. possibilities and prospects. *Indian J Agroforestry.* 8: 58–65.

Surendhiran, D., Sirajunnisa, A.R., Anandan, M., Anbarasu, M., Mahin, B. S., Mohammed, YA., Gulam, MA and Md. Sadiq Mohiuddinc (2016). Direct conversion of lipids from marine microalga *C. salina* to biodiesel with immobilised enzymes using magnetic nanoparticle. *J. Environ Chem Engineer.* 4:1393–1398.

Tran, N.H., Bartlett, J.R., Kannangara, G.S.K., Milev, A.S., Volk, H and M.A. Wilson (2010). Catalytic upgrading of biorefinery of oil from Microalgae. *Fuel.* 89: 265-74.

Ullah, F., Nosheen, A., Hussain, I and A. Bano (2009). Base catalyzed transesterification of wild apricot kernel oil for biodiesel production. *Afr J Biotech.* 8:3289–93.

Unpaprom, Y., Tipnee, S and R. Ramaraj (2015). Biodiesel from Green Alga *Scenedesmus acuminatus. IJSGE.*4: 1-6.

Vicente, G., Martýnez, M and J. Aracil (2003) Integrated biodiesel production: a comparison of different homogeneous catalysts systems, *Bioresour Technol:* 300.

Vincenzo, B., Maria, C.E., Roberto, E., Matteo, L., Rosa, T., Francesco, R and S.D. Martino (2015). A novel and robust homogeneous supported catalyst for biodiesel production, *Fuel.* 171: 1-4.

Wen, L., Wang, Y., Lu, D., Hu, S and H. Han (2010). Preparation of KF/CaO nanocatalyst and its application in biodiesel production from Chinese tallow seed oil. *Fuel.*89: 2267-2271.

Wen, Z., Yu, X., Tu, S.T., Yan, J and E. Dahlquist (2010). Biodiesel production from waste cooking oil catalyzed by TiO2–MgO mixed oxides. *Bioresour Tech.*101: 9570-9576.

Xu, H., Miao, X and Q. Wu (2006). High quality biodiesel production from a microalga *Chlorella protothecoides* by heterotrophic growth in fermenters. *J Biotechnol.* 126:499–507.

Yu, X., Wen, Z., Li, H., Tu, S.T., Yan, J (2011). Transesterification of *Pistacia chinensi*s oil for biodiesel catalyzed by CaO-CeO$_2$ mixed oxides. *Fuel.* 90: 1868-1874.

Zhang, X.L., Yana, S, Tyagi, R.D and R.Y. Surampalli (2013). Biodiesel production from heterotrophic microalgae through transesterification and nanotechnology application in the production. *Renew. Sustain. Ener. Review.* 26: 216-233.

8

Glycosyl Hydrolases for Biomass Processing

Juliet Victoria, Suruchi Rao, Arvind Lali and Annamma Odaneth

DBT-ICT Centre for Energy Biosciences, Institute of Chemical Technology, (University under Section 3 of UGC Act- 1956, Elite Status and Centre of Excellence - Govt. of Maharashtra, TEQIP Phase II Funded), Mumbai, India
E-mail: a.dbtceb@gmail.com; arvindmlali@gmail.com

Introduction

Research into the green technologies that can feed the world's requirement for food, feed and fuel have now been researched for over eight decades. Natural renewable resources currently evaluated as candidates include solar, wind, water and biomass. Methodologies that can convert available resources into usable form, are however limited and fail to meet the requirements of being green, clean and also economically viable.

Biomass, in particular, provides sugars as precursors that can be converted by fermentation into an astonishingly wide variety of organic material employed in energy and feed (Popp *et al.*, 2016). Digestible components from biomass particularly, starch from corn grain and simple sugars from sugarcane and beets are currently being used directly for this purpose (Ho, Ngo and Guo, 2014). Non-digestible materials include wood and agricultural processing byproducts and waste, municipal solid waste, animal wastes, waste from food processing and aquatic plants and algae (Dusselier, Mascal and Sels, 2014).

Derived from the non-edible fraction of plant matter, termed Lignocellulose or LBM (biomass), which is the primary building block of plant cell walls. All plant cells exhibit thick cell walls that consist of polysaccharides and the aromatic polymer lignin evolved to form complex composite material which is highly resistant to pathogenic attack. Primarily, plant polysaccharides consist of cellulose, the β-1,4 linked homopolymer of glucose; hemicellulose, a heterogeneous, branched polysaccharide made up of a β-1,4 linked polymers including xylan, glucoronxylan, xyloglucan, glucomannan, and arabinoxylan backbones with heterogeneous side chains and pectin, a typically minor component in cell walls consisting of a complex set of polysaccharide polymers enriched in α-linked galacturonic acid or galacturonic acid and rhamnose monomers. Lignin is a heterogeneous, branched, alkyl-aromatic polymer comprising three phenyl-propanoid monomers linked by myriad C–O and C–C bonds that are likely formed through radical coupling reactions during cell wall synthesis (Schneider *et al.*, 2016).

These complex structures are not easily available as energy sources for most life forms, however some saprophytes and detritivores utilize enzymes and proteins to release this trapped carbon. Cellulose microfibrils in plants are packed into multiple forms of tightly bound, crystalline lattices termed polymorphs (Klemm *et al.*, 2005). Wolfenden *et al.*, first estimated that the uncatalyzed half-life of O- glycosidic linkages such as those found in cellulose, chitin, and other polysaccharides are more stable than DNA or peptide bonds, respectively, to uncatalyzed hydrolysis at neutral pH with a half-life of an astounding five million years and sometimes. Chemically intact cellulose has been found in fossilized plants that are significantly older.

The microbial degradation of the plant cell wall is a fundamental biological process and one that is of considerable industrial significance (Karaki *et al.*, 2016). Cellulose is a major reservoir of carbon and energy where plants sink photosynthetically fixed carbon into their cell walls as a means to provide resistance against herbivore and pathogen attack (Singh and Shukla, 2016). Insolubility of carbohydrate substrates from plant cell walls compel microorganisms to use extracellular enzymes, free or in complex, so as to convert the polysaccharides into soluble products that are capable of being transported into the cells. However, the capacity to deconstruct these structural carbohydrates is restricted only to a selected number of microorganisms. For effective biocatalysis, cellulolytic organisms produce a multitude of cellulases, hemicellulases, lytic polysaccharide monooxygenases (LPMOs), and other enzymes that target various parts of the plant cell wall matrix, which have been classified into families in the Carbohydrate Active Enzyme database (https://http://www.cazy.org/) on the basis of their sequence similarity (Lombard *et al.*, 2014). The enzyme consortium produced by organisms in response to the complex substrate so as to effect efficient degradation of plant cell wall polysaccharides reflects the composite nature of cell wall matrix polysaccharides, which requires the synergistic action of enzymes to degrade this complex structure (Cragg *et al.*, 2015).

Today, the most prevalent method of deriving fuel precursors (primarily glucose and xylose) from LBM involves a process combining chemical treatment with biocatalysis (Lee, Hamid and Zain, 2014). The chemical treatment (Pretreatment/

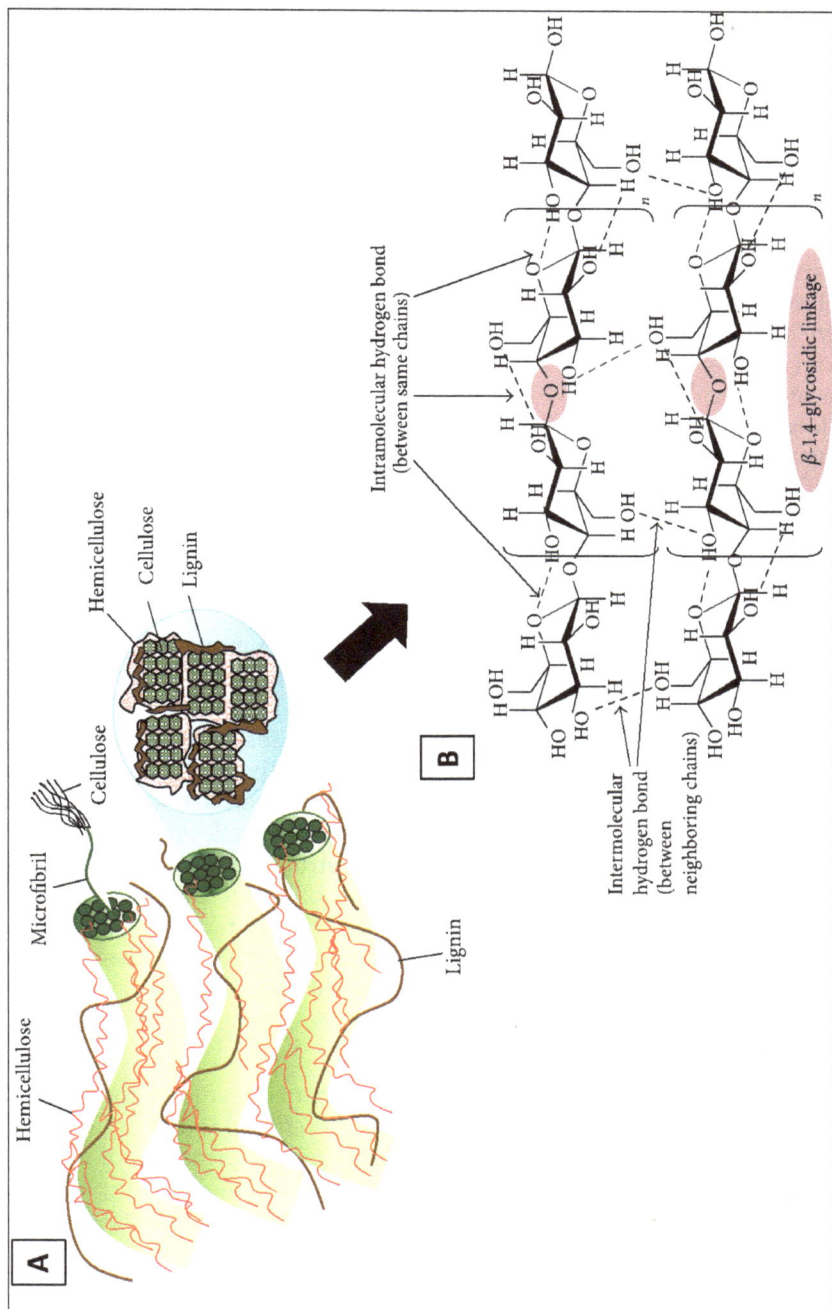

Figure 8.1: Structure of Plant Cell Wall Represented by (A) a Cross-Section of Micro-Fibril Showing the Cell Wall Consisting of Strands of Cellulose Embedded in a Matrix of Hemicellulose and Lignin, (B) Chemical Structure of Individual Cellulose Chains and the Inter- and Intramolecular Hydrogen Bonding Holding together Chains of β-1,4 Linked Glucose Monomers (Adapted from Lee, Hamid and Zain, 2014).

Fractionation) and Biocatalysis (Saccharification) represent two areas that currently are major bottlenecks and require significant improvements. The pretreatment process, which fractionates the intercalating layers of polymeric cellulose, hemicellulose, and lignin in LBM lays the foundation for the biocatalysis step. Diverse biomass substrates present a challenge for pretreatment as toxic compounds are rapidly generated during the processes which impede subsequent biocatalytic conversions, affecting the overall economics of the process. The biocatalytic activity loss on account of the above phenomenon also result in high enzyme dosage requirements that prove to be cost prohibitive. These factors define the prerequisites for the biocatalytic reaction. The complex hierarchical structure of lignocellulosic biomass needs fractionation into key components, namely cellulose, hemicellulose and lignin (Hollinshed *et al.*, 2014). This allows overcoming issues of irreversible, unproductive binding of the enzyme; poisoning of the enzyme and also inhibitions resulting from competing oligosaccharides and monomers in the biocatalytic step. Mechanisms of glycosyl hydrolases that define recognition, binding and subsequent catalysis followed by desorption from the insoluble substrate define the success rate of the interfacial solid-liquid hydrolysis technology that sequentially depolymerises polysaccharides into their respective monomers. In this chapter, the diversity, mechanism and influence of glycoside hydrolases on the saccharification of cellulose has been discussed.

The Enzyme System

Diversity in biomass composition present in nature necessitates a repertoire of enzymes for its synthesis, degradation as well as modification. Such enzymes are grouped as carbohydrate active enzymes which include glycoside hydrolases (GHs), polysaccharide lyases (PLs) or glycosyl transferases (GTs). A dedicated database for all these enzymes known as Carbohydrate Active Enzymes Database (CAZy) had been created in 1988, which is a manually curated list of the primary enzyme classes known to act on carbohydrates (Murphy *et al.*, 2011). All enzymes that degrade, modify or create glycosidic bonds are structured as families based on their structural similarity of the catalytic and carbohydrate-binding modules in the database (http://www.cazy.org). Till date, 135 GH families have been characterized and include almost every enzyme responsible for biomass destruction. Other enzymes responsible for the breakdown are Polysaccharide Lyases (PLs), Carbohydrate Esterases (CEs) and Auxiliary Activities (AAs) that have been recently included into the CAZY database.

Glycoside hydrolases are however the most prominent enzymes in this arsenal for hydrolyzing the glycoside bond. They are defined as enzymes that catalyze the hydrolysis of O-, N- and S-linked glycosides, leading to the formation of a sugar hemiacetal or hemiketal and the corresponding free aglycon. Due to multiple types of possible glycoside linkages, Glycosyl Hydrolases generally constitutes ~ 1-2 per cent of genome from soil and marine organisms; recent study reveals that ~40 per cent of the genomes of all sequenced bacteria encode at least one cellulase gene. Amongst these GHs, cellulases comprise around ~ 27 per cent of the characterized GHs.

Catalytic Mechanism for GHs

Glycosyl Hydrolases works by mechanism similar to that of acid-base catalysis. The catalytic mechanisms for GHs were proposed long ago, since then several efforts have been made to study the details about the action of GH enzymes. One striking feature of this breakdown is the conformational changes undergone by the free sugar residue at the -1 position which is directed by the glycosyl hydrolases and the intrinsic properties of the sugar moiety. The sugar is stressed out from its lowest energy chair conformation to a variety of different boat, skew boat, and half-chair forms which is specific for a given enzyme and substrate (Davies *et al.*, 2012). These distortions, position the glycosidic bond and favor the "electrophilic migration" of the anomeric center along the reaction coordinate catalysis.

Retaining glycosyl hydrolases release their products in soluble form with simultaneously maintaining configuration of the sugar at the point of hydrolysis (*i.e.* at the newly exposed anomeric center). They follow a double-displacement mechanism in which the covalent linkage between sugar residue is distorted and broken to allow a bond formation with the enzyme generating a transient state. For glycosyl hydrolases active site contains two amino acids, *i.e.* glutamic acid at 1st position and glutamic acid or aspartic acid on 2nd position; separated by a distance of 5.5°A. Retaining mechanisms fall into two general types (Davies and Williams 2016), where the nucleophilic residue derives either from the enzyme (typically aspartate or glutamate, or occasionally tyrosine) or the substrate (2-acetamido group leads to formation of an oxazoline intermediate).

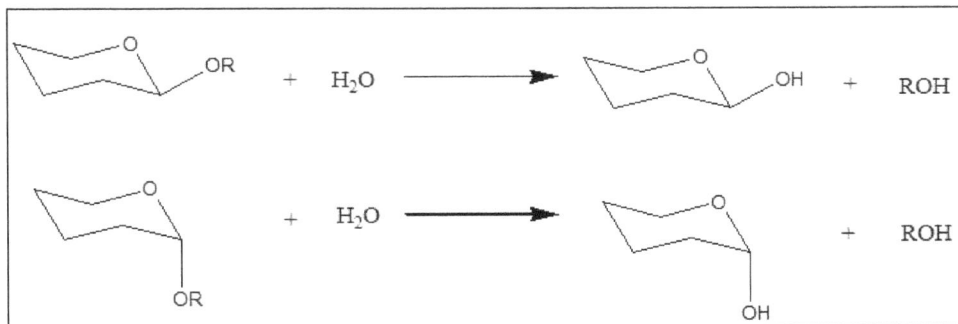

Figure 8.2: Retaining Mechanism for Hydrolysis of Glycosidic Bonds.

On the other hand, Inverting cellulases follow direct displacement mechanism through oxocarbenium ion-like transition state. Two active site residues are 11°A apart from each other. During cleavage water molecule is accommodated below pyranoside ring. One acid facilitates base catalytic assistance to water attack and other residue (Bronsted acid) gives acid catalytic assistance to cleavage of glycosidic bond. Nucleophile and acid/base of active site are located 5-6°A.

Non-catalytic Substrate Recognition

Multiplicity of functions associated with several carbohydrate-active proteins has evolutionarily pushed these molecules to acquire non-catalytic, substrate

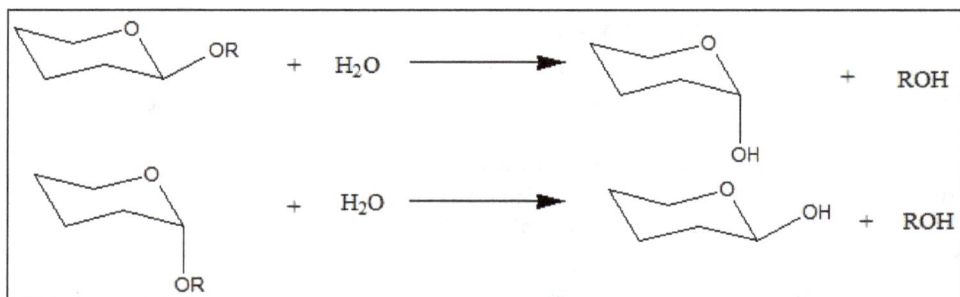

Figure 8.3: Inverting Mechanism for Hydrolysis of Glycosidic Bonds.

recognizing and binding modules that interact very specifically with cognate carbohydrates. These independently folding structures termed carbohydrate binding modules (CBMs) are auxiliary domains meant for recognition of the heterogeneous and complex carbohydrate arrangement (Boraston *et al.*, 2004).

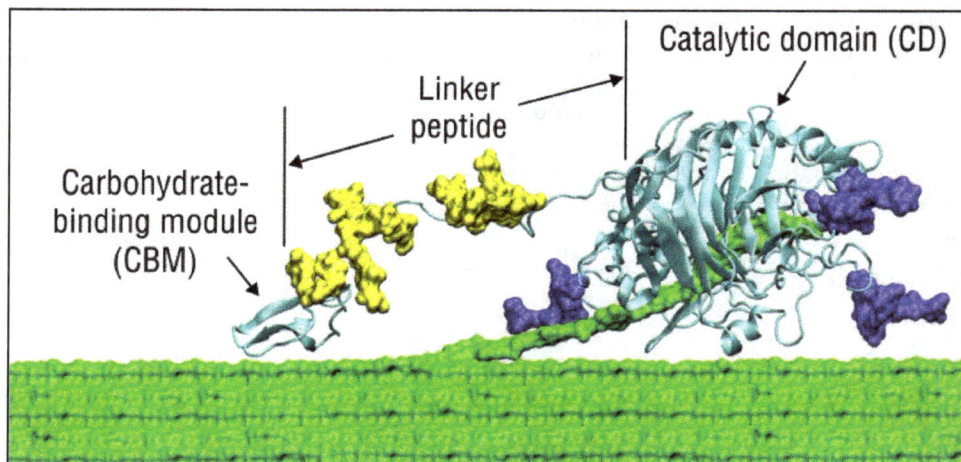

Figure 8.4: Modular Arrangement of Cellobiohydrolase: A Glycosyl Hydrolases (Credit: NREL).

Due to their association with enzymes involved in many life processes such as metabolism, energy storage, structural scaffold, immunological detection; CBMs are found within proteins that recognize a range of polysaccharides including cellulose, chitin, β-glucans, starch, glycogen, inulin, pullulan, xylan, and many other different sugars such as arabinofuranose, mannan, fucose, lactose, galactose, polygalacturonic acid, β-D-galactosyl-1,4-β-D-N-acetylglucosamine and lipopolysaccharides. The primary function of CBMs is to recognize and bind specifically to carbohydrates, resulting in diverse functions, such as enhanced hydrolysis of insoluble substrates, enhanced proximity of catalytic domain to the target substrate, anchoring associated proteins to cell surface and in some instances even disruption of polysaccharide structure. These characteristics are retained particularly in cellulases involved in biomass saccharification; for example, in extracellular fungal enzyme systems, the enzymes may be composed of a single catalytic module or of multiple domains which bind to polysaccharide backbone via CBM.

These CBM moieties are connected to catalytic domains via a linker peptide. Its length varies between 30-100 amino acids depending on source and predominantly contains proline, serine and threonine. This motif is susceptible to proteolysis when exposed to aqueous microenvironment, hence some linker peptides have been glycosylated to provide protection against proteolysis.

Diversity of Cellulolytic GHs

Fungal

The genetic diversity in fungal genome is tremendous and is capable of secreting copious amount of enzymes responsible for degrading plant cell wall. Fungi produce enzymes that populate 91 of 137 CAZy families, and thus cover all critical activities needed for efficient conversion of natural biomass. Fungal cellulases belong to 9 families (5, 6, 7, 8, 9, 12, 45, 48 and 61) which majorly consists of endoglucanases and cellobiohydrolases (CBH) responsible for cleaving C-O bond. Along with these, enzymes for hemicellulose degradation are more diverse containing exoglycosidases needed to hydrolyze arabinoxylans as well as glucogalactomannans and are included in GH families 1, 2, 3, 5, 8, 10, 11, 12, 16, 26, 43, 51, 54, 61, 62, 67 and 74. Xylanases are also placed in GHs and are classified in GH families 3, 10, 11 and 39 (Payne *et al.*, 2015; Jovanovic *et al.*, 2009).

These are modular enzymes with functionally distinct modules, a carbohydrate binding domain or module (CBD/CBM) and a catalytic domain (CD), connected by a serine – and threonine – rich polylinker with varying chain length. The function of CBD is to target the catalytic domain of cellulases or hemicellulases to the insoluble hydrogen bonded carbohydrate polymers, thereby increasing the effective enzyme concentration on polysaccharide surface (Boussaid and Saddler 1999).The presence of hyphal extension in cellulolytic filamentous fungi aids in cellulosic substrate penetration and cellulase secretion in the confined cavities within cellulosic particles. The production of "free" cellulases, with or without CBMs, may therefore suffice for the efficient hydrolysis of cellulose under these conditions. These soluble enzymes, secreted in the environment are called "noncomplexed" systems. These noncomplex systems facilitate ease of extraction and purification, thus important for commercial production of cellulases.

CBHs are known to possess a tunnel-shaped catalytic site where the cellulose chain is threaded and the enzyme moves along the cellulose chain cleaving it sequentially in a processive manner (Divne *et al.*, 1998). The catalytic site of Endoglucanases (EG) is groove-shaped and can accommodate a long cellulose chain (Davies and Henrissat, 1995), whereas a pocket-type active site is found in β-glucosidases for accommodation of smaller substrates. LPMOs, on the other hand, have a flat catalytic site (Vaaje-Kolstad *et al.*, 2010), which aligns with the crystalline surface of a cellulose elementary fibril. These structural features further determines the processive (in case of CBH) and non processive (in case of EG) nature of fungal cellulolytic enzymes (Horn *et al.*, 2012). In addition to the active site morphology, non complex cellulases with one catalytic unit and single CBM attached may be restricted in digesting only the surface of the crystalline cellulose microfibril leading to tapered morphology of cellulose microfibril.

One of the most extensively studied fungi *T. reesei* is an industrial power house of cellulolytic enzymes. It secretes atleast five different endoglucanases (EG I-V, or Cel 7B, Cel5A, Cel 12A, Cel 61A and Cel 45A) (Sajith *et al.*, 2016), two types of cellobiohydrolases (CBH-I or Cel 7A and CBH II or Cel 6A) of which the CBH-I constitutes half of the secreted proteins as well as number of xylanases and atleast one β-xylosidases. In addition to *Trichoderma*, nature has wide array of fungi which are cellulolytic these are *Aspergillus, Penicillium, Fusarium, Stachybotrys, Cladosporium, Alternaria, Acremonium, Ceratocystis, Myrothecium, Humicola,* etc. Species of *Aspergillus* and *Penicillum* are known to secrete β-glucosidases at a comparatively higher levels than *T. reesei*. Since β-glucosidases secreted by *T. reesei* remains bound to mycelium, it is not recovered efficiently during industrial production of cellulases resulting in low levels in concoction. *Aspergillus* and *Penicillum* are therefore known as major producers of β-glucosidases.

Bacterial

In contrast to fungi, bacteria lack the ability to effectively penetrate cellulosic materials, therefore had found an alternative mechanism for degrading cellulose. In presence of other micro-organism with limited ATP availability for cellulases synthesis, bacteria have evolved to produce "complexed" cellulases system known as cellulosomes. Cellulosomes are multi enzyme complexes first identified in 1983 in anaerobic, thermophilic spore forming bacterium *Clostridium thermocellum* (Bayer *et al.*, 2004). Cellulosomes are protuberances produced on cell wall of cellulolytic bacteria while growing on cellulosic materials. These cellulosomes are stable complexes that are firmly bound to bacterial cell wall but are flexible enough to bind tightly to microcrystalline cellulose. Electron microscopy indicated that cellulosomes are compact "fist" like structures that open up when attaching to micro crystalline cellulose, allowing for local spreading of the catalytic domains. Thus, complexed Cellulosomes enzymes with multiple CBMs and catalytic domains are likely to fully digest single microfibrils or regions of microfibrils leading to splaying of remaining undigested cellulose microfibrils thereby increasing the total substrate surface area available to enzymatic digestion. Cellulosomes also eliminate wasteful expenditure of energy of micro-organism by continuously producing copious amount of free enzymes. Synergism is optimized by the correct ratio between components, eliminating non-productive adsorption (Maki *et al.*, 2009).

However, complex and noncomplex cellulases exhibits striking difference in the mechanism of biomass saccharification. Complex cellulases were reported to show higher activity on pure cellulose while noncomplex cellulases were active on pretreated biomass. In case of pretreated biomass, Cellulosomes are likely to become inactivated more easily because of reduced access to exposed cellulose, whereas free cellulases can more readily diffuse to other regions of biomass for productive engagement. Hence noncomplex free cellulases are preferred than the complex Cellulosomes for saccharification of pretreated biomass.

The most common cellulolytic bacteria are *Acetivibrio cellulolyticus, Bacillus spp., Cellulomonas* spp., *Clostridium* spp., *Erwinia chrysanthemi, Thermobispora bispara, Ruminococcus albus, Streptomyces* spp., *Thermonospora* spp. and *Thermobifida fusca.* Apart from these, recently; *Streptomyces abietis, Kallotenue papyrolyticum, Ornatilinea apprima, Bacteroides luti, Alicyclobacillus cellulosilyticus, Anaerobacterium chartisolvens, Caldicellulosiruptor changbaiensis, Herbinix hemicellulosilytica* and *Pseudomonas coleopterorum* have also been discovered for possessing cellulolytic properties.

Amino-acid sequences of more than 15000 GH from both bacterial and fungal sources have been identified (Cantarel *et al.*, 2008) and sequence-based classifications for the catalytic domains available on the CAZy (CArbohydrate-Active EnZymes) database has led to the definition of more than 100 different families. The database provides a series of regularly updated sequence based classification of the enzymes at catalytic domain level. This hierarchical system of families, clans and folds to rationalize structural and mechanistic information helps reliable prediction of catalytic residues, mechanism and folds within family sequences. The unique and unexpectedly small numbers of structural folds that make up these proteins emphasize the fact of their being adapted and fine tuned to achieve the diverse and quite specific functions mediated by these enzymes. Figure 8.5 represents the different types of fold that GH usually adapts to (Henrissat *et al.*, 1995). The Tim-barrel, which has eight b/a motifs folded into a barrel structure, is one of the most common protein folds. A common origin, whereby diversification has resulted by gene duplication, followed by fusion and diversification has been proposed for all $(\alpha/\beta)_8$ barrel domians. Almost 50 per cent of known glycosidases have the $(\alpha/\beta)_8$ – barrel fold for their catalytic domains.

Currently, commercial production of cellulolytic glycosyl hydrolases, both native and heterologous has been carried out in hosts including *Bacillus subtilis, Trichoderma reesei, Aspergillus niger and others*. However, these systems continue to result in preparations that are ineffectively being used for cellulose hydrolysis resulting in a very high contribution of the enzyme costs in the final economics of the saccharification process. Strategies to overcome this barrier may originate from innovations in enzyme production methods and improvements in process design for use of the produced enzymes during cellulose saccharification, thereby resulting in increased sugar yields and reduced production costs.

Technologies for Enzymatic Cellulose Hydrolysis

As in nature, biological/biocatalytic conversion of biomass to sugars involves the cooperative action of a multitude of enzymes owing to the complexity of the substrate. Cellulose hydrolysis is catalyzed by endocellulases, exocellulases, and beta-glucosidases. Endocellulase randomly cleave the internal glycosidic linkages in the cellulose chain reducing the size of the polymer and introducing new chain ends. Exocellulases act in a processive manner either from the reducing or the non-reducing ends generating cellobiose majorly (Kostylev and David, 2012). The concerted action of endocellulases and exocellulases are majorly responsible for the solubilization of the polymer to cello-oligosaccharides and cellobiose, which are then cleaved by beta-glucosidase to glucose (Kumar *et al.*, 2008). Hemicellulose which acts

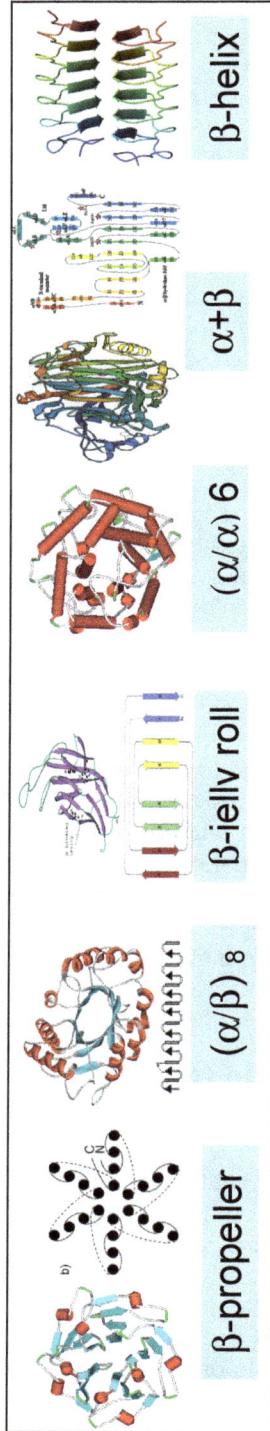

Figure 8.5: Different Types of Folds for Glycosyl Hydrolases.

like a shield for cellulose is a complex polysaccharide with different glyco- units and glycosidic bonds. The hemicelluloses polymer has side chain and hence they exist in various forms such as xylan, xyloglucan, arabinoxylan, mannan, galactan etc. Hence in order to hydrolyze glycosidic bonds, ester bonds and remove the chain's substituents or side chains enzymes like glycoside hydrolases, arabinofuranosidase, carbohydrate esterases, mannanase, polysaccharide lyases, endo-hemicellulases and present in nature (Sweeney and Xu 2012). In addition, alternative enzymatic partners are involved in the oxidation of cell wall components which aid the depolymerisation of the biomass. Lytic polysaccharide monooxygenases (LPMO) are metalloenzymes which uses molecular oxygen and an electron donor to degrade cellulose. LPMOs work in concert with cellulases and enhance cellulose degradation by catalyzing the oxidative cleavage of crystalline region of cellulose. Oxidative cleavage of the cellulose polymer improves the digestibility of the inaccessible polymer and avoids non-productive binding of enzymes.

Despite the plethora of enzymes, naturally, the depolymerisation of polysaccharides is slow and ineffective process. In order to improve the bioconversion of lignocellulose to biofuel, enzymes must be robust, tolerant to product inhibition, have high catalytic rate, adsorption capabilities, and thermal stability (Maki *et al.*, 2009). Research is therefore focused at mining, discovering, cloning and expressing new glycosyl hydrolases from environments rich with cellulolytic organisms. Engineering these to have improved functionality and stability and then formulating them into potent cocktails that work for different biomass varieties and pretreatments is crucial. An important aspect of the development of technology for cellulose hydrolysis is also the processing strategy for developed enzymes so that near theoretical yields and conversions are achieved. The commonly used process for utilization for lignocellulosic biomass includes pretreatment, saccharification and fermentation (Verardi *et al.*, 2012). Pretreatment is a basic pre-requisite however the remaining three processes namely enzyme production, saccharification, fermentation can be operated individually or in combination, depending on which they are classified as Consolidated Bioprocessing (CBP), Simultaneous Saccharification and Fermentation (SSF) and Separate Hydrolysis and Fermentation

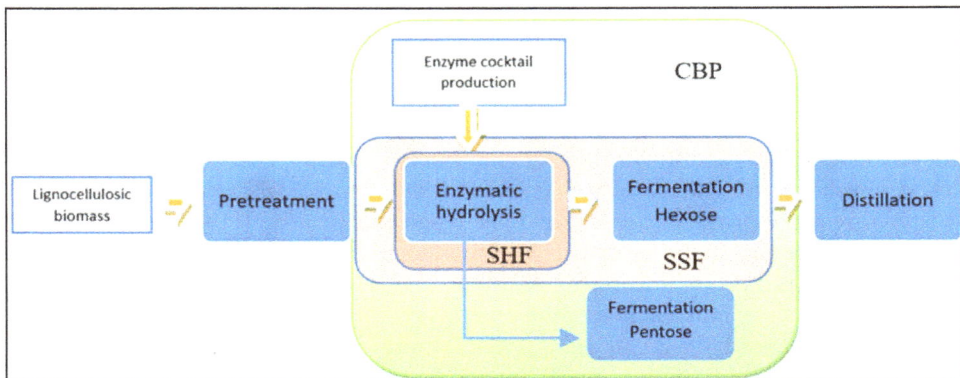

Figure 8.6: Different Process Configuration for Conversion of Cellulosic Biomass to Ethanol.

(SHF) (Lynd *et al.*, 2002). SSF has been extensively researched as it overcomes the limitation of feedback inhibition of cellulase (Galbe and Zacchi 2002) by converting the released sugars to ethanol. An extension to the SSF process is SSCF wherein both hexose and pentose sugars are utilized for production of ethanol. The system can either employ a single organism capable of utilizing both hexoses and pentoses (Christine *et al.*, 2016) or use two cultures in combination in a single reactor (Ire, Ezebuiro, and Ogugbue 2016). Co-fermentation of glucose and xylose is impeded since glucose is a preferred substrate and higher affinity of glucose transporters hinder xylose uptake (Kim, Block, and Mills 2010).

Similarly, CBP aims to improve the overall process economics by selecting a single organism which can produce the enzymes and also perform saccharification and fermentation (Brethauer and Studer 2015). CBP offers the advantage of low capital cost as all of the reactions are happening within a single bioreactor and it also avoids external addition of expensive enzymes. Additionally, simultaneous conversion of the sugars to ethanol reduces end-product inhibition by sugar which is a major concern during saccharification. Certain microbes like *Clostridium thermocellum*, *Neurospora crassa*, *Fusarium oxysporum* and *Paecilomyces sp.* can be used for this purpose (Kumar and Reetu 2015). Despite these advantages, CBP compromises between the optimal reaction temperature of saccharification and fermentation leading to sub-optimal condition operation. Attempts to engineer strains that can ferment sugars at higher temperature (48°C - 50°C) are being carried out. The feasibility of CBP will be fully established only when a microorganism or microbial consortium is developed that satisfies all the above requirements (Menon and Rao 2012).

Separate Hydrolysis and Fermentation (SHF) is a mode, wherein enzymatic hydrolysis and fermentation are performed separately giving higher reaction rates and greater degree of freedom for optimizing reaction parameters (Wilson, 2009). The performance of enzymes used is mainly attributed to the structural features of the substrates involved and the concoction of the enzyme used. The effectiveness of a multi-cellulase system as in nature rather than a single cellulase is a valid proof that these enzymes have very specific recognition patterns and work in close synergy to identify different substrates, thereby allowing them to function sequentially and effect complete saccharification of the cellulose and/or hemicellulose (Zhang, 2004b). The optimal design of the multi-enzyme cocktail, therefore dictates the performance of the hydrolysis stage. This approach is usually associated with rapid hydrolysis followed by 'cellulase slowdown' phenomenon. Loss of enzyme activity due to irreversible binding of enzyme to substrate and product inhibition become major contributors of this enzyme inefficiency and affect the economics of the large-scale usage of these enzymes in saccharification (Himmel ME., 1999). To increase the productivity of the enzymes in such a case, complete hydrolysis has been centered on better pretreatment, modification of reaction conditions and intermittent dosing of enzymes to the hydrolysis reaction (Jørgensen H, 2007). All the above strategies have limited success and therefore, recombinant enzymes with accessory proteins (LPMO's and AAs) are being designed for varied substrates and for harsh environments generated by the pre-treatment process. These developments have resulted in better enzyme cocktails that yield higher production efficiency and greater stability (Schülein, 2000). Cellic® CTec3, the latest on the shelf from Denmark-based Novozymes is claimed to be 1.5 times better than the previous Novozymes' Cellic® CTec2 and requires five times less enzyme dosage to make the same amount of ethanol.

These advances in cellulase cocktail design combined with innovations in enzyme production surely pave the way for effective enzymes that can be used. Off-site or commercial enzyme preparations are generally produced using mutants of *T. reesei*, as it produces large amounts of extracellular cellulolytic enzyme. Onsite productions, on the other hand, have a varied set of celluloytic organisms and aid in decentralizing the process, making the enzyme available at site and thus reduces the transportation cost. Additionally, on site enzyme preparation gives the flexibility of enzyme production suitable for a specific pretreatment and/or feedstock as and when required. Since enzyme preparations are freshly prepared, it does not require any further processing to maintain the stability of the enzymes. These factors together pool in to reduce the CAPEX and OPEX costs for enzyme production as both processes could share location, infrastructure and utilities. However, on site enzyme production requires the strain to possess a variety of cell wall degrading enzymes together with high expression and activity and the productivity of enzyme production maybe compromised due to the presence of inhibitory compounds in the substrate. The potential application of this strategy is still at a nascent stage and very elusive.

Apart from the above-mentioned approaches, reaction engineering is an important strategy for improving the hydrolysis performance and reducing the overall costs associated with enzyme usage.

Table 5.1: Various Approaches Practised for Cellulose Hydrolysis

Strategy	Outcome	Reference
Membranes with ionic liquids (IL)	Conversion: 45 per cent Time: 70 hr	Christian *et al.*, 2013
Integrated Membrane Reactor	Conversion: 53 per cent Time: 50 hr	Gan *et al.*, 2002
Immobilized cellulase With IL	Conversion: 13 per cent Time: 8 hr	Paetrice *et al.*, 2010
Cross-linked cellulase	Conversion: 52 per cent Time: 4 hr	Sharma *et al.*, 2001
Polyethylene glycol (PEG) addition	Conversion: 82 per cent Time: 48 hr	Johan *et al.*, 2007

Membrane reactors (integration of bioreactors with a separation unit) improve the rate of hydrolysis by over 40 per cent in comparison to batch processes. Different configurations and designs have been tested for improving the operational stability and product removal using membrane reactors. Product removal can also be achieved by segregating the enzyme hydrolysis reactor and ultra-filtration membrane module (Andric *et al.*, 2010). Using these configurations, hydrolysed sugars can be continuously or semi-continuously removed from the reaction system. However, these are also not free from enzyme inhibitions and stall after a few hours of operations. Further improvisations to the process can be made by using membranes to effectively remove the solubilised cellulose as oligosaccharides as quickly as they are formed and further subjected to complete hydrolysis with chemical/enzymatic methods. This allows the reaction to be completed in a very short time as the solubilisation is a faster reaction and works independent of the hydrolysis reaction (Lali *et al.*, 2014). Compartmentalization of the saccharification process into two stages of solubilisation and hydrolysis also helps in recycling the enzymes much more efficiently leading to longer half lives and therefore better economic viability of the process.

Conclusion

The ever-increasing repertoire of glycosyl hydrolases presents to us the question of is there a "one size fits all" solution for the breakdown of cellulose from diverse lignocellulosic biomass? Attempts to redesign expression of enzymes by fungal systems have been undertaken extensively to achieve the above goal and have could reduce the enzyme usage costs by tenfold. However, there is a lot more desired if the saccharification step should be economically viable. To meet the desired targets for enzyme usage, various aspects of process intensification for cellulase production and application, improved enzyme activity and stability while reducing enzyme usage per ton of biomass processed need to be worked upon. The road ahead has several hurdles, however, when the Glycosyl Hydrolase puzzle is solved; it opens a new chapter in bioeconomics of human civilization where sugars dominate the feed, food and energy landscape.

References

Andriæ, P., Meyer, A. S., Jensen, P. A., and Dam-Johansen, K. (2010). Reactor design for minimizing product inhibition during enzymatic lignocellulose hydrolysis: I. Significance and mechanism of cellobiose and glucose inhibition on cellulolytic enzymes. *Biotechnology advances*, 28(3), 308-324.

Andriæ, P., Meyer, A. S., Jensen, P. A., and Dam-Johansen, K. (2010). Reactor design for minimizing product inhibition during enzymatic lignocellulose hydrolysis: II. Quantification of inhibition and suitability of membrane reactors. *Biotechnology advances*, 28(3), 407-425.

Bayer, E. A.; Belaich J-P.; Shoham, Y. and Lamed R. (2004). "The cellulosomes: Multienzyme machines for degradation of plant cell wall polysaccharides". *Annual Review of Microbiology*, 58, 521-554.

Bech, L., Herbst, F. A., Grell, M. N., Hai, Z., and Lange, L. (2015). On-Site Enzyme Production by Trichoderma asperellum for the Degradation of Duckweed. *Fungal Genomics and Biology, 2015*.

Boraston, A.B.; Bolam, D. N.; Gilbert H. J. and Davies G. J. (2004). "Carbohydrate-binding modules: fine-tuning polysaccharide recognition". *Journalof Biochemistry*, 382, 769-781.

Boussaid, A. and Saddler, J. N. (1999). "Adsorption and activity profiles of cellulases during the hydrolysis of two Douglas fir pulps". *Enzyme Microbial Technology*, 15, 138–143.

Brethauer, S. and Michael, H. S. (2015). "Biochemical Conversion Processes of Lignocellulosic Biomass to Fuels and Chemicals – A Review". *Chimia International Journal for Chemistry*, 69 (10), 572–581.

Christian, A.; Kristof, T.; Helene, W.; Antje, C. S. and Matthias, W. (2013). "Membrane-based recovery of glucose from enzymatic hydrolysis of ionic liquid pretreated cellulose". *Bioresource Technology*, 149, 58–64.

Christine, Samantha, Amanda Silva, De Sousa, Suzane Rodrigues, Robson Tramontina, Roberto Ruller, Fabio Márcio, *et al.*, 2016. "Bioethanol Production by Recycled Scheffersomyces Stipitis in Sequential Batch Fermentations with High Cell Density Using Xylose and Glucose Mixture." *Bioresource Technology*, 219.

Cragg, S. M.; Beckham, G. T.; Bruce, N. C.; Bugg, T. D.; Distel, D. L. *et al.* (2015). "Lignocellulose degradation mechanisms across the Tree of Life". *Current Opinion in Chemical Biology*, 29, 108-119.

Davies, G. and Henrissat B. (1995). "Structure and mechanisms of glycosyl hydrolases". *Structure*, 3(9), 853-859.

Davies, G. J.; Planas A. and Rovira C. (2012). "Conformational Analyses of the Reaction Coordinate of Glycosidases". *Accounts of Chemical Research*, 45 (2), 308–316.

Davies G. J. and Williams S. J. (2016). "Carbohydrate Active Enzymes: sequences, shapes, contortions and cells". *Biochemical Society Transactions*, 44 (1), 79-87.

Divne, C.; Stanlberg, J.; Teeri, T. T. and Jones, T. A. (1998). "High resolution crystal structure reveal how a cellulose chain is bound in the 50 A long tunnel of cellobiohydrolase I from Trichoderma ressei". *Journal of Molecular Biology*, 275, 309-325.

Dusselier, M.~/ Mascal, M. and Sels, B. F. (2014). "Top chemical opportunities from carbohydratebiomass: a chemist's view of the Biorefinery". *Topics in Current Chemistry*, 353, 1-40.

Gan, Q.; Allen, S. J. and Taylor, G. (2002). "Design and operation of an integrated membrane reactor for enzymatic cellulose hydrolysis". *Biochemical Engineering Journal*, 12, 223–229

Galbe, M., and G. Zacchi. (2002). "A Review of the Production of Ethanol from Softwood." *Applied Microbiology and Biotechnology*, 59 (6), 618–628.

Henrissat B, Callebaut I, Fabrega S, Lehn P, Mornon JP, Davies G. (*1995*) Conserved catalytic machinery and the prediction of a common fold for several families of glycosyl hydrolases. *PNAS; 92:7090-94*

Ho, D. P.~/ Ngo, H. H. and Guo, W. (2014). "A mini review on renewable sources for biofuel". *Bioresource Technology*, 169, 742-749.

Horn, S. J.; Vaaje-Kolstad, G.; Westereng, B. and Eijsink V. (2012). "Novel enzymes for the degradation of cellulose". *Biotechnology for Biofuels*, 5, 45.

Hollinshead, W.~/ He, L. and Tang, Y. J. (2014). "Biofuel production: an odyssey from metabolicengineering to fermentation scaleup". *Frontiers in Microbiology*, 5, 344(1-8).

Imam, J.~/ Singh, P. K. and Shukla, P. (2016). "Plant Microbe Interactions in Post Genomic Era: Perspectives and Applications". *Frontiers in Microbiology*, 7, 1488.

Johan, B.; Ragna, P. and Folke, T. (2007). "Enhanced enzymatic conversion of softwood lignocellulose by poly(ethylene glycol) addition". *Enzyme and Microbial Technology*, 40, 754–762.

Jovanovic, I.; Magnuson, J. K.; Collart, F.; Robbertse, B.; Adney, W. S.; Himmel, M. E. and Baker, S. E. (2009). "Fungal glycoside hydrolases for saccharification of lignocellulose: outlook for new discoveries fueled by genomics and functional studies". *Cellulose*, 16, 687–697.

Karaki, N.; Aliawish, A.; Humeau, C.; Muniglia L. and Jasniewski J. (2016). "Enzymatic modification of polysaccharides: Mechanisms, properties, andpotential applications: A review". *Enzyme and Microbial Technology*, 90, 1-18.

Kim, J.; David E. B. and David A M. (2010). "Simultaneous Consumption of Pentose and Hexose Sugars/: An Optimal Microbial Phenotype for Efficient Fermentation of Lignocellulosic Biomass." *Applied Microbiology and Biotechnology*, 88, 1077–1085.

Klemm, D.; Heublein, B.; Fink, H. P. and Bohn A. (2005). "Cellulose: fascinating biopolymer and sustainable raw material". *Angewandte Chemie International Edition*, 44 (22), 3358-3393.

Kostylev, M and Wilson, D. (2012). " Synergistic interactions in cellulose hydrolysis". *Biofuels*, 3(1), 61-70.

Kumar, J. and Saini, R. (2015). "Lignocellulosic Agriculture Wastes as Biomass Feedstocks for Second-Generation Bioethanol Production/: Concepts and Recent Developments." *3 Biotech*, 5, 337–353.

Kumar, R.; Singh, S and Singh, O. V. (2008). "Bioconversion of lignocellulosic biomass: biochemical and molecular perspectives". *Journal of Industrial Microbiology and Biotechnology*, 35, 377–391.

Lali, A. M., Nagwekar, P. D., Varavedekar, J. S., Wadekar, P. C., Gujarathi, S. S., Valte, R. D.,. and Odaneth, A. A. (2014). *U.S. Patent No. 8,673,596*. Washington, DC: U.S. Patent and Trademark Office.

Lee, H. V.~/ Hamid, S. B. and Zain, S. K. (2014). "Conversion of lignocellulosic biomass to nanocellulose: structure and chemical process". *The Scientific World Journal*, 631013(1-20).

Lombard, V.; Ramulu H. G.; Drula E.; Coutinho P. M. and Henrissat B. (2014). "The carbohydrate-active enzymes database (CAZy) in 2013". *Nucleic Acids Research*, 42 (1), 490-495.

Lynd, Lee R, Paul J Weimer, Willem H van Zyl, and Isak S Pretorius. (2002). "Microbial Cellulose Utilization: Fundamentals and Biotechnology." *Microbiology and Molecular Biology Reviews*, 66 (3), 506–577.

Maki, M.; Leung, K. T.and Qin, W. (2009). "The prospects of cellulase-producing bacteria for the bioconversion of lignocellulosic biomass". *International Journal Biological Sciences*, 5 (5), 500-516.

Murphy C.; Powlowski, J.; Wu, M.; Butler, G. and Tsang, A. (2011). "Curation of characterized glycoside hydrolases of Fungal origin". *Database*, 2011, bar020.

Payne, C. M.; Knott, B. C.; Mayes, H.B.; Hansson, H.; Himmel, M. E.; Sandgren, M.; Stahlberg, J. and Beckham, G. T. (2015). "Fungal cellulases". *Chemical reviews*, 115 (3), 1308-1448.

Popp, J.; Haranqi-Raskos, M.; Gabnai, Z.; Balogh, P.; Antal, G. and Bai, A. (2016). "Biofuels and Their CoProducts as Livestock Feed: Global Economic and Environmental Implications". *Molecules*, 21(3), p. 285(1-26).

Sajith, S.; Priji, P.; Sreedevi, S. and Benjamin, S. (2016). "An Overview on Fungal Cellulases with an Industrial Perspective". *Journal of Nutrition and Food Sciences*, 6, 461.

Sharma, A.; Khare, S. K. and Gupta, M. N. (2001). "Hydrolysis of rice hull by cross-linked Aspergillus niger cellulase". *Bioresource Technology*, 78, 281-284.

Schneider, R.; Hanak, T.; Persson, S. and Voigt C. A. (2016). "Cellulose and callose synthesis and organization in focus, what's new?" *Current Opinion in Plant Biology*, 34, 9-16.

Sweeney M.D., Xu Feng. 2012. " Biomass Converting Enzymes as Industrial Biocatalysts for Fuels and Chemicals: Recent Developments" *Catalysts*, 2 (2), 244-263.

Tomes, D., Lakshmanan, P., and Songstad, D. (Eds.). (2010). Biofuels: global impact on renewable energy, production agriculture, and technological advancements. Springer Science and Business Media.

Vaaje-Kolstad, G., Westereng, B., Horn, S. J., Liu, Z., Zhai, H., Sørlie, M. and Eijsink, V. G. H. (2010). "An oxidative enzyme boosting the enzymatic conversion of recalcitratnt polysaccharides". *Science*, 330, 219–222.

Verardi, Alessandra, Isabella De Bari, Emanuele Ricca, and Vincenza Calabrò. (2012) "Hydrolysis of Lignocellulosic Biomass/: Current Status of Processes and Technologies and Future Perspectives." *Bioethanol*, 290.

9

Enzyme Catalysis in Bioproduct Development

T.C.K. Sugitha, U. Sivakumar and K. Ramasamy

Department of Agricultural Microbiology,
Tamil Nadu Agricultural University, Coimbatore – 641 002

Introduction

Biomass conversion has emerged as a potential renewable energy reserve and is regarded as an important route for catalytic transformation of biomass into a variety of bioproducts. Three major approaches have been widely explored in lignocellulosic biomass conversion process: physical (*e.g.* various pre-treatment/conversion technologies such as milling pressures high temperatures etc.,), chemical (*e.g.* strong acid treatment) and biological process. Physical and chemical processes are generally more efficient in terms of total conversion, but they are energy consuming and generate more by-products. In biological processes, such as fermentation and enzymatic catalysis, peculiar enzymes and microorganisms are used to break down lignocellulosic substrates and thus a variety of few commodity chemicals can be obtained(Zhang and Lynd, 2004; Pedersen and Meyer, 2010).

The enzyme-based biological processes can be performed under mild conditions with high specificity for the target products. Nevertheless, such biological processes generally suffer from unsolvable problems such as low efficiencies, specific conditions for reactions and limited scale of production (Shafiee and Topal, 2009; Anwar *et al.*, 2014). Enzyme-assisted biomass conversion can therefore constitute an alternative green approach, which can potentially reduce costs, byproducts. This will enhance the reaction efficiency (under optimum conditions) and specificity. The use of one-pot simple protocols mediated by multiple enzymes without any need for isolation steps (cascade reactions) and/or continuous flow processes can offer significant benefits when compared to physico-chemical biomass conversion processes. These include decreased reactor volume, decreased unit operations, shortened cycle times, increased volumetric and space-time yields (Ricca *et al.*, 2011).

In general, the biomasss undergoes depolymerization and hydrolysis of cellulose to monomer glucose which is regarded as a necessary initial step for the process to move further. Later glucose is catalytically degraded into various intermediates, fine chemicals and fuels. The various catalytic approaches were *viz.*, homogeneous/heterogeneous and acid/bases based catalysis for the conversion of carbohydrates or carbohydrate-derived chemicals. Recent research has paid much heed in accelerating the conversion of lignocellulosic biomass in the presence of multifunctional enzymes as catalysts *viz.*, amylase, cellulase, xylanases, laccase, pectinases and lipase etc. So a thorough understanding is needed regarding the catalytic properties of these enzymes, structure-activity relationship, activation energy and kinetics in order to derive maximum efficiency of these catalysts. The present chapter will highlight recent advances during enzymatic reaction in conversion of lignocellulosic substrates to derived products, substrate specificity, enzyme kinetics, models and integrated enzymatic catalysis for deconstruction of biomass.

1. What is Enzyme Catalysis?

Catalysis is defined as a phenomenon in which the rate of the reaction is altered, and the substance used to accelerate remains unchanged regarding its quantity and chemical properties. The substance used to change the speed of the reaction is termed as a 'catalyst'. Enzymes are catalysts which are responsible for increasing the rate of reaction. The catalysis in which enzymes act as a catalyst is called enzyme catalysis.

2. Why Enzyme Catalysis is Used in Biomass Conversion?

☆ A single molecule of catalysis can transform a million molecules of the reactant per second. Hence it is highly efficient.

☆ Biochemical catalysts are unique in nature *i.e.* the same catalyst cannot be used in more than one reaction.

☆ The efficiency of any catalyst is maximum at its optimum temperature. The activity declines at either side of the optimum temperature.

☆ Biochemical catalysis is highly dependent upon the pH of the solution. Any catalyst functions best at an optimum pH range of 5-7.

☆ The activity of the enzymes usually increases due to the presence of a co–enzyme or an activator such as Na^+, Co^{2+}. The rate of the reaction increases due to the presence of a weak bond which exists between the enzyme and a metal ion.

3. Biochemical Perspective of Enzyme Catalysis

In any reaction, the reactant molecules must contain sufficient energy to cross a potential energy barrier, which is called as the **activation energy**. All molecules possess varying amounts of activation energy depending on their collision but, generally, only a few have sufficient energy for the reaction to occur. The lower the potential energy barrier to reaction, the more reactants possess sufficient energy and, hence, faster the reaction. The various parameters involved in enzyme catalysis is are detailed below:

a) Transition State

In general, chemical reactions are adiabatic (without transfer of heat and matter), nonadiabatic, or in the intermediate regime. All catalysts, including enzymes, function by forming a transition state, with the reactants, of lower free energy. It was noted that the catalysed reaction pathway goes through various transition states TS_{c1}, TS_{c2} and TS_{c3}, with standard free energy of activation ΔG_c^*, whereas the uncatalysed reaction goes through the transition state TS_u with standard free energy of activation ΔG_u^*. The catalytic effect lowers the standard free energy of activation from ΔG_u^* to ΔG_c^* and has no influence on the overall change in free energy (*i.e.* the difference between the initial and final states) or the related equilibrium constant.

b) Free Energy Barrier

The enzyme systems involve the motion of many atoms. The free-energy profile is projected onto a single collective reaction coordinate. According to Zhang and Lynd (2004), the transition state is identified with the configuration at the top of the free-energy barrier. On the basis of fundamental reaction rate theory, enzymatic catalysis can be analyzed in terms of free-energy barrier and the coefficient of transmission relative to the reaction. Since the coefficient of transmission is a pre-factor, whereas the free-energy barrier is in the exponential phase of the overall rate expression, the most important contributions to enzymatic catalysis arise from lowering of the free-energy barrier rather than from dynamical effects on the transmission coefficient. As per Schiffer, (2013) the free energy barrier can be calculated by generating the free energy profile along a collective reaction coordinate, as depicted in Figure 9.1.

c) Active Sites and Substrate Interaction

The structures of enzymes determine the energies available for binding with their substrates *viz.*, optimal non-covalent ionic and/or hydrogen bonding forces plus a good 3-dimensional fit. The enzyme has specific binding sites also known as specific sites for the attachment of substrates and to modify the same. The specificity of enzyme binding depends upon the absence of unsolvated or unpaired charges, the presence of sufficient hydrogen bonds and minimal steric repulsion. In case of cellulose, carboxy methyl cellulose and p-nitro phenol are considered to be the substrate. Here the complex cellulose is converted into its monomeric form, D-glucose. The three types of active sites found in glycosyl hydrolases are 1) pocket or crater 2) cleft or groove and 3) tunnel. The geometry of the active site depends

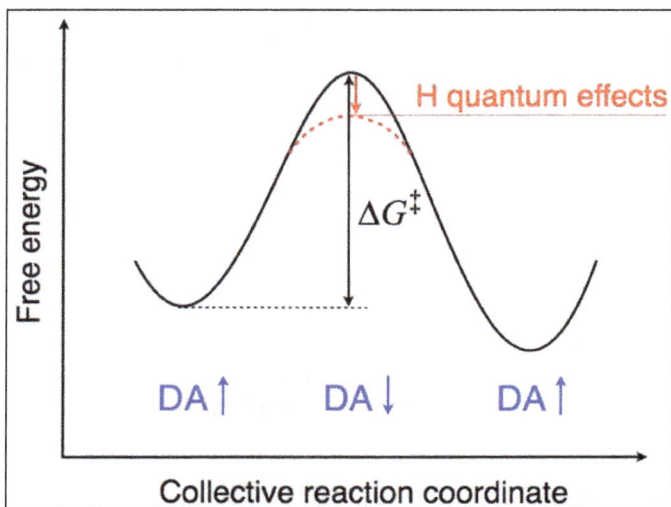

Figure 9.1: Schematic Depiction of an Adiabatic Hydrogen Transfer Reaction.
(*Source*: Adopted from catalytic efficiency of enzymes by Schiffer, 2013:A theoretical analysis/Biochemistry).

The free energy is plotted along the collective reaction coordinate, with the free energy barrier denoted by $\Delta G^{\ddagger}$. The dashed red curve and red arrow indicate the decrease in the free energy barrier when the nuclear quantum effects of the transferring hydrogen are included. DA denotes the hydrogen donor-acceptor distance, which is typically larger at the reactant and product (*i.e.*, minima) and smaller at the transition state (*i.e.*, top of the barrier) because the donor and acceptor must get closer for the hydrogen to transfer.

on the endo or exo specificity of the enzyme. Substrates bind to the open cleft in endoglucanases and xylanases. This causes the twisting of cellulose strands along the chain facilitating endo mode of action. On the other hand, the active sites of exoglucanases and cellobiases are perfectly enclosed substrate binding tunnels to cleave the cellulose chains from its non-reducing end. According to Kurasin and Valjamae (2011), the cleft like active site typically exhibit endocellulolytic activity binding along the length of the cellulose molecule and hydrolysing β1-4 glycosidic bonds where as the tunnel like active sites shows exocellulolytic activity.

d) Protein Folding Patterns

The protein folding patterns determine the conformation of active sites and enzyme activity. Cellulases with endomode of action possess large number of protein folds (Leonid *et al.*, 2011). Many enzymes involved in bioconversion process are multi-domain proteins having acccessory domains like carbohydrate binding modules in-addition to its catalytic domain (Fontes and Gilbert, 2010). The bacterial cellulosome complex contains a scaffolding protein with cohesions and dockerins on which the enzyme subunits are placed periodically. However the cellulosome compositions are not similar in all the strains. Cellulosomes increases the surface area accessible to enzyme attack by separating cellulose microfibrils (Michael *et al.*, 2013).

e) Specific Sctivity and Enzyme Kinetics

Specific activity is determined based on enzyme-substrate interaction and affinity. The more the affinity the specific activity is also more. Moreover, the specific activity is also decided by the purity of the enzyme. Usually specific activity is inversely proportional to purification fold. The concentration of substrate versus enzyme activity determines the kinetics of enzymes in lignocellulosic biomass conversion. The kinetics is explained elaborately below in this chapter.

f) Inhibition of Enzyme Activity during Biomass Bioconversion

The enzyme tends to get inhibited or loses its property by denaturation of native conformation of the enzyme. Glucose, cellobiose, ethanol and butanol are possible non competitive inhibitors, 6 gluconalactone is a mixed inhibitor and acetone is a non competitive activator during lignocellulosic biomass conversion (Holtzapple *et al.*, 1990). The types of inhibition and changes in K_m and V_{max} values are given in the Table 9.1. Certain phenolic compounds also inhibit cellulsae activity (Eduardo *et al.*, 2010).

Table 9.1: Types of Inhibition and their Respective K_m and V_{max} Values

Type of inhibition	Competitive	Non Competitive
K_m	Decreases	unchanged
V_{max}	unchanged	increases

g) Enzyme Tunnelling: Is it a Puzzle during Enzyme Catalysis?

Hydrogen tunnelling plays a major role in the active sites of the enzyme in case of many enzyme mediated reactions and enzyme kinetics. Enzymes posseses the property of quantum hydrogen tunnelling in which protons vanish from one position and rematerialise in another position. The biochemical reactions involving the transfer of electrons and hydrogen nuclei may occur through quantum mechanical tunnelling. According to Marcus theory the collective reaction coordinate for electron transfer is comprised of the nuclear motions that influence the relative stabilities of the two charge-transfer states. Electron tunnelling can occur when fluctuations of the nuclei lead to a configuration for which the two states are degenerate. The probability of hydrogen tunnelling depends on the width and height of the hydrogen-transfer barrier, as determined by the configuration of the heavy atoms. Enzyme motion for reactions involving quantum mechanical tunnelling can be discussed in terms of two distinct types of effects (Klinman and Kohen, 2013). First, the motion can alter the probability of achieving configurations with degenerate quantum states and, second, the motion can alter the probability of tunnelling at this degenerate configuration.

4. Rate-Limiting Factors of Enzymatic Hydrolysis during Biomass Conversion

The hydrolysis of lignocellulosic biomass mediated by enzymes, involve the use of cellulase and hemicellulase enzymes to convert cellulose and hemicelluloses into hexoses (glucose, galactose and mannose) and pentoses (xylose and arabinose) respectively. In the hydrolysis of xylan, which is the dominant hemicelluloses in many biomass, three major enzymes including endo-β-1-4 xylanse, catalyse the hydrolysis of β-1-4 bonds between D xylose residues of heteroxylans and xylooligosaccharides, exoxylanase, which releases xylobioses and β-xylosidases which degrade xylooligosaccharides to xylobiose and xylose are involved (Saha and Bothast,1999). The removal and solubilisation of xylan and lignin enhances the accessibility of cellulase to cellulose (Xu *et al.*, 2011). The choice of combinations of enzymes depends on the biomass feedstock and type of pretreatment technology.

4.1. A Glimpse of Biochemical Process during Lignocellulosic Substrates

a) Enzyme Substrate Complex

In enzymatic hydrolysis, the first step is the formation of an enzyme-substrate complex, which involves the mass transfer of enzyme from bulk aqueous phase to cellulose/hemicelluloses surface and the formation of enzyme substrate complex following enzyme adsorption. The subsequent hydrolysis of cellulose/hemicelluloses/lignin has two possible modes of action, with respect to the location of enzyme and substrate.

a) One mode of action focuses on the movement of substrate and incorporates three steps. According to O'Dwyer *et al.* (2007), the reactant molecules are first transferred to the active site of the enzyme-substrate complex. Then the reaction is catalyzed by the enzyme, followed by the release of soluble products to the bulk aqueous phase.

b) The other mode of action refers to the location of enzyme, which neither moves to the next reaction site along the surface of cellulose or desorbs and then re-adsorbs onto cellulose. In cellulose hydrolysis, the conversion of cellulose to cellobiose and glucose involves two heterogeneous reactions, while the degradation of cellobiose to glucose is considered to be a homogeneous react

b) Determinants of Hydrolysis Rate

The major determinants deciding rate of hydrolysis and kinetics of cellulolytic enzymes during lignocellulosic biomass conversion are: crystallinity index, degree of polymerization, lignin content, particle size, pore volume, and accessible surface area (Zhang and Lynd, 2004; Del and Luque, 2015).

i) Fine Structural Features of Lignocellulosic Biomass

The structural features of lignocellulosic substrate, exerts great influence in deciding the rate of hydrolysis. It is not possible to obtain a discrete population of particles with identical fine-structure due to fine-structure variability of cellulose. There exists a great degree of variability in the size and shape of individual particles (Walker *et al.*, 1990, Weimer *et al.*, 1990). The interrelationships among the various structural features of a cellulose moiety determine the structure-utilization relationship. For instance, the same structural discontinuities that contribute to increased pore volume, also serve to lower the average degree of crystallinity. Hence it is difficult to change (*e.g.*, by physical or chemical treatments) one fine-structure feature without simultaneous modification of others. Thus, as noted by several workers (Weimer *et al.*, 1995), studies pertaining to identify of the major structural features that determine the rate of hydrolysis/utilization often have been over interpreted, due to a failure to measure or consider changes in other structural determinants.

ii) Cellulose Crystallinity Index

Crystallinity is a major determinant of cellulose hydrolysis at both microbial and enzymatic levels. Relative crystallinity index (RCI) is the parameter used to estimate the degree of interchain bonding. In any enzymatic reaction with cellulose, the initial hydrolysis rate is predominantly affected by crystallinity index. As a general phenomenon, crystallinity increases during the course of cellulose hydrolysis which results in more rapid removal of amorphous material. On the other hand pre-treatments of biomass reduce crystallinity and it usually enhance the hydrolysis of cellulose by fungal cellulases (Fochar *et al.*, 1981). Amorphous celluloses are degraded 5 to 10 times more rapidly than the highly crystalline celluloses by both fungal enzymes (Gama *et al.*, 1994) and ruminal bacteria (Weimer *et al.*, 1991). For the bacteria *Cellulomonas*, *F. succinogenes* (Weimer *et al.*, 1993), *Clostridium cellulolyticum* (Desvaux *et al.*, 2000; Del and Luque, 2015), and members of the ruminal microflora (Garleb *et al.*, 1991), no significant changes in relative crystallinity index were observed during growth on cellulose.

iii) Lignin Content

Both lignin content and crystallinity index of cellulosic biomass have remarkable impact on ultimate sugar yield (Chang and Holtzapple, 2000). The overall conversion efficiency of carbohydrates was reported to be mainly dependent on the amount of residual lignin in pretreated biomass (Laureano-Perez *et al.*, 2005). During pretreatment, hydrogen bonds that contribute to the high degree of cellulose crystallinity can be broken down, which is beneficial for enhancing the initial hydrolysis rate.

iv) Pore Size and Surface Area

The available surface area is regarded as a potential determinant of hydrolytic rate. The pore structure of cellulosic substrates can accommodate particles proportionate to the size of a cellulolytic enzyme (Grous *et al.*, 1986, Thompson *et al.*, 1992). Hence, a good correlation has been observed between total surface area and the rate of substrate hydrolysis. According to Gama *et al.* (1994), cellulolytic enzymes do not penetrate the pore structure of purified celluloses. However, effective cellulolysis requires synergism among several enzymes, *viz.*, cellulases, hemicellulase, xylanase, laccase etc. The combined size is normally larger than the micropores that would accommodate a single enzyme. The external surface area, including the macropores, is the actual effective contact area between enzymes and cellulose during the start up of the reaction. The pore structure is an important determinant of hydrolysis in natural biomass materials than in purified celluloses, which have relatively lower porosity and smooth surfaces.

Cellulose hydrolysis is affected by the inability of enzymes to access additional substrate if hydrolytic cleavage of bonds is moderately rapid. This shows that a part to most cellulose chains is being buried within the microfibrils. The coverage of some surface chains by the "footprint" of enzyme molecules (particularly those in a complexed form) that covers many bonds (Gilkes *et al.*, 1992) physically restricts the binding of additional enzyme molecules to neighbouring sites on the fibre. But, continued hydrolysis of cellulose requires both excision and removal of hydrolytic products from the site of attack to reveal the underlying cellulose chains to the enzymes.

v) Fragmentation of Cellulose

Fragmentation also increases reaction rate with time due to increased surface area. The importance of fragmentation is well understood from the studies of Weimer *et al.* (1991), in which the comparisons of the rate of weight loss of various cellulose allomorphs by mixed ruminal microflora and pure cultures as well. The most slowly utilized allomorphs (cellulose II and cellulose III$_{II}$) hydrogen bonds between adjacent sheets, in addition to the intrachain and inter-chain hydrogen bonds common to all crystalline allomorphs. This would impede the fragmentation of microfibrils into individual chains. While employing whole cells, the rates of cellulose hydrolysis have been found to be constant or to decline with increasing substrate conversion.

vi) Physical and Chemical Conditions of the Environment

As a microbial process, cellulose utilization by enzymes is subject to physical and chemical conditions in the environment. Among the environmental parameters, affecting the rate and extent of cellulose utilization, temperature plays a predominant role followed by pH and redox potential. Rates of degradation were linear, suggesting that the process was limited by microbial activity rather than by the amount of substrate (Weimer, *et al.*, 1991; Sugitha, 2007).

In case of reactions involving bacteria, most anaerobic cellulolytic bacteria, like other fermentative microbes, prefer to grow within a fairly narrow pH range. However, pH fluctuations permit cellulose hydrolysis to occur even at lower pH values than those supporting the growth of cellulolytic population in some habitats. For instance, substantial cellulose hydrolysis can occur by rumen bacteria at pH below 6.0. According to Maurino *et al.* (2001), once the bacteria adheres to cellulose, synthesizes a glycocalyx, and initiate bacterial growth at a higher pH Cellulose removal in some anaerobic mixed cultures is observed at pH as low as 4.5 (Chyi *et al.*, 1994) and has an optimum near pH 5.0.

5. Quantitative Description during Hydrolysis of Lignocellulosic Biomass

Quantitative description of lignocellulosic hydrolysis is of potential value in two contexts: (i) for structuring and testing fundamental understanding and (ii) for designing and evaluating engineered systems based on quantitative models.

Cellulase enzyme systems are composed of multiple proteins that interact by performing complementary functions and also by forming multiprotein complexes in some cases. Such interactions will exhibit different behaviour depending on the cellulase system components whether they are present in combination or in isolation. Cellulosic substrates are typical insoluble macroscopic particles with characteristic distribution of sizes, shapes, and, composition. In general, cellulosic substrates contain significant amount of lignin to which many cellulase components bind. Presence of hemicellulose, both in naturally occurring and pretreated substrates interferes the access of cellulase components to 1,4-β-glucosidic bonds. Hence it may require enzymatic activities distinct from those involved in cellulose degradation for hydrolysis (Yang et al., 2011).

During the course of enzymatic hydrolysis, kinetic properties like adsorption capacity and affinity, substrate reactivity, chemical properties (fractional composition of different components), and physical properties (size, shape, density, and rigidity) exhibits a remarkable change. Cellulolytic organisms influence the rate of cellulose hydrolysis by determining the rate of cellulase production which is in turn determined by the interaction of multiple complex processes. All these factors varies and totally depends on the nature of substrate and types of enzyme systems.

5.1. Kinetics of Enzymes during Lignocellulosic Substrate Catalysis

The kinetics of enzymatic hydrolysis of lignocellulosic substrates is very complex. It is due to multiple hydrolytic enzyme activities encompasses in the process and heterogeneous nature of substrate (Kadam et al., 2004). For successful modelling of cellulose kinetics, a thorough fundamental understanding of physical and chemical properties of the reacting substrate and its relevant enzyme, a complete idea on the rate limiting factors are required (Bansal et al., 2009). As a potential improvement over the assumptions of number of kinetic models, Bansal et al. (2009) classified all the models into four categories. They are: empirical models, Michaelis Menton based models, adsorption based models and soluble substrate based models.

Empirical models have limitations on their applicability to conditions outside those used for model development. They are of not much useful in terms of disclosing the mechanism of enzymatic catalysis during hydrolysis. Likewise, soluble substrate based models are not applicable due to the interference of insoluble lignocellulosic substrate. Adsorption based models and Michaelis Menton based models are discussed briefly in this book chapter.

5.1.1. Adsorption Based Kinetic Models Approach

a) What is Adsorption?

During bioconversion of lignocellulosic biomass, the carbohydrate binding modules (CBMs) of cellulase enzymes readily adsorb to accessible sites on a lignocellulosic substrate particle to form a complex held together by specific, non-covalent interactive forces. In some cases, catalytic domains of cellulase system components may specifically adsorb to cellulose independently of CBMs. Cellulase may also adsorb to lignin, which is thought to be nonspecific (Ooshima *et al.*, 1991). The prerequisite for cellulose hydrolysis is the formation of enzyme-substrate complex. Such complexes are a central feature of most quantitative and conceptual models for cellulose hydrolysis. Klyosov (1990) showed a strong correlation between values of the adsorption equilibrium constant and hydrolysis rates.

The adsorption of cellulase(s) to cellulose can be quantitatively described based on the concentration of a cellulose-enzyme complex. It is expressed as a function of a vector of variables relevant to cellulase adsorption. The variables include, total amount of cellulase, total amount of substrate, substrate-specific and enzyme-specific parameters that influences adsorption (*e.g.*, affinity and capacity), and parameters that describe the physical and chemical environment (*e.g.*, temperature and ionic strength). In any enzymatic reaction for biomass bioconversion, experimental determination of the concentration of cellulose enzyme complex, $[CE]$, is usually carried out by taking the difference between total cellulase present and unadsorbed cellulase, This can be explained as follow for a substrate containing only cellulose:

$$[CE] = [E_T] - [E] \tag{1}$$

where,

$[E_T]$ is the total concentration of binding sites on the enzyme and $[E]$ is the concentration of binding sits on the enzyme not adsorbed to cellulose (Mansfield *et al.*, 1999).

b) Equilibrium in Adsorption Based Models

In many adsorption models equilibrium assumption is often justified by the time required for adsorbed cellulase to reach a constant value. It is shorter relative to the time required for hydrolysis. Most studies revealed that adsorbed cellulase reaches a constant value in $\leq$90 mins. However few researchers have found $\leq$30 min to be sufficient (Singh *et al.*, 1991, Strobel *et al.*, 1995). But complete hydrolysis of cellulose usually requires a day or more. The simplest representation of adsorption equilibrium is given by means of an equilibrium constant, K_d:

$$[K_d] = [E][C]/[CE] \tag{2}$$

where,

$[C]$ is the concentration of accessible binding sites on cellulose not adsorbed to enzyme. K_d, $[E]$, $[C]$, and $[CE]$ are expressed as micromoles per liter. The use of units other than micromoles per liter for K_d is considered below. As an alternative to equilibrium models, some models (Nidetzky *et al.*, 1994) employ a dynamic description of adsorption such as

$$d[CE]/dt = k_f[E][C] - k_r[CE] \tag{2}$$

where,

where $k_r/k_f = K_d$.

Studies involving various cellulases and cellulose samples indicate that once a cellulase-cellulose complex is formed, the enzyme remains bound to the cellulose (substrate) for a significant period (*e.g.*, 30 min or more) as mentioned above. During this time period, hundreds of catalytic events may occur. For instance, surface diffusion rates of *Cellulomonas fimi* cellulase on microcrystalline cellulose have been measured by Jervis *et al.* (1997). According to Ooshima *et al.* (1991), soluble sugars have no influence on adsorption behaviour. On the other hand, inhibition of adsorption by unidentified compounds in protein-extracted lucerne fibers has been reported by Stutzenberger and Lintz, (1986).

c) Cellulose Adsorption Parameters

The adsorption parameters for the same enzyme-substrate system differ by an order of magnitude. The maximum binding capacities (enzyme per unit substrate) and affinity constants $(1/K_d)$ for isolated cellulase components, cellulase mixtures and complexes as well as several substrates are presented in Table 9.2. Avicel and pretreated hardwood showed highest binding affinity (liters per gram of cellulase) by over 100-fold for the complexed type of cellulose, *C. thermocellum* system.

Table 9.2: Summary of Cellulase Adsorption Parameters

Organism, Substrate and Temp.	Binding Capacity			Binding Affinity		Reference
	mg/g	µmol/g	µmol/µmolb	Liters/g	Liters/µmol	
Components						Gilkes *et al.*, 1992
Cellulomonas fimi (30°C)						
CenA BMCCCex BMCC	144	**3.1**	5.0×10^{-4}	**41**	1.89	
	184	**3.6**	5.8×10^{-4}	**33**	1.71	
Thermobifida fusca (50°C)						
E3, Avicel	26	**0.4**	6.5×10^{-5}	3.1	**0.2**	Bothwell *et al.*, 1997
E3, BMCC	741	**11.4**	1.8×10^{-3}	1.5	**0.1**	
E4, Avicel	31	**0.34**	5.5×10^{-5}	0.85	**0.077**	
E4, BMCC	875	**9.7**	1.6×10^{-3}	0.49	**0.044**	
E5, Avicel	31	**0.67**	1.1×10^{-4}	4.8	**0.22**	
E5, BMCC	556	**12**	1.9×10^{-3}	2.8	**0.13**	
Trichoderma reesei						
CBHI, Avicel (4°C)	48	**0.74**	1.2×10^{-4}	0.69	**0.044**	Medve *et al.*, 1997
CBHI, Avicel (25°C)	57	**1.1**	1.8×10^{-4}	5.4	**0.28**	Stalberg *et al.*,1991
CBHI, Avicel (25°C)	15	**0.29**	4.7×10^{-5}			Tomme *et al.*,1988
CBHI, Avicel (50°C)	25	**0.48**	7.8×10^{-5}	1.7	**0.09**	Bothwell *et al.*,1997
CBHI/BMCC (50°C)	239	**4.6**	7.5×10^{-4}	5.4	**0.28**	Bothwell *et al.*,1997
CBHII, Avicel (4°C)	28	**0.52**	8.4×10^{-5}	1	0.053	Medve *et al.*, 1997
CBHII, Avicel (25°C)	11.3	**0.24**	3.9×10^{-5}			Tomme *et al.*,1988

Contd...

Table 9.2–Contd...

Organism, Substrate and Temp.	Binding Capacity			Binding Affinity		Reference
	mg/g	µmol/g	µmol/µmolb	Liters/g	Liters/µmol	
Trichoderma viride (30°C)						
ExoI	6.6	0.11	1.8×10^{-5}	5	0.3	Beldman et al.,1987
ExoIII	63	1	1.6×10^{-4}	6.9	0.43	
EndoI, Avicel	130	2.5	4.0×10^{-4}	0.88	0.04	
EndoIII, Avicel	26	0.45	7.3×10^{-5}	12	0.68	
EndoV, Avicel	110	1.8	2.9×10^{-4}	0.89	0.05	
Multicomponent mixtures and complexes						
Trichoderma reesei						
Avicel (50°C)	92	1.92	3.1×10^{-4}	1.04	0.05	Steiner et al.,1988
Avicel (40°C)	56	1.21	2.0×10^{-4}	3.21	0.015	
Cellulose a (40°C) in pretreated wood	81	1.68	2.7×10^{-4}	1.82	0.087	Ooshima et al., 1990
C. thermocellum						
Avicel b(60°C)	17.5	0.0083	1.4×10^{-5}	246	517	Bernadez et al., 1993
Pretreated wood(60°C)b,c	317	0.15	2.5×10^{-5}	344	722	

a: Dilute-acid-pretreated wood prepared at 220°C; an average molecular weight of 48,000 is assumed. b: Calculated quantities based on a specific activity of 2.4 µmol/mg/min and a molecular mass of 2.1×10^{6} Da. c: Dilute-acid-pretreated wood prepared at 220°C.

5.1.2. Soluble Substrates

For reactions involving soluble substrates, the ratio of enzyme-binding capacity to substrate reactive sites is ≤0.001 for most described cellulase-cellulose systems. Because of this difference, it is unusual for substrate to be in excess during enzymatic hydrolysis of cellulose. There exists a saturation of cellulosic substrates with cellulose. On contrary, during engineered bioprocess, free cellulase activity is present throughout the course of hydrolysis (Ooshima et al., 1990, Yu et al., 1995), which is indicative of accessible substrate sites not being in excess.

5.1.3. Langmuir-type Isotherm

Adsorption of cellulases onto insoluble substrates is an essential part of the kinetics of lignocellulosic catalysis. A material balance on accessible cellulose binding sites is important for cellulosic substrates and it is usually not included in enzyme kinetics of soluble substrates. The equation is as follow:

$$\sigma[S_T] = [C] + [CE] \tag{4}$$

where,

where σ is the binding capacity of the substrate, corresponding to the density of accessible binding sites on cellulose expressed in micromoles per gram. $[S_T]$ is the concentration of substrate (grams per liter). Substrate is usually not in excess during enzymatic hydrolysis of cellulose. So [C] is not higher than [CE]. Hence equation (4) may be substituted into equation (2) and solved for [CE] to give the Langmuir equation as below:

$$[CE] = \frac{\sigma[S_T][E]}{K_d + [E]} \tag{5}$$

The first attempt to model the kinetics of cellulose catalysis at high substrate and enzyme concentrations was by Wald *et al.* (1984). A Langmuir equation of the general form of equation 5 is the most commonly used relationship to describe cellulase adsorption. It has been used to describe cellulase adsorption in studies involving individual components (Woodward *et al.*, 1998), multicomponent non complexed systems (Wald *et al.*, 1984), and complexed systems and for describing adsorption to lignocellulosic materials (Bernadez *et al.*, 1993), lignin (Ooshima *et al.*, 1990), and purified cellulose (Ooshima *et al.*, 1983).

Substitution of both enzyme and substrate material balances (equations 1 and 4) into the equilibrium constant (equation 2) gives a quadratic equation in [CE], as noted by several authors (Steiner *et al.*, 1988, Wald *et al.*, 1984, Bothwell *et al.*, 1995):

$$[CE]^2 - [CE]\{\sigma[S_T] + [E_T] + K_d\} + \sigma[S_T][E_T] = 0 \tag{6}$$

Equation (6) has two roots. The physically meaningful root being the one that satisfies the condition that $0 \leq [CE] \leq [E_T]$ and $\sigma[S_T]$. The values of [CE] exhibit the following important features. (i) Values for [CE] approach a constant value when either $[S_T]$ is increased and $[E_T]$ is held constant or $[E_T]$ is increased at constant $[S_T]$, representing saturation with either substrate or enzyme. Such dual saturation has been conclusively demonstrated for both noncomplexed and complexed cellulases. (ii) The value of $[S_T]$ achieves an arbitrary extent of saturation in the concentration of [CE] (*e.g.*, half the maximum concentration) is a function of $[E_T]$, with higher $[S_T]$ required at higher $[E_T]$. This behavior is consistent with the adsorption study by Bernardez *et al.* (1993) as well as the kinetic studies by Wald *et al.* (1984), Lynd and Grethlein (1987), and Bernardez *et al.* (1994).

According to Bansal *et al.* (2009), the languir isotherm model can also be given as,

$$E_b = (E_{max}K_{ad}E_f S)/1 + K_{ad}E_f) \tag{7}$$

E_B is the adsorbed enzyme concentration (mg cellulose/L), E_{max} is the maximum adsorption capacity in the unit of mg cellulose per gram cellulose, K_{ad} is the disassociation constant, E_f is the free enzyme concentration and S is the substrate concentration. A simplified expression of the equation (7) is: [adsorbed enzyme]=constant × [free enzyme] × [substrate]. Langmuir isotherm is invalid for cellulolytic enzymes adsorption onto lignocellulosic substrates but provides a good fit for cellulosic substrates.

Cellulase enzyme components adsorb onto lignin as well as cellulose in a lignocellulosic biomass. It is evident from the reports of Sutcliffe and Saddler, (1986) and Chernoglazov *et al.* (1988) that binding of cellulose to lignin can decrease the rate of hydrolysis by several times. This can stop hydrolysis altogether before cellulose is exhausted.

5.1.4. Michaelis-Menton Based Kinetic Models Approach

The simplest enzymatic reaction is considered to be irreversible, and no product inhibition exists during the reaction. Thus, the Michaelis Menton equation is for most purposes not useful for understanding cellulose enzyme systems acting on crystalline cellulose. The reaction scheme is

$$E+S \underset{k_{-1}}{\overset{K_1}{\rightleftharpoons}} ES \xrightarrow{K_2} E+P \tag{8}$$

where,

E is the enzyme, S is the substrate, ES is the enzyme-substrate complex, P is the product, k_1 is the forward rate constant for the formation of enzyme-substrate complex, k_{-1} is the dissociation rate constant of the enzyme-substrate complex, and k_2 is the rate constant of product formation. From E+P, the product disassociates and the enzyme is available for next round of reaction. A typical kinetic model for homogeneous enzymatic reaction follows the Michaelis Menton equation, which is described as,

$$v = (v_{max}[S]/(K_M+[S]) \tag{9}$$

where,

v is the conversion rate of substrate, v_{max} is the maximum conversion rate of substrate, [S] is the substrate concentration, v_{max} is the maximum conversion rate of substrate and K_M is Michaelis constant.

The quasi-steady state assumption employed in the derivation of equation (10) cannot be directly applied in enzymatic hydrolysis of insoluble lignocellulosic substrates since it is a heterogeneous reaction system (Bansal *et al.*, 2009). But Michelis –Menton based based models can also be applied in simulating enzymatic reaction for both pure cellulose (Bezera and Dias, 2004; Caminal *et al.*, 1985) and lignocellulosic substrates (Brown *et al.*, 2010).

The identification of inhibition pattern of kinetic models for enzymatic hydrolysis of cellulosic biomass is important, as product inhibition can limit the bioproduct development from biomass. Competitive inhibition is the primary concern in the Michelis –Menton based model and its fits very well according to Kadam *et al.* (2004). The non-competitive inhibition pattern which is the consequence of non-preferential and irreversible binding of enzymes to lignin can be expressed by the HCH-I model as given below:

$$V=k[S][E]/(\alpha+\ddot{o}[S]+/[E]) \tag{10}$$

where,

[S] is the substrate concentration, [E] is the enzyme concentration, ö is the fraction of the substrate surface that is accessible to be hydrolyzed, and k, α and / are parameters that represent the degree of substrate reactivity. As reviewed by Bansal *et al.* (2009), the decreasing rate of enzymatic hydrolysis with increasing conversion can be attributed to enzymatic deactivation, biphasic composition, of cellulose, decrease in substrate reactivity and accessibility, fractal or spatially constrained reacting environment, decrease in synergism of cellular components and interfering by lignin. All these factors should be taken under consideration to simulate enzymatic catalysis at high conversion levels.

5.2. Limitations of Kinetic Models

Most of the lignocellulosic kinetic models developed for hydrolysis lack in the involvement of more than one substrate state variable and more than one hydrolyzing activity. Zhang and Lynd (2006) proposed a functionally based model taking into account multiple substrate variables and more than one solubilising activity, for cellulose kinetics using pure substrate. According to Wang *et al.* (2011) the fraction of accessible β-glucosidic bonds and degree of polymerization are highly correlated with substrate reactivity, thus influencing the hydrolysis rate. The degree of synergy between endoglucanse and exoglucanse predicted by the model increases as the extents of two substrate parameters are raised.

6. Rates of Enzymatic Hydrolysis

The previous section on enzyme kinetics dealt with how the concentration of the cellulose-cellulase complex varies with respect to the relevant variables. This section deals primarily with the proportionality factor, k' which is a function of temperature and also a function of additional state variables such as substrate conversion.

According to Lynd *et al.* (2002), models for the rate of enzymatic catalysis are based on the mathematical product of the concentration of the enzyme substrate complex and a proportionality factor relating this concentration to the reaction rate as follow:

$$r_C = k + [CE] \tag{11}$$

where,

r_C is the cellulose hydrolysis rate (substrate units/[volume × time]) and k is the rate constant, a proportionality factor between $[CE]$ and r_C (units as needed for dimensional consistency).

During enzymatic reaction, hydrolysis of one or more soluble oligosaccharides occurs. The general form of the equation for rate of reaction for oligosaccharides, including cellobiose, is as follows: overall rate of reaction of G_j = rate G_j formed from cellulose hydrolysis + rate G_j formed from reaction of oligosaccharides of length $>j$ – rate G_j reacts to form oligosaccharides with chain length $<j$, or

$$rG_j + fC \rightarrow G_j r_C + \sum_i rG_{j,i} \tag{12}$$

where,

G_j represents a soluble glucooligosaccharide of chain length j, r_{Gj} is the overall rate of formation of G_j (substrate units/volume/time]), $f_C{\rightarrow}_{Gj}$ is the fraction of cellulose hydrolyzed directly to G_j, and $r_{Gj,i}$ is the rate of formation of intermediate G_j by the ith reaction (+ if produced, − if consumed). If there are multiple soluble intermediates, equation (8) must be written for each intermediate.

6.1. Specific Activity of Cellulase

During biomass bioconversion, the cellulase-cellulose complex is expressed in terms of units of enzyme. Here the parameter 'k represents the specific activity (rate/unit of enzyme) of the cellulase system. The experimental data for specific activities for both multicomponent mixtures or complexes and components thereof with respect to Avicel is presented below in Table 9.3. The highest specific activities reported for multicomponent mixtures or complexes are higher and in the case of C. *thermocellum* (Morag *et al.*, 1992, Ahsan *et al.*, 1997). The larger benefit of multiple protein components for the C. *thermocellum* system compared with T. *reesei* may be because most of the catalytically active components of the C. *thermocellum* cellulosome lack a cellulose-binding domain and are thus dependent on the noncatalytic dockerin protein for cellulose binding.

7. Consolidated Bioprocessing of Biomass

Nowadays bioconversion of biomass has exhibited a trend toward increasing consolidation over time. This trend is most evident in the case of production of ethanol from lignocellulosic substrates, which has received the most attention among biomass-derived fermentation products. However, it is likely to be applicable to other products as well. Prior to the mid-1980s, simultaneous saccharification and fermentation is currently viewed as an economically attractive near-term goal for process development in bioprocess technology (Lynd *et al.*, 2005). It offers potential cost benefits relative to SSF. This can be viewed as an alternative *Cellulose Hydrolysis Paradigms*. Over the last decade, the development of microorganisms capable of converting xylose and other hemicellulose-derived sugars to ethanol at high yields has been one of the most significant advances in the fields of both biomass conversion and metabolic engineering. Both of the strategies presented above with respect to cellulose conversion have been successfully applied to xylose conversion.

For quantitative modelling of enzymatic hydrolysis of cellulose during simultaneous saccharification and fermentation in continuous well-mixed reactors, South *et al.* (1995) found it necessary to integrate reaction rates over the time individual particles, or segments of the particle population, spend in the reactor according to the following equation:

$$\chi(\tau) = \int_0^\infty E(t,\tau) \times \chi(t)d$$

(13)

where,

τ is the mean residence time, $\chi(\tau)$ is the substrate conversion as a function of τ, $E(t,\tau)$ is the particle residence time distribution equals to $e^{(t/\tau)}/\tau$, and

$$\chi(\tau) = \int_0^t \frac{r[S(t)]}{S_{in}}dt$$

Table 9.3: Specific Hydrolysis Rates for Cellulosic Substrates of High Crystallinity in Cell-free Systems

Organism	Cellulase Preparation	Substrate and Assay	Sp Act (µmol/min/mg)	Reference
Isolated component				
Clostridium cellulolyticum	CelA, expressed in E. coli and chromatographically purified	Avicel; 37°C, RS	0.11	Fierobe et al., 1991
	CelC, same	Avicel; 37°C	0.017	Fierobe et al., 1993
	CelE, same	Avicel; 37°C	0.06	Gaudin et al., 2000
	CelF, same	Avicel; 45°C	0.17	Reverbel-Leroy et al.,1997
Clostridium thermocellum	Ion exchange-purified celA expressed in E. coli	Avicel; 60°C, RSa, 1 h	0.083	Schwarz et al., 1986
		Avicel (summarizes data for 8 components from various studies)	0.0052-0.083	Ahsan et al.,1997
Neocallimastix patriciarum	Immunoaffinity-purified CELA expressed in E. coli	Avicel; 40°C, RS, 30-60 min	9.7	Denman et al., 1996
	Crude recombinant CELA	Avicel; 40°C, RS, 30-60 min	1.7	Xue et al., ., 1992

Source: Y. Zhang and L. R. Lynd, 2004.

8. Quantitative Structure-Activity Relationship (QSAR) Approach

Biomass pretreatment is needed to break down the recalcitrant structure of the plant cell wall for subsequent enzymatic hydrolysis and fermentation. However, the pretreatment processes generate inhibitors from the degradation of cellulose, hemicellulose and lignin, many of which significantly reduce the microbial growth and fermentation productivity. According to Tu (2005), the molecular structure and functional groups in carbonyls will govern the reactivity of carbonyl compounds and the reactivity of carbonyls will potentially dominate their inhibitory effects on enzymatic conversion during bioconversion.

9. Future Thrust

Bioconversion of biomass has shown great potential of applications in the production of biofuel and other value added products such as xylitol, lactic acid, ethanol, vanillin, methanol, butanol, acetobutane etc. Reese and Mandels (1971) wrote in 1971: "The dream of cellulase investigators is to develop a commercially feasible process for converting waste cellulose to glucose. The recent successful commercial practice of enzymatically converting starch to glucose gives new hope to this dream. The vast majority of studies investigating lignocellulosic hydrolysis and cellulose/laccase enzyme systems have proceeded within the context of an enzymatically oriented intellectual paradigm. But the major challenge is the efficient release of sugars from recalcitrant biomass in a reasonable time. To realize this goal, factors that hinder the conversion of cellulosic biomass need to be well understood, which require computer simulations with experimental results. Today, biotechnology is pioneering the progress towards both the microbial and enzymatic cellulose hydrolysis paradigms. Research efforts on modelling the conversion process have been focussed on the kinetic behaviour of cellulose, xylan and lignin present in the biomass. The following points can be considered in developing a more efficient multi-enzyme complex with more efficient conversion rate at cheaper rate.

i) For the microbial paradigm, achieving the organism development milestones with regard to both the native and recombinant cellulolytic strategies.

ii) For the enzymatic paradigm, cellulases, laccases, xylanases, amylases etc. with higher specific activity than current commercial preparations are highly desirable and are likely to be required. Developing such enzymes can be approached either by using new heterologous expression systems that naturally have high specific activity or by using protein engineering to create new, improved enzyme systems.

iii) Allele mining of thermophilic microbes/archaea for cellulolyitc complex to withstand adverse environment of high temperature during bioprocessing. Incorporation of both these paradigms into advanced technology in a convergent manner for high-specific-activity cellulases and consolidated bioprocess enabling microorganisms.

iv) Enzyme tunnelling and its impact in cellulase and laccase enzymes need to be investigated for enhancing the efficiency and conversion rate.

v) In the development of kinetic models, the impact of pre-treatment conditions such as temperature, chemical concentration or pH, residence time as well as the interaction between lignin and other carbohydrates on the reaction rate of each major component need to be considered. A deeper insight of pretreatment mechanisms of recalcitrant biomass is necessary for building robust models.

vi) A severity factor that combines all pre-treatment conditions into one variable can be integrated into kinetic or empirical models to simplify the inputs for saving efforts in process optimization.

vii) Non-kinetic models that exclude mathematical formulas such as artificial neural network and fuzzy logic based inference systems, tend to be more efficient than kinetic models.

viii) Definitive quantitative evaluation of enzyme–microbe synergy is an important objective for future research, and could provide further evidence supporting the desirability of consolidated bioprocessing.

ix) Despite all these advancements, the enzymatic decrystallization process is still critical and poorly understood in many cases. Emerging technologies such as cascading techniques should be further investigated for more efficient biocatalytic conversions

10. Conclusion

Since commercially feasible cellulose conversion processes featuring enzymatic hydrolysis are available, sustainable resource supply, energy security, and global climate change have emerged as dominant issues. Despite the great effort that has been devoted to the biomass and bioprocess technology, there exists biotechnological and nanotechnological approaches for developing practical processes for the conversion of lignocellulose to fuels and commodity chemicals that are both promising and relatively unexplored. Synergistic activity of enzyme systems needs to be explored to attain efficient bioconversion. Inclusion of multiple substrate state variables and more than one hydrolyzing activity into a functionally based model development will provide vivid understanding on enzymatic catalysis during lignocellulosic biomass.

References

Ahsan, M. M., M. Matsumoto, S. Karita, T. Kimura, K. Sakka, and K.Ohmiya. 1997. Purification and characterization of the family J catalytic domain derived from the *Clostridium thermocellum* endoglucanase Cell *J. Biosci. Biotechnol. Biochem.* 61:427–431.

Anwar, Z., M. Gulfraz and M. Irshad. 2014. Agro-industrial lignocellulosic biomass a key to unlock the future bioenergy: A brief review. *Journal of Radiation Research and Applied Sciences*, http://dx.doi.org/10.1016/j.jrras.2014.02.003.

Bansal, P., M. Hall, M.J. Realff, J.H. Lee and A.S. Bommarius. 2009. Modelling cellulose kinetics on lignocellulosic substrates. *Biotechnol.* Adv.,27:833-848.

Beldman, G., A. G. J. Voragen, F. M. Rombouts, M. F. Searle-van Leeuwen, and W. Pilnik. 1987. Adsorption and kinetic behavior of purified endoglucanases and exoglucanases from *Trichoderma viride*. *Biotechnol. Bioeng.*, 30:251–257.

Bernardez, T. D., K. Lyford, D. A. Hogsett, and L. R. Lynd. 1993. Adsorption of *Clostridium thermocellum* cellulases onto pretreated mixed hardwood, Avicel, and lignin. *Biotechnol. Bioeng.* 42:899–907.

Bezera, R.M.F. and A.A. Dias. 2004. Discrimination among eight modified MIchelis – Menten kinetics models of cellulose hydrolysis with a large range of substrates/enzyme ratios. *Appl. Biochem.Biotechnol.*,112:173-184.

Bothwell, M. K., S. D. Daughhetee, G. Y. Chaua, D. B. Wilson, and L. P. Walker. 1997. Binding capacities for *Thermomonospora fusca* E3, E4, and E5, the E3 binding domain, and *Trichoderma reesei* CBHI on Avicel and bacterial microcrystalline cellulose. *Biores. Technol.* 60:169–178.

Brown, R.F., F.K. Agbogbo and M.T. Holtzapple. 2010. Comparison of mechanistic models in the initial rate enzymatic hydrolysis of AFEX-treated wheat straw. *Biotechnol. Biofuels,* 3(6):1-13.

Caminal, G., Lopez-Santin and J. Solac. 1985. Kinetic modelling of enzymatic hydrolysis of pretreated cellulose. *Biotechnol. Bioeng.*,27:1282-1290.

Chang, V.S and M.T. Holtzapple. 2000. Fundamental factors affecting biomass enzymatic activity. *Appl. Biochem.Biotechnol.*,84:5-37.

Chernoglazov, V. M., O. V. Ermolova, and A. A. Klyosov. 1988. Adsorption of high-purity endo-1,4-_-glucanases from *Trichoderma reesei* on components of lignocellulosic materials: cellulose, lignin, and xylan. *Enzyme Microb.Technol.*, 10:503–507

Chyi, Y. T., and R. R. Dague. 1994. Effects of particulate size in anaerobic acidogenesis using cellulase as a sole carbon source. *Water Environ. Res.*66:670–678.

De1, S. and R. Luque. 2015. Integrated enzymatic catalysis for biomass deconstruction: a partnership for a sustainable future Sustainable Chemical Processes (2015) 3:4, DOI 10.1186/s40508-015-0030-9

Denman, S., G.-P. Xue, and B. Patel. 1996. Characterization of a *Neocallimastix patriciarum* cellulase cDNA (*celA*) homologous to *Trichoderma reesei* cellobiohydrolase II. *Appl. Environ. Microbiol.* 62:1889–1896.

Desvaux, M., E. Guedon, and H. Petitdemange. 2000. Cellulose catabolism by *Clostridium cellulolyticum* growing in batch culture on defined medium.*Appl. Environ. Microbiol.* 66:2461–2470.

Eduardo, X., K. Youngmi, M. Nathan, D. Bruce and L. Michael. 2010. Inhibition of cellulose by phenols. *Enzyme Microbial. Technol.*, 46:170-176.

Fie´robe, H.-P., C. Bagnara-Tardif, C. Gaudin, F. Guerlesquin, P. Sauve´, A.Be´laÿ¨ch, and J.-P. Be´laÿ¨ch. 1993. Purification and characterization of endoglucanase

C from *Clostridium cellulolyticum*. Catalytic comparison with endonuclease A. *Eur. J. Biochem.* 217:557–565.

Fie´robe, H.-P., C. Gaudin, A. Be´lay¨ch, M. Loutfi, E. Faure, C. Bagnara, D.Baty and J.-P. Be´lay¨ch. 1991. Characterization of endoglucanase A from *Clostridium cellulolyticum. J. Bacteriol.* 173:7956–7962.

Focher, B., A. Marzetti, M. Cattaneo, P. L. Beltrame, and P. Carniti. 1981. Effects of structural features of cotton cellulose on enzymatic hydrolysis. *J. Appl. Polym.* Sci. 26:1989–1999.

Fontes, C.M., and H.J. Gilbert, 2010. Cellulosomes: highly efficient monomachines designed to deconstruct plant cell wall the complex carbohydrate. *Annual review sof biochem.,* 79:655-681.

Gama, F. M., J. A. Teixeira, and M. Mota. 1994. Cellulose morphology and enzymatic reactivity: a modified solute exclusion technique. *Biotechonol. Bioeng.* 43:381–387.

Garleb, K. A., L. D. Borquin, J. T. Hsu, G. W. Wagner, S. J. Schmidt, and G. C. Fahey, Jr. 1991. Isolation and chemical analysis of nonfermented fiber fractions of oat hulls and cottonseed hulls. *J. Anim. Sci.* 69:1255–1271.

Gaudin, C., A. Be´lay¨ch, S. Champ, and J.-P. Be´lay¨ch. 2000. CelE, a multidomain cellulase from *Clostridium cellulolyticum*: a key enzyme in the cellulosome? *J. Bacteriol.* 182:1910–1915.

Gilkes, N. R., E. Jervis, B. Henrissat, B. Tekant, R. C. Miller, R. A. J.Warren, and D. G. Kilburn. 1992. The adsorption of a bacterial cellulose and its two isolated domains to crystalline cellulose. *J. Biol. Chem.* 267:6743–6749.

Grous, W. R., A. O. Converse, and H. E. Grethlein. 1986. Effect of steam explosion pretreatment on pore size and enzymatic hydrolysis of poplar (*Populus tremuloides*). *Enzyme Microb. Technol.* 8:274–280.

Holtzapple, M., M. Cognata and Y.Shu. 1990. Inhibition of T. *reesei* cellulase by sugars and solvents. *Biotechnol. Bioeng.,*36:275-287.

Jervis, E. J., C. A. Haynes, and D. G. Kilburn. 1997. Surface diffusion of cellulases and their isolated binding domains on cellulose. *J. Biol. Chem.*272: 16–23.

Kadam, K.L., E.C. Rydholm and J.D. McMIllan. 2004. Development of a kinetic model for enzymatic saccharification of lignocellulosic biomass. *Biotechnol. Programme.*20: 698-705.

Klinman, J.P and A. Kohen. 2013. Hydrogen Tunneling Links Protein Dynamics to Enzyme Catalysis. *Annu Rev Biochem.,* 82: 471–496.

Klyosov, A. A. 1990. Trends in biochemistry and enzymology of cellulose degradation. *Biochemistry* 29:10577–10585.

Kurasin, M and P. Valjamae. 2011. Processivity of cellobiohydrases is limited by the substrate. *J. Biol. Chem.*286: 169-177.

Laureano-Perez, L., F. Teymouri, H. Alizadeh and B.E. Dale. 2005. Understanding factors that limit enzymatic hydrolysis of biomass. *Appl.Biochem. Biotechnol.,*124(1):1081-1099.

Leonid, O.S., J.C. Brian, P. Mircea and B.Z.Igor. 2011. Cellullases:ambiguous non homologous enzymes in a genomic perspectives. *Trends Biotechnol.*,29(10): 473-479.

Lynd, L. R. P. J. Weimer, W. H. van Zyl, and I. S. Pretorius. 2002. Microbial Cellulose Utilization: Fundamentals and Biotechnology. *Microbiol. Mol. Biol. Rev.*, 66(3):. 506–577.

Lynd, L. R., and H. E. Grethlein. 1987. Hydrolysis of dilute acid pretreated mixed hardwood and purified microcrystalline cellulose by cell-free broth from *Clostridium thermocellum. Biotechnol. Bioeng.* 29:92–100.

Lynd, L.R.,W. H van Zyl, J. E. McBride and M. Laser. 2005. Consolidated bioprocessing of cellulosic biomass: an update Current Opinion in Biotechnology 2005, 16:577–583

Mansfield, S. D., C. Mooney, and J. N. Saddler. 1999. Substrate and enzyme characteristics that limit cellulose hydrolysis. *Biotechnol. Prog.* 15:804–816.

Medve, J., J. Sta°hlberg, and F. Tjerneld. 1997. Isotherms for adsorption of cellobiohydrolase I and II from *Trichoderma reesei* on microcrystalline cellulose. *Appl. Biochem. Biotechnol.* 66:39–56.

Michael, G.R., S.D Bryon, O.B. John, R.D. Stephen, A.B. Edward, T.B. Gregg and E.H. Michael. 2013. Fungal cellulases and complexed cellulosomal enzymes exhibit synergistic mechanism in cellulose deconstruction. *Energy Environ. Sci.*,6:1858-1867.

Morag, E., E. A. Bayer, and R. Lamed. 1992. Affinity digestion for the near-total recovery of purified cellulosome from *Clostridium thermocellum. Enzyme Microb. Technol.* 14:289–292.

Mourin˜o, F. M., R. Akkarawongsa, and P. J. Weimer. 2001. Initial pH as a determinant of cellulose digestion rate by mixed ruminal microorganisms in vitro. *J. Dairy Sci.* 84:848–859.

Nidetzky, B., W. Zachariae, G. Gercken, M. Hayn, and W. Steiner. 1994. Hydrolysis of cellooligosaccharides by *Trichoderma reesei* cellobiohydrolases: Experimental data and kinetic modeling. *Enzyme Microb. Technol.*16:43–52.

O'Dwyer, J.P., L.Zhu, C.B. Granda and M.T. Holtzapple. 2007. Enzymatic hydrolysis of lime pre-treated corn stover and investigation of HCH-I model; inhibition pattern, degree of inhibition,validity of simplified HCH-I model. *Bioresource Technol.*,98:2969-2977.

Ooshima, H., D. S. Burns, and A. O. Converse. 1990. Adsorption of cellulose from *Trichoderma reesei* on cellulose and lignaceous residue in wood pretreated by dilute sulfuric acid with explosive decompression. *Biotechnol. Bioeng.* 36:446–452.

Ooshima, H., M. Kurakake, J. Kato, and Y. Harano. 1991. Enzyme activity of cellulase adsorbed on cellulose and its change during hydrolysis. *Appl.Biochem. Biotechnol.* 31:253–266.

Ooshima, H., M. Sakata, and Y. Harano. 1983. Adsorption of cellulose from *Trichoderma viride* on cellulose. *Biotechnol. Bioeng.* 25:3103–3114.

Pedersen, M and A. S. Meyer. 2010. Lignocellulose pretreatment severity-relating pH to biomatrix opening, New Biotechnol., 27(6), 739–750.

Reese, E. T., and M. Mandels. 1971. Enzymatic degradation.*In* N. M. Bikales and L. Segal (ed.), Cellulose and cellulose derivatives. Wiley Interscience, New York, N.Y. pp: 1079–1094

Reverbel-Leroy, C., S. Page's, A. Be´laý¨ch, J.-P. Be´laý¨ch, and C. Tardif. 1997. The processive endocellulase CelF, a major component of the *Clostridium cellulolyticum* cellulosome: purification and characterization of the recombinant form. *J. Bacteriol.* 179:46–52.

Ricca E., B. Brucher and J.H. Schrittwieser, 2011. Multi-enzymatic cascade reactions: overview and perspectives. *Adv. Synth. Catal.*, 353:2239–62.

Saha B.C. and Bothast R.J. 1999. Production of 2,3-butanediol by newly isolated *Enterobacter cloacae*. *Appl Microbiol Biotechnol.*,52(3):321-6.

Schiffer, S.H. 2013. Catalytic Efficiency of Enzymes: A Theoretical Analysis | *Biochemistry*, 52(12). dx.doi.org/10.1021/bi301515j

Shafiee, S. and E. Topal. 2009.When will fossil fuel reserves be diminished?, *Energy Policy*, 37(1), 181–189.

Singh, A., P. K. R. Kumar, and K. Schugerl. 1991. Adsorption and reuse of cellulases during saccharification of cellulosic materials. *J. Biotechnol.* 18:205–212.

South, C. R., D. A. L. Hogsett, and L. R. Lynd. 1995. Modeling simultaneous saccharification and fermentation of lignocellulose to ethanol in batch and continuous reactors. *Enzyme Microb. Technol.* 17:797–803.

Stahlberg, J., G. Johansson, and G. Petterson. 1991. A new model for enzymatic hydrolysis of cellulose based on the two-domain structure of cellobiohydrolase-1. *Bio/Technology* 9:286–290

Steiner, W., W. Sattler, and H. Esterbauer. 1988. Adsorption of *Trichoderma reesei* cellulase on cellulose: experimental data and their analysis by different equations. *Biotechnol. Bioeng.* 32:853–865.

Strobel, H. J., F. C. Caldwell, and K. A. Dawson. 1995. Carbohydate transport by the anaerobic thermophilic *Clostridium thermocellum* LQRI. *Appl. Environ. Microbiol.* 61:4012–4015.

Stutzenberger, F. and G. Lintz. 1986. Hydrolysis products inhibit adsorption of *Trichoderma reesei* C30 cellulases to protein-extracted lucerne fibres. *Enzyme Microb. Technol.* 8:341–344.

Sugitha, T.C.K. 2007. Anaerobic fermentation of coffee processing wastes. Ph.D Thesis submitted to Tamil Nadu Agricultural University.

Sutcliffe, R., and J. N. Saddler. 1986. The role of lignin in the adsorption of cellulases during enzymatic treatment of lignocellulosic material. *Biotechnol.,Bioeng.* Symp. Ser. 17:749–762.

Suvajittanont, W., J. McGuire, and M. K. Bothwell. 2000. Adsorption of *Thermomonospora fusca* E5 cellulase on silanized silica. *Biotechnol. Bioeng.,*67:12–18.

Thompson, D. N., H. C. Chen, and H. E. Grethlein. 1992. Comparison of pretreatment methods on the basis of available surface area. *Biores. Technol.*39:155–163.

Tomme, P., H. Van Tilbeurgh, G. Pettersson, J. Van Damme, J. Vanderkerckhove, J. Knowles, T. Teeri, and M. Claeyssens. 1988. Studies of the cellulolytic system of *Trichoderma reesei* QM 9414. Analysis of domain function in two cellobiohydrolases by limited proteolysis. *Eur. J. Biochem.,*170:575–581.

Tu, M. 2015. Carbonyl Inhibition of Biofuels Production from Biomass Hydrolysates: A Quantitative Structure-Activity Relationship (QSAR) Approach. Paper presented in Conference on Society for Industrial Microbiology and Biotechnology during August 02 - 06, 2015 at Philadelphia, PA

Wald, S., C. R. Wilke, and H. W. Blanch. 1984. Kinetics of the enzymatic hydrolysis of celluose. *Biotechnol. Bioeng.* 26:221–230.

Walker, L. P., D. B. Wilson, and D. C. Irwin. 1990. Measuring fragmentation of cellulose by *Thermomonospora fusca* cellulase. *Enzyme Microb. Technol.* 12:378–386.

Wang, Z.,J.Xu and J.J. Cheng. 2011. Modelling biochemical conversion of lignocellulosic material for sugar production: a review. *BioResources,* 6(4):5282–5306.

Weimer, P. J. 1993. Effects of dilution rate and pH on the ruminal celluloytic bacterium *Fibrobacter succinogenes* S85 in cellulose fed continuous culture. *Arch. Microbiol.* 160:288–294.

Weimer, P. J., A. D. French, and T. A. Calamari. 1991. Differential fermentation of cellulose allomorphs by ruminal cellulolytic bacteria. *Appl. Environ.Microbiol.* 57:3101–3106

Weimer, P. J., J. M. Hackney, and A. D. French. 1995. Effects of chemical treatments and heating on the crystallinity of cellulose and their implications for evaluating the effect of crystallinity on cellulose biodegradation. *Biotechnol. Bioeng.* **48:**169–178.

Weimer, P. J., J. M. Lopez-Guisa, and A. D. French. 1990. Effect of cellulose fine structure on kinetics of its digestion by mixed ruminal microorganisms invitro. *Appl. Environ. Microbiol.* 56:2421–2429.

Woodward, J., M. K. Hayes, and N. E. Lee. 1988. Hydrolysis of cellulose by saturating and non-saturating concentrations of cellulase: implications for synergism. *Bio/Technology* 6:301–304.

Xu, J., Z. Wang and J.J. Cheng. 2011. Bermuda grass as feed stock for biofuel production: a review. *Bioresource Technol.,*102:7613-7620

Xue, G.-P., K. S. Gobius, and C. G. Orpin. 1992. A novel polysaccharide hydrolase cDNA (*celD*) from *Neocallimastix patriciarum* encoding three multifunctional catalytic domains with high endoglucanase, cellobiohydrolase and xylanase activities. *J. Gen. Microbiol.* 138:2397–2403.

Yang, B., Z. Dai, S.Y. Ding and C. E. Wyman. 2011. Enzymatic hydrolysis of cellulosic biomass. *Biofuels*, 2(4), 421–450.

Yu, A. H. C., and J. N. Saddler. 1995. Identification of essential cellulose components in the hydrolysis of steam-exploded birch substrate. *Biotechnol.Appl. Biochem.*,21:185–202.

Zhang, Y.H.P. and L. R. Lynd. 2004. Toward an aggregated understanding of enzymatic hydrolysis of cellulose: Noncomplexed cellulase systems, *Biotechnol. Bioeng.*, 88(7), 797–824.

Zhang, Y.H.P. and L.R. Lynd. 2006. Afunctionally based model for hydrolysis of cellulose by fungal cellulose. *Biotechnol. Bioeng.*,94: 888-898.

10

Enzymatic Biorefinery for the Sequential Production of Biodiesel, Bioethanol, Biohydrogen, Biomethane using Leather Solid Wastes and Treatment Plant Sludges

P. Shanmugam

Senior Principal Scientist, Environmental Science and Engineering,
CSIR-Central Leather Research Institute, Adyar, Chennai
e-mail: pashanmugam@yahoo.com

1. Introduction

Leather manufacturing process results in substantial level of solid wastes of different kinds. One ton of wet salted yields only 200kg of leather but over 600kg of solid wastes. During the pre-tanning processes of leather manufacturing, the extraneous matter such as hair along with epidermis, globular proteins, blood, fat and adipose layer are removed. Particularly during fleshing operation, the adipose layer of the hides and skins are removed. As much as 70,000 tons of fleshings is generated in India per annum. About 65 per cent of the weight of the fleshings is contributed by water. Protein and Fat are the chief constituents of fleshings which contribute to about 15 per cent and 20 per cent respectively. Solid waste generated from leather making and were termed fleshing (LF) and this waste was generated around 15 per cent of the raw hide or skin processed. The management of leather solid waste becomes a cause of serious concern in many of the developing countries. The raw hide and skins from buffalos, cows, goat, pig and sheep were the principal raw material used for the manufacture of leather.

There are around 2300 tanneries in India using 100000 tones of raw hides and skins, producing around 60000 tones of solid waste per annum. This generates huge quantity of solid wastes needs effective management system before contaminating the ground water due to open dumping and percolation of contaminants during rainy seasons. The rapid growth of civilization and urbanization requires an increased demand for leather production. These solid wastes are yet to find its safe disposal. There are many solid waste management technologies such as land filling, incineration, composting, etc., but triggers primary secondary and tertiary environmental impacts such as ground water contamination, odor problem, green house gas emissions etc., There are many biomethanisation plants installed in India for the safe and odour free management of solid wastes coupled with biogas and power generation by CLRI, Chennai. The residual solid wastes after biogas recovery invites further disposal problem such as E.Coli, pungent odor problems, etc. These biogas plants are used for generating methane and power from the leather organic solid wastes. There are the two stage high rate biomethanisation technologies implemented through overseas technology that was absorbed by CLRI. But the residual solid waste after digestion is enormous and the production of methane again contributes green house gas emissions. Alternatively, the biodiesel, bioethanol, and biohydrogen and biomethane recovery followed by fertilizer recovery from the same solid wastes and treatment plant sludge has been experimented in the recent past and found an interesting results as a "biorefinery" concept of "Waste into Wealth" a total solution to achieve the *Zero Solid waste Discharge* (ZSD) with multiple fuel recovery. This clean/green energy production from leather solid wastes and tannery effluent treatment plant sludges reduces the coal/CNG consumption, GHG emissions, ground water contaminations, toxic gas emissions, depletion of natural resources besides achieving the environmental sustainability. Consequently, a mobile pilot biorefinery plant (10 L) capacity was designed and built in CLRI operated successfully. Consequently, The Union Ministry of Environment Forests and climate change (MoEF and CC), Govt. of India have evaluated this proposal submitted by the author of this paper and sanctioned Rs 1.15 Crore to demonstrate this project onsite at Ranipet common effluent treatment plant (Ranitec) for 2 tons per day capacity of leather fleshing wastes and treatment plant sludge. This paper describes the concept of enzymatic biorefinery and its feasibility in recovering multiple biofuels from fat, protein and carbohydrates rich solid wastes using the enzymes such as lipase and protease.

2. Aims and objectives

1. To produce multiple biofuels (biodiesel, bioethanol, biohydrogen, and biomethane) as a source of renewable green/clean energy coupled with an efficient leather solid waste management.

2. To produce by products like glycerol after biodiesel

3. To maximize the production of lipase from the leather fleshing waste and to use it again for enzymatic hydrolysis and trans-esterification fat rich solid wastes.

4. To achieve zero solid discharge technology using leather fleshing and CETP sludges.

5. To scale up and develop a pilot plant studies for the four stage sequential production of biodiesel, bioethanol, and biohydrogen and biomethane from the complete conversion of leather solid waste and treatment plant sludges.

3. Concept of Biorefinery

This biorefinery is similar to petroleum refinery. In petroleum refinery crude oil is used as raw material, whereas in this biorefinery plant leather solid wastes and treatment plant sludge are used as raw material. The aim is to extract the fat content present in leather solid wastes, which are detrimental to biogas plants and other problems such as odour, ground water contaminations and convert the animal (tannery) fleshing in to oil and further in to biodiesel. The biodiesel making has also got an another by product called glycerine. Consequent upon the biodiesel recovery from fat, the residual protein and carbohydrates will be used for bioethanol recovery with little addition of molasses, and yeast. After the ethanol is recovered the residual waste is used in the hydrogenesis process to recover the biohydrogen and the residual waste after biohydrogen will be used for biomethane recovery using the solventogenesis, hydrogenesis and methanogenesis respectively. The residual waste after biomethanisation will be used in a filter press to separate the solid cake from the digestate and use it as an agricultural fertilizer.

4. Solid Wastes from Leather Manufacturing Process

Leather manufacturing process generates huge quantities of by-products and solid wastes higher than that of finished leather (Maire and Lipsett 1980). Highly polluting solid wastes containing varied quantities of protein are currently being wasted due to their non-utilization. One metric ton of wet salted hides yield 200 kg of leather, along with about 250 kg of tanned solid waste and about 350 kg of non-tanned waste and 100 kg is lost in the wastewater. In India, approximately 1,50,000 tons of tannery offals in the form of raw hide trimmings, limed animal fleshings, green animal fleshings, hide splits and chrome shavings were available during leather processing which are not utilized or underutilized, thus creating a solid waste disposal problem in tanneries (Muralidhara Rao, 1994).Leather process has been playing a key role in the growth of tanning industry in India. It played an important role by providing trained manpower and appropriate technologies for the production of varieties of finished leathers in the 70's when the Government of India banned the export of raw and semi processed leathers in India. The division again played a lead role in late 90's by providing unit specific cleaner processing options for reducing pollution loads in tannery effluents when the Apex Court of the country closed down more than 300 tanneries in the state of Tamil Nadu during 1996.

Leather is made from raw hide or skin in three steps:

a. Pre-tanning
b. Tanning
c. Post tanning and Finishing

The **leather manufacturing process** is divided into three sub-processes: preparatory stages, tanning and crusting. All true leathers will undergo these sub-processes. A further sub-process, surface coating may be added into the sequence. The list of operations that leathers undergo vary with the type of leather. The preparatory stages are when the hide/skin is prepared for tanning. During the preparatory stages many of the unwanted raw skin components are removed. Many options for pretreatment of the skin exist. Not all of the options may be performed. Preparatory stages may include:

☆ **Preservation**: the hide/skin is treated with a method which renders it temporarily unputrescible.

☆ **Soaking**: water for purposes of washing or rehydration is reintroduced.

☆ **Liming**: unwanted proteins and "opening up" is achieved.

☆ **Unhairing**: the majority of hair is removed.

☆ **Fleshing**: subcutaneous material is removed.

☆ **Splitting**: the hide/skin is cut into two or more horizontal layers.

☆ **Reliming**: the hide/skin is further treated to achieve more "opening up" or more protein removal.

☆ **Deliming**: liming and unhairing chemicals are removed from the pelt.

☆ **Bating**: proteolytic proteins are introduced to the skin to remove further proteins and to assist with softening of the pelt.

☆ **Degreasing**: natural fats/oils are stripped or as much as is possible from the hide/skin

☆ **Frizing**: physical removal of the fat layer inside the skin.

☆ **Bleaching**: Chemical modification of dark pigments to yield a lighter coloured pelt.

☆ **Pickling**: lowering of the pH value to the acidic region. Must be done in the presence of salts. Pickling is normally done to help with the penetration of certain tanning agents, *e.g.*, chromium (and other metals), aldehydic and some polymeric tanning

☆ **Tanning**: Tanning is the process that converts the protein of the raw hide or skin into a stable material which will not putrefy and is suitable for a wide variety of end applications. The most commonly used tanning material is chromium, which leaves the leather, once tanned, a pale blue colour (due to the chromium), this product is commonly called "wet blue"

☆ **Crusting**: Crusting is when the hide/skin is thinned, retanned and lubricated.

☆ **Wetting back**: semi-processed leather is rehydrated.

☆ **Sammying**: 45-55 per cent (m/m) water is squeezed out the leather.

☆ **Splitting**: the leather is split into one or more horizontal layers.

☆ **Shaving**: the leather is thinned using a machine which cuts leather fibres off.

☆ **Neutralisation**: the pH of the leather is adjusted to a value between 4.5 and 6.5.

☆ **Retanning**: additional tanning agents are added to impart properties.

☆ **Dyeing**: the leather is coloured.

☆ **Fatliquoring**: fats/oils and waxes are fixed to the leather fibres.

☆ **Filling**: heavy/dense chemicals that make the leather harder and heavier

☆ **Stuffing**: fats/oils and waxes are added between the fibres.

☆ **Stripping**: superficially fixed tannins are removed.

☆ **Whitening**: the colour of the leather is lightened.

☆ **Fixation**: all unbound chemicals are chemically bonded/trapped or removed from the leather

☆ **Setting**: area, grain flatness are imparted and excess water removed.

☆ **Drying**: the leather is dried to various moisture levels (commonly 14-25 per cent).

☆ **Conditioning**: water is added to the leather to a level of 18-28 per cent.

☆ **Softening**: physical softening of the leather by separating the leather fibres.

☆ **Buffing**: abrasion of the surfaces of the leather to reduce nap or grain defects

5. Biodiesel Recovery

Biodiesel is produced by transesterification of large, branched triglycerides in tosmaller, straight chain molecules of methyl esters, using an alkali or acid or enzyme ascatalyst. There are three stepwise reactions with intermediate formation of diglycerides and monoglycerides resulting in the production of three moles of methyl esters and one mole ofglycerol from triglycerides. The overall reaction is: Alcohols such as methanol, ethanol, propanol, butanol and amyl alcohol are used inthe transesterification process. Methanol andethanolare used most frequently, especially methanol because of its lowcost,andphysical and chemical advantages. They can quickly react withtriglycerides and sodium hydroxide is easily dissolved in these alcohols. 3:1. In practice, the ratio needs to be higher to drive the equilibrium to a maximum ester yield.

The leather solid waste contains 20 per cent of Fatty material which can be used for biodiesel recovery. The biodiesel recovery has been attempted from the Jatropha plants, slaughter house waste tallow etc. But the present study, the transesterification process of free fatty acids converts the fatty substances in the leather solid waste in to biodiesel. The recovered biodiesel can be mixed with a small (10 per cent) addition of actual diesel to be used as an automotive fuel. The research on biodiesel recovery process optimization, efficient storage and utilization are focused in this research proposal.

5.1. Lipase Enzyme

Lipases (Triacylglycerol acyl hydrolase, E.C.3.1.1.3) are hydrolytic enzymes which will act on the interface between their hydrophobic lipid substrate and the hydrophilic space medium (oil-water interface) cleaving water insoluble glycerides into fatty acids (Jaeger and Eggert 2002). Because of their important attributes they turn to be one of the most important biocatalyst and have industrial applications. To quote some, they are involved in biodiesel production, organic synthesis and some bio transformations. In view of these applications a large scale production of lipase has become essentially important.

Lipases from microbial origin has gained an enormous momentum in their industrial production.Many bacterial, yeast and mould species were identified to produce lipases with different enzymological properties and specification.

5.2. Lipase Based Enzymatic Transesterification

The high cost of biodiesel, compared to petroleum-based diesel, is a major barrier to its commercialization. It has been reported that 60-90 per cent of biodiesel cost arises from the cost of the feedstock oil. The production of biodiesel is an important step in reducing and recycling waste. A fresh vegetable oil and its waste differ significantly in water and free fatty acids (FFAs) contents, which are around 2000 ppm and 10-15 per cent, respectively. This renders conventional alkaline catalyst unsuitable, and limits the use of chemical catalysts to the slower acidic ones. When a biocatalyst (enzyme) namely lipase is used, the issues faced during the chemical based transesterification will be resolved. The basic mechanism of lipase is it converts all the FFAs contained in oil to their respective fatty acids methyl esters (FAMEs). However, the cost of the lipase associated with its preparation processes, including purication from culture broth and immobilization on a carrier, seems to be one of the main obstacles to industrial application.

To overcome these drawbacks, the study has been focused on a whole-cell immobilized biocatalyst for biodiesel production, which enables the direct use of lipase-producing microorganism. Among several microorganisms, *Aspergillus* species were reported to be promising hosts due to the following reasons:

1. They can be spontaneously immobilized within porous biomass support particles (BSPs) during cultivation
2. They can produce a large amount of proteins
3. Intensive research studies have made many advances, including those in molecular biology and the development of improved promoters for the high-level expression of heterogonous genes. These advantages led us to develop *Aspergillus* species whole-cell biocatalysts.

However, despite the advantages of using enzymes, biodiesel production plants using lipases are not yet achieved its industrial-scale operations in production. The reason for this is that there are some challenges that are yet to be resolved before biocatalysts can be made feasible for biodiesel production, such as their higher cost, biodiesel productivity, and inhibition by reactants and products.

This study explores the possible application of lipase enzyme for leather fleshing waste disposal through revenue generating methods.

6. Bioethanol Recovery

Although biodiesel fuels are produced chemically and enzymatically there is a need to utilize the biodiesel production waste in order to reduce the waste discharged from the process, and glycerol is essentially generated as the by-product. Currently, the waste generated in the form of glycerol during biodiesel production is utilized by the cosmetic industry, but a further increase in the production of biodiesel fuels would raise the problem of efficiently treating this waste stream. The biological production of ethanol from glycerol is attractive because ethanol can be used as a raw material and a supplement to petro-fuel in vehicles and industrial applications. It has been observed that the yields of H_2 and ethanol decreased with an increase in the concentrations of biodiesel wastes and a high salinity of the medium with biodiesel wastes would be one of the causal factors for the inhibition of product formation. A pH of 4.5 has been reported to be the most favourable for ethanol production [9] Present work deals with optimization of conditions for recovery of fat from fleshings by chemical treatment and chemical trans esterification for FAME formation. The residual glycerol is subject to anaerobic digestion with pre-fermented strictly anaerobic seed sludge. The effect of process parameters on ethanol production are studied and optimal conditions provided for ethanol production accordingly. Further, a continuous Stirred Tank Reactor can be operated for studying optimization of hydraulic (HRT and OLR) and process related parameters for ethanol production with different ratios of substrate and seed sludge.

The residue after recovering fat for biodiesel from the leather solid wastes can be used with small addition of molasses/spent wash from distilleries in an anaerobic fermentation process to recover bioethanol. This bioethanol recovery can be performed in the anaerobic CSTR process and purified to be used as an automotive fuel. The residual waste will have optimum C/N ratio required for further processes.

7. Biohydrogen Recovery

It has been perceived that the production of biohydrogen becomes difficult for the proteinaceous solid wastes such as leather wastes. This is attributed to lower C/N ration of such high proteinaceous solid wastes. However the co-digestion becomes an attractive option to increase the biohydrogen recovery. The previous process of co-digestion of leather solid waste with a minor addition of spent wash/molasses from distilleries for bioethanol recovery increases the C/N ratio required for maximizing the hydrogen yield. The hydrogen as one of the clean/green fuel of the future having the calorific value three fold higher than methane. Therefore it is advantageous to have produced more hydrogen rather than methane by routing the biochemical mechanism in to biomethanisation. The fermentative hydrogen recovery has been attempted with the author of this proposal for his PhD in UK using the leather solid wastes and treatment plant sludges. But the sequential production of hydrogen after ethanol recovery is expected to increase the hydrogen yield from the previous level.

8. Biomethane Recovery

The residual waste from the previous biohydrogen processes expected to have an optimum C/N ratio that could be very much useful for biomethanisation. The two stage biomethanisation digesters of hydrolyser followed by methaniser are ameliorated to biohydrogeniser and biomethaniser producing hydrogen and methane. The recovered methane and hydrogen will be used to produce the hydrogen/methane fuel cell. This is not only an attractive option to increase the renewable energy generation but also a good environmental protection measures from ground water contamination, odor problem, GHG emissions and depletion

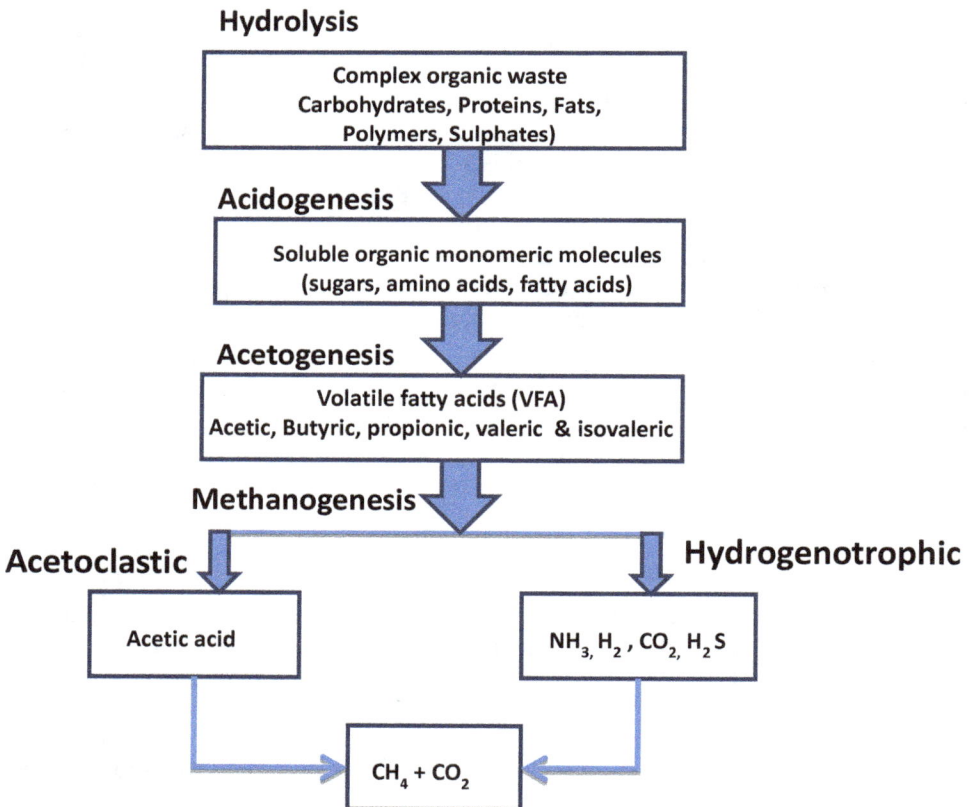

Figure 10.1: Biochemical Pathways of Biomechanization.

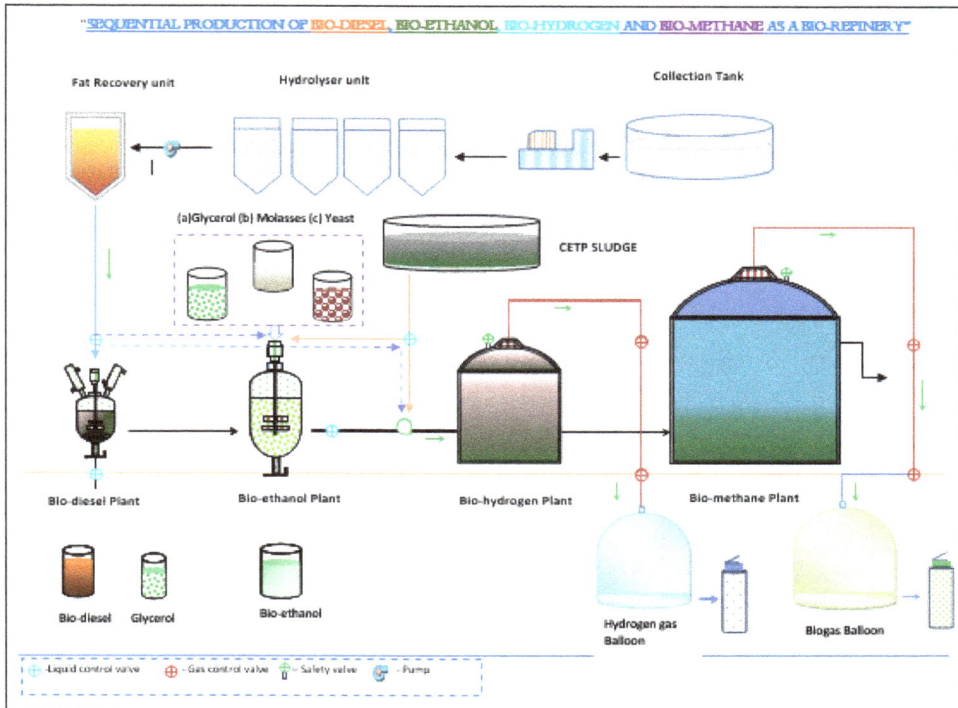

Figure 10.2: Process Flow Diagram of the Biorefinery Project.

of natural resources. Adoption of this Biorefinary concept in to other industrial solid waste is feasible.

9. Conclusion

☆ The *Aspergillus* species as whole cell biocatalyst is economical and suggested as whole-cell biocatalyst for application in biodiesel production.

☆ The maximum fat recovery obtained was about 80 per cent from 100g of fleshing at an optimum pH of 8.0 and an optimum ammonium sulphate concentration of 0.2g/g of fleshing.

☆ In the case of enzymatic transesterification, it was attempted first of its kind with fleshing fat oil and yielded 85 per cent recovery of biodiesel.

☆ The produced biodiesel was confirmed with FT-IR analysis. The bands obtained from the FT-IR spectrum were analyzed and confirmed as FAME (Biodiesel)

☆ The FT-IR peak at 1743 cm^{-1} refers to the C=O stretch which is typical for ester bonds were observed in the spectrum obtained proving the presence of methyl esters.

☆ Based on the above observations the biodiesel produced is confirmed both qualitatively and quantitatively.

☆ Pre-hydrolysis using crude lipase increased the biogas yield about 60 per cent over control (without adding lipase) and gave qualitatively better methane composition.

☆ The lipase enzyme confirmed the maximized gas production with as VFA level of 5446.8 mg. l^{-1} against the control (without lipase).

☆ Ammonia was considerable lower than the reactor without lipase with 642 mg.l^{-1} compared to 826 mg. l (in control reactor) which proves the enhanced gas production in the lipase catalyzed reactor.

References

Aloy, M. (1987). Methane gas from tannery effluent. *Leather*. 189, 53–54.

Andres Donoso-Bravo., Maria Fdz-Polanco. (2013). Anaerobic co-digestion of sewage sludge and grease trap: Assessment of enzyme addition, *Process Biochemistry*, 48,936–940

Angelidaki, I., Ahring, B.K. (1995). Establishment and characterization of an anaerobic thermophilic (55 °C) enrichment culture degrading long-chain fatty acids. *Appl. Environ. Microbiol.* 61, 2442–2445.

Assefa, G, Erikson, D and Frostell, B 2005. 'Technology assessment of thermal treatment technologies using ORWARE energy conversion and management', vol.46, pp.797-819.

Bajza, I and Vrcek, V 2001, 'Thermal and enzymatic recovering of proteins from untanned leather waste', vol. 21, pp.79-84.

Beccari, M, Carucci, G, Majone, M and Torrisi, L 1999, 'Role of Lipids and Phenolic Compounds in the Anaerobic Treatment of Olive Oil Mill Effluents', *Environmental Technology*, vol. 20, no.1, pp. 105-110.

Borja.R, Martýn. A, Sanchez.E., Rincon.B, Raposo.F,2005. Kinetic modelling of the hydrolysis, acidogenic and methanogenic steps in the anaerobic digestion of two-phase olive pomace (TPOP). *Process Biochemistry* 40.1841–1847.

Brooks, AA and Asamudo, NU 2011, 'Lipase production by strains of Aspergillus species isolated from contaminated body creams. *Journal of Toxicology and Environmental Health Sciences*', vol. 3, no. 11, pp. 311-316.

Broughton, M, Thiele, J, Birch, E and Cohen, A 1998, 'Anaerobic batch digestion of sheep tallow', *Water Research*, vol. 32, no. 5, pp. 1423-1428.

Cammarota, MC, Teixeira, GA and Freire, DMG 2001. 'Enzymatic pre-hydrolysis and anaerobic degradation of wastewaters with high fat contents', *Biotechnology Letters*, vol.23, pp.1591–5.

Chan, OC., Liu,W-T., Fang HHP. (2000).Study of microbial community of brewery-treating granular sludge by denaturing gradient gel electrophoresis of 16S rRNA gene. *Water Sci Tech*; 43(1):77–82.

Cirne, D.G., Paloumet, X., Bjornsson, L., Alves, M.M., Mattiasson, B., 2006. Evaluation of biological strategies to enhance hydrolysis during anaerobic digestion of complex. *Waste*, 32,965–975.

Demirel. B., and Scherer. P., 2008 "The roles of acetotrophic and hydrogenotrophic methanogens during anaerobic conversion of biomass to methane: a review," *Reviews in Environmental Science and Biotechnology*, vol. 7, no. 2, pp. 173–190.

Fang, HP and Chui, HK, 1993, 'Microstructural Analysis of Anaerobic Granules', *Biotechnology*, vol. 7, no. 7, pp. 407-410.

Ferris,MJ., Muyzer,G., Ward, DM. (1996). Denaturing gradient gel electrophoresis profiles of 16S rRNA-defined populations inhabiting a hot spring microbial mat community. *Appl. Environ Microbiol*; 62:340–6.

Giovannoni, SJ., DeLong EF., Olsen, GJ., Pace, NR. (1988). Phylogenetic group-specific oligonucleotide probes for identification of single microbial cells. *J Bacteriol*; 170:720–6.

Hashimoto, A. G., Chen, Y. R.; Varel, V. H. (1981). Anaerobic fermentation of beef cattle manure: final report. Roman L. Hruska U.S. Meat Animal Research Center, U.S.Department of Agriculture Clay Center. Nebraska.

Hashimoto, AG, Chen, YR and Varel, VH 1981. 'Anaerobic fermentation of beef cattle manure: final report', Roman L. Hruska U.S. Meat Animal Research Center, U.S. Department of Agriculture Clay Center. Nebraska.

Henze, M. H., Jonsen, J. C., Arven, E. (1997). Waste Water Treatment. *Springer-Verlog*, Berlin, Germany.

Heuer,H., Krsek, M., Baker, P., Smalla, K., Wellington, EM. (1997). Analysis of actinomycete communities by specific amplification of genes encoding 16S rRNA and gel-electrophoretic separation in denaturing gradients. *Appl. Environ Microbiol*; 63:3233–41.

Karakashev, D., Batstone, D. J., and Angelidaki, I. 2005. Influence of Environmental Conditions on Methanogenic Compositions in Anaerobic Biogas Reactors. *Applied and Environmental Microbiology*, 71(1), 331–338.

Kirk, DW, Chan, PCY and Marsh, H 2002, 'Chromium behaviour during thermal treatment of MSW fly-ash', *Journal of Hazardous Material*. B vol.90, pp.39-49.

Liu,W.T., Marsh,T.L., Cheng, H., Forney, L.J.,(1997).Characterization of microbial diversity by determining terminal restriction fragment length polymorphisms of genes encoding 16S rRNA. *Appl. Environ Microbiol*; 63:4516–22.

Mladenovska, Z, Dabrowski, S, Ahring, BK, 2003. 'Anaerobic digestion of manure and mixture of manure with lipids: biogas reactor performance and microbial community analysis.' *Water Science and Technology*, vol. 48, pp. 271–278.

Mshandete, Anthony, Kivaisi, Amelia, Rubindamayugi, Mugassa and Mattiasson, bo. 2004, 'Anaerobic batch co-digestion of sisal pulp and fish wastes'. *Bioresource technology*, vol. 95, no. 1, pp. 19-24.

Muyzer G, de Waal EC and Uitterlinden AG. 1993.Profiling of complex microbial populations by denaturing gradient gel electrophoresis analysis of polymerase chain reaction amplified genes coding for 16S rRNA. *Appl Environ Microbiol* 59: 695–700.

Nealson, K,H., 1997 "Sediment bacteria: who's there, what are they doing, and what's new?" *Annual Review of Earth and Planetary Sciences*, vol. 25, pp. 403–434.

Owen, WF, Stuckev, DC, Healv, JB, Young, LY and Mccagrv, PL 1979. 'Bioassay for monitoring Bio Chemical Methane Potential and Anaerobic Toxicity', vol. 13, no. 5, pp. 485-492.

Peter Weiland., 2010. Biogas production: current state and perspectives. *Appl Microbiol Biotechnol*, 85:849–860

Pimentel, M.C., Krieger, N., Coelho,L.C., Fontana, J.O., Melo, E.H., Ledingham, W.M., Lima Filho, J.L. (1994). Lipase from a Brazilian strain of Penicillium citrinum. *Appl. Biochem Biotechnol*; 49:59–74.

Plugge, C.M., van Lier, J.B., Stams, A.J.M., 2010. Syntrophic communities in methane formation from high strength wastewaters. In: Insam, H. *et al.* (Eds.), Micobes at Work. Springer-Verlag, Berlin, Heidelberg, pp. 59–77

Pokorny D, Friedrich J, Cimerman A, 1994, 'Effect of nutritional factors on lipase biosynthesis by *Aspergillus niger*', Biotechnology letters, vol. 16, no. 4, pp.363-366

Prabhudessai, V, Salgaonkar, B, Braganca, J and Mutnuri, S, 2014. 'Pretreatment of Cottage Cheese to Enhance Biogas Production.', *BioMed research international*, vol. 2014, pp. 1-6

Rajesh Banu, J, Esakki Raj, S and Logakanthi, S 2005, 'Biomanagement of petro chemical sludge using an exotic earthworm Eudrilus eugineae.' *Journal of Environmental Biology*, vol. 26, no. 1.

Raposo,F., Fern´andez´, V.C,Rubia, M.A., Borja, R., Beline, F., Cavinato,C., Demirer,G., Fern´andez,B., Fern´andezPolanco,M., Frigon,JC., Ganesh,R., Kaparaju,P., Koubova,J., M´endez, R., Menin,G., Peene, A., Scherer, P., Torrijos,M., Uellendahl,H., Wierinck, I, and V. deWilde, 2011. Biochemical methane potential (BMP) of solid organic substrates: evaluation of anaerobic biodegradability using data from an international inter laboratory study, *J Chem Technol Biotechnol*; 86: 1088–1098.

Riesner,D., Steger,G., Zimmat, R., Owen,RA., Wagenhofer,M., Hillen,W., Vollabach,S., Henco,K.,(1989). Temperature-gradient gel electrophoresis of nucleic acids: analysis of conformational transitions, sequence variations, and protein-nucleic acid interaction. *Electrophoresis*; 10377–89.

Selvam, G, and Kumar, AG 2015. 'Performance of Acidogenic Microbial Treatment Process of Tannery Wastewater in Batch Digester and Packed Bed Reactor', *Biosciences Biotechnology Research Asia*, vol. 12, pp. 355–359.

Shah, FA, Mahmood, Q, Shah, MM, Pervez, A, and Asad, SA, 2014, 'Microbial Ecology of Anaerobic Digesters/: The Key Players of Anaerobiosis', *The Scientific World Journal*, vol. 2014, pp. 21

Shanmugam, P., Horan, N.J. (2009). Optimising the biogas production from leather fleshing waste by co-digestion with MSW. *Bioresource Technology* 100; 4117–4120.

Shanmugam, P., Horan, N.J. (2009). Simple and rapid methods to evaluate the methane potential and biomass yield for range of mixed solid wastes. *Bioresource Technology* 100, 471–474.

Simeonova, LS and Dalev, PG 1996, 'Utilization of a leather industry waste', *Waste Management*, vol.16, no. 8, pp. 765-769.

Sousa, DZ, Smidt, H, Alves, MM, Stams, AJ, 2007. 'Syntrophomonas zehnderi sp. nov., an anaerobe that degrades long-chain fatty acids in co-culture with Methanobacterium formicicum'. *International Journal of Systematic Evolutionary Microbiology*. vol.57, no.3, pp. 609–615.

Watanabe, T, Asakawa, S, Nakamura,A, Nagaoka,K, and Kimura,M 2004,'DGGE method for analyzing 16S rDNA of methanogenic archaeal community in paddy field soil,' *FEMS Microbiology Letters*, vol. 232, no. 2, pp. 153–163.

Weiss A, Jerome V, Freitag R and Mayer H. 2008.Diversity of the resident microbiota in a thermophilic municipal biogas plant. *Appl Microbiol Biotechnol* 81: 163–173.

Ying Meng., Sang Li., Hairong Yuan., Dexun Zou., Yanping Liu., Baoning Zhu., Xiujin Li., 2015. 'Effect of lipase addition on hydrolysis and biomethane production of Chinese food waste', *Bioresource Technology* 179 (2015) 452–459.

11

Beta Glucosidase: Functional Role and Involvement in Cellulolysis

J. Beslin Joshi, T.C.K. Sugitha and Sivakumar Uthandi

Biocatalysts Lab. Department of Agricultural Microbiology,
Tamil Nadu Agricultural University, Coimbatore – 641 002

1. Introduction

Increase in worldwide demand for energy and depletion of petroleum sources have led to resurgence in the development of alternative energy that can displace fossil fuel. Many countries have initiated extensive research in development of fuels from a sustainable and renewable energy resources (Himmel *et al.,* 2007). Among the different substrates used in biofuel production, lignocellulosic biomass (LCB) are considered as cost-effective, environmentally sustainable and alternative to non-renewables (Behera *et al.,* 2014). Biofuels produced LCB can easily be blended with petrol/diesel to reduce environmental pollution in developing countries like India. Bioethanol production from LCB is achieved by the breakage of 1,4-β-D-glycosidic linkages in cellulose to release glucose monomers for subsequent fermentation by yeast. This is achieved by the action of cellulases and of which beta-glucosidase (BGL) is critical enzyme to complete the hydrolysis cycle by converting cellobiose to glucose units. However, there exists a long gap between the quantity of ethanol required and produced. Though there are many potential cellulase producers, the amount of BGL produced is quite less for efficient biomass conversion and further they are sensitive to increased concentration of glucose. Consequently, the enzymatic hydrolysis process gets inhibited due to glucose accumulation. It is vivid that the BGL is one of the bottlenecks in bioconversion of biomass to ethanol through enzymatic route. In order to overcome this problem, β-glucosidase is supplemented externally for complete hydrolysis of cellobiose and cellodextrins into glucose to increase the saccharification efficiency. For efficient biomass hydrolysis, glucotolerant BGL which is active at high concentration of glucose is the need of hour. With this consideration, the present chapter throws light on the catalytic role of BGL in cellulolysis during biomass conversion.

2. Importance of BGLs

In general BGLs are associated with diverse physiological roles and depends on the biological system in which these enzymes occur. Nowadays beta-glucosidases are gaining momentum due to their biotechnological potential for industrial applications.

☆ Beta-glucosidase is involved in cellulase induction (due to its transglycosylation activities) and cellulose hydrolysis in cellulolytic microorganisms.

☆ In plants system, the enzyme is involved in beta-glucan synthesis during cell wall development, pigment metabolism, fruit ripening, and defense mechanisms.

☆ In humans and other mammals, BGL is involved in the hydrolysis of glucosyl ceramides. The defects in BGL activity in humans are associated with Gaucher's disease, a non-neuropathic lysosomal storage disorder (Lieberman *et al.*, 2007).

☆ Hydrolysis of soybean isoflavone glycosides is an important application of BGL. Isoflavones are known to prevent certain cancers, lowers risks of cardiovascular diseases and improve bone health (Izumi *et al.*, 2000).

☆ β-Glucosidase is used in food industry (production of wine, juice and colouring agents) (Hang 1995; Martino *et al.*, 1994), glycoconjugates synthesis (Hirofumi *et al.*, 1991; Toshio and Susumi 1985), pharmacology, cosmetic, and synthesis of high value added biomolecules (Saibi *et al.*, 2007).

☆ Some of the other biotechnological/industrial processes include the production of fuel ethanol from agricultural residues, the release of aromatic compounds from flavourless precursor, etc.

3. Classification of BGLs

BGLs are ubiquitous and heterogeneous group of hydrolytic enzymes. They are classified based on the criteria of their substrate specificity or their protein structure or nucleotide sequence identity (NSI).

A. Based on Substrate Specificity

BGLs are grouped into three classes according to Rajan *et al.* (2004):

(i) aryl- β-D-glucosidases (having strong affinity for aryl- β--D-glucosides)

(ii) cellobiases (hydrolyze only disaccharides)

(iii) broad specificity glucosidases (exhibit activity on many substrate types and are the most commonly found β-glucosidases)

B. Based on the Sequence Homology

According to (Cantarel *et al.*, 2009; Krisch *et al.*, 2010), BGLs have been divided into two subfamilies:

(i) BGA (β-glucosidases and phospho-β-glucosidases from bacteria to mammals)

(ii) BGB (β-glucosidases from yeasts, molds and rumen bacteria).

C. Based on Amino Acid Sequence and Structural Similarity

According to Henrissat and Davies (1997) β-glucosidases are classified based on their aminoacid and structural similarity. The enzymes with overall amino acid sequence similarity and well conserved sequence motifs are grouped under a single family. International Union of Biochemistry and Molecular Biology (IUBMB) classified GH families based on the structural features of the enzymes as it is more informative than substrate specificity. Based on the structure of a family member, other members in the family can be elucidated by bioinformatics tools and system biology approach.

Till now, 133 glycoside hydrolase (GH) families are listed in Carbohydrate Active enZYme (CAZY) website (http://www.cazy.org) (Cantarel *et al.*, 2009; Cairns and Esen 2010). Though most of the β-glucosidases belongs to GH1 and GH3, β-glucosidases are also found in families 5, 9 and 30 of glycosyl hydrolases (Cantarel *et al.*, 2009; Opassiri *et al.*, 2007). Among them, 62 β-glucosidases are from archaebacteria, plants and mammals. The 6-phosphoglycosidases and thioglucosidases belonging to GH1 and GH3 families consist of nearly 44 beta-glucosidases and hexosaminidases of fungus, bacteria and yeast origin (Singhania *et al.*, 2013). X-ray crystallographic study revealed the characteristic structure of GH1 family, a classical $(\beta/\alpha)_8$ barrel folds in *Phanerochaete chrysosporium* (Nijikken *et al.*, 2007) and human cytosolic β-glucosidase (Tribolo *et al.*, 2007).

4. Catalytic Mechanism of β-glucosidases

In any enzyme, the active site (containing binding site and catalytic site) determines its specific activity. β-Glucosidases exhibit pocket or crack catalytic structure (Tribolo *et al.*, 2007) and perform catalysis in 2 steps *via* glycosylation and deglycosylation (Figure 11.1). The catalytic active site conserved among all β-glucosidases and the overall catalytic reaction is performed by 2 glutamate residue in the active site (Davies and Henrissat 1995; Wang *et al.*, 1995).

1. Glycosylation step: The anomeris carbon is attacked by the nucleophillic glutamate (conserved as 'I/VTENG' motif) forming glucose–enzyme intermediate product.

2. Deglycosylation step: Acid/base catalyst glutamate residue (conserved as a 'TFNEP' motif), activate the nearby water molecule act as a nucleophile and breaks the glycosidic bond to release glucose (Litzinger *et al.*, 2010).

X-ray crystallographic structure of human cytosolic β-glucosidase revealed that the acid/base catalyzing residue was located on strand 4 and nucleophilic residue on strand 7 (Tribolo *et al.*, 2007). The mechanism of action of β-Glucosidases from GH9 family is different where nucleophilic attack on the anomeric carbon to displace the aglycone in a single step is activated by water molecule (Qi *et al.*, 2008).

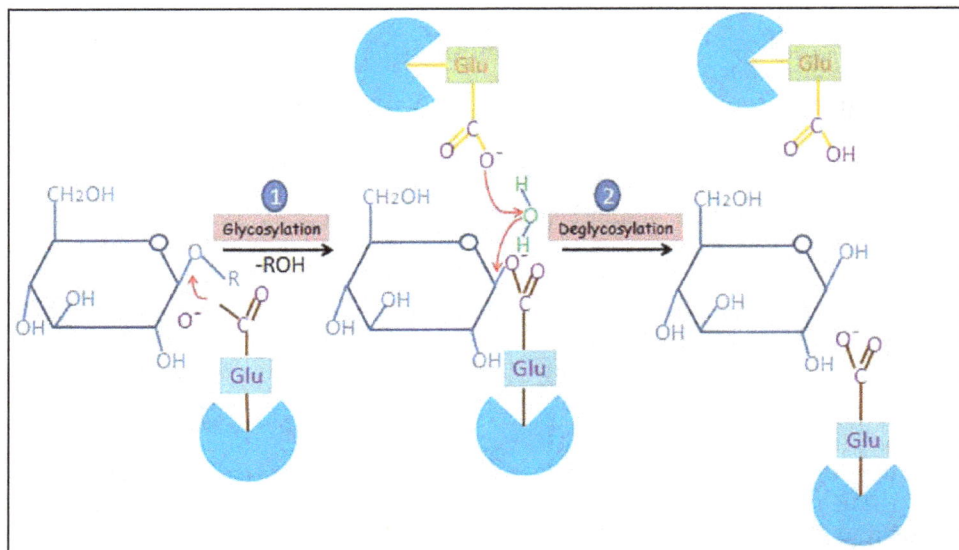

**Figure 11.1: Catalytic Mechanism of β-Glucosidases
(Adopted from Singh *et al.*, 2016).**

5. BGL Activities and Substrate

The interaction of beta-glucosidases with their substrates, especially in respect of the glycone moiety is not much understood. There exists a tremendous diversity in terms of substrate range and subtle differences in substrate specificity. BGL activities are measured using artificial substrates such as pNPG (para-nitrophenyl beta-D-glucopyranoside) or methyl umbelliferyl beta-D-glucoside (MUG) or cellobiose. High catalytic activity of BGLs has been reported with artificial substrates like pNPG and MUG.

The kinetics of the beta-glucosidase depends on the configuration of its substrate. Cellobiose requires a conformational change for catalysis. BGL has a very rigid structure in S1 substrate binding site which will accommodate glucose of cellobiose but the second glucose of cellobiose will change the conformation using rotation of the r-bond of the glucoside so as to fit in the substrate binding site (Nam *et al.*, 2010). But nitrophenol in pNPG follows the same binding patterns as of glucose. This is the reason behind low Kcat/Km of beta-glucosidase towards cellobiose than the substrate pNPG.

For improving the efficiency of saccharification process, supplementation of enzymes is very important to carry out the reactions coherently and synergistically. Hence knowledge on the structure of lignocelluloses matrix structure and mechanism of action of hydrolytic enzymes, hemicellulases such as xylanases and β-glucosidases is a mandate.

6. β-glucosidase: A Key in Cellulase Cocktail

Beta glucosidases (beta-D-glucoside glucohydrolase, EC 3.2.1.21) are biologically important enzyme that catalyzes the transfer of glycosyl moieties between oxygen nucleophiles. Cellulose degradation requires a multi-enzymatic system (cellulases) composed of three enzymes: the endoglucanases, which randomly attack cellulose in amorphous zones releasing cellooligomers; the cellobiohydrolases, which liberate cellobiose from reducing and non-reducing ends and finally the β-glucosidases which hydrolyze cellobiose and other cello oligosaccharides into glucose units. Among the 3 enzymes, β-glucosidase is the rate-limiting enzyme because it completes the final step of hydrolysis by converting the cellobiose (an intermediate product of cellulose hydrolysis) to glucose (Alef and Nannipieri 1995). During saccharification of lignocellulosic biomass, instead of glucose as end product, cellooligosaccharides and cellobiose are produced due to low levels of β-glucosidase, resulting in inefficient ethanol fermentation (Figure 11.2).

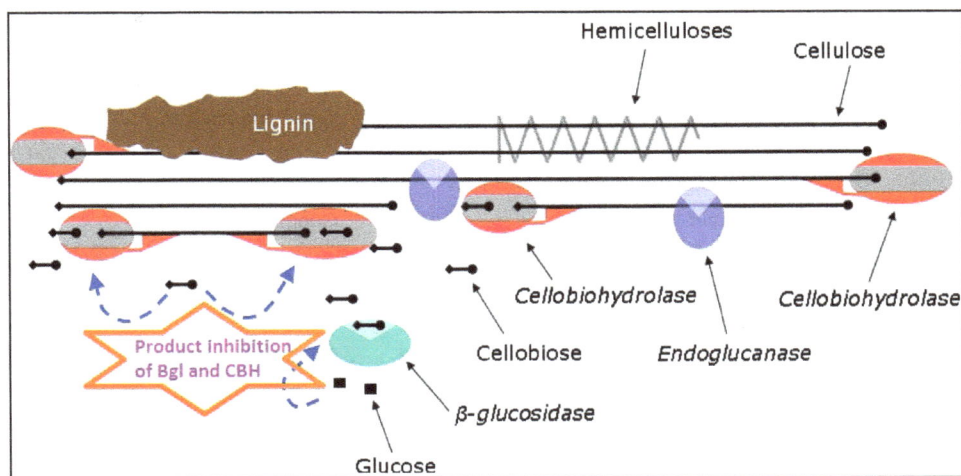

Figure 11.2: Overview of Enzymatic Conversion of Cellulose (Adapted from Jorgensen *et al.*, 2007 with slight modification).

7. Microorganisms Producing β-Glucosidase

β-Glucosidase are widely distributed in all living organisms as it is involved in several biological process. Microbial sources have been widely exploited for beta-glucosidase production by both solid-state fermentation (SSF) and submerged fermentation. The list of microbes producing β-glucosidase is given in the Table 11.1. Fungi are the main source of β-glucosidase producers followed by bacteria

Table 11.1: List of Microrganisms Producing β-glucosidases

Group	Microorganism	References
Fungi	*Debaryomyces pseudopolymorphus*	Villena *et al.* (2006)
	Trichoderma atroviride	Kovacs *et al.* (2008)
	Penicillium pinophillum	Joo *et al.* (2010)
	Penicillium citrinum	Ng *et al.* (2010)
	Periconia sp.	Harnpicharnchai *et al.* (2009)
	Penicillium decumbens	Chen *et al.* (2010)
	Penicillium echinulatum	Martins *et al.* (2008)
	Stachybotrys sp.	Amouri and Gargouri (2006)
	Humicola isolens	Souza *et al.* (2010)
	Fomitopsis palustris	Okamoto *et al.* (2011)
	Fomitopsissp	Deswal *et al.* (2011)
	Aspergillus niger	Singhania *et al.* (2011)
	Aspergillus tubingensis	Decker *et al.* (2001)
	Aspergillus oryzae	Riou *et al.* (1998)
	Aspergillus aculeatus	Kawaguchi *et al.* (1996)
	Phanerochaete chrysosporium	Lymar *et al.* (1995)
	Pyrococcus Furiosus	Kempton and Withers (1992)
	Thermoascus aurantiacus	Parry *et al.* (2001)
	Trichoderma reesei	Saloheimo *et al.* (2002)
	Trichoderma Koningii	Wood and McCrae (1982)
	Volvariella volvacea	Li *et al.* (2005)
	Neurospora crassa	Yazdi *et al.* (2003)
	Botrytis cinerea	Gueguen *et al.* (1995)
	Fusarium oxysporum	Christakopoulos *et al.* (1994)
	Thermomyces lanuginosus	Lin *et al.* (1999)
Bacteria	*Vibrio Cholerae*	Park *et al.* (2002)
	Azospirillum irakense	Faurc *et al.* (1999)
	Clostridium thermocellum	Ait *et al.* (1982)
	Alicyclobacillus sp.	Lauro *et al.* (2006)
	Bacillus polymyxa	Gonzalez *et al.* (1990)
	Agrobacterium sp.	Pouwels *et al.* (2000)
	B.circulans	Paavilainen *et al.* (1993)
	Pseudomonas sp.	Her *et al.* (1999)
	Paenibacillus sp.	Shipkowski and Brenhley (2005)
Yeast	*Candida peltata*	Saha and Bothast (1996)
	Saccharomycopsis fibuligera	Machida *et al.* (1988)
	Pichia anomala	Kohchi and Toh-e (1985)
	Debaryomyces sp.	Villena *et al.* (2006)
Actinomycetes	*Streptomyces* sp.	Ozaki and Yamada (1992)

and yeast. Some of the major fungal species producing β-glucosidase are the filamentous fungi such *Aspergillus niger* (Gunata and Vallier 1999), *A. oryzae* (Riou *et al.*, 1998), *Penicillium brasilianum* (Krogh *et al.*, 2010), *P. decumbens* (Chen *et al.*, 2010), *Phanerochaete chrysosporium* (Tsukada *et al.*, 2006), *Paecilomyces sp.* (Yang *et al.*, 2009) etc. *Penicillium decumbens* 114-2, a filamentous fungus was found to produce more balanced extracellular lignocellulolytic enzyme with more of β-glucosidase of high specific activity.

BGL from filamentous fungal sources could be preferred as it offers an advantage of production *via.* solid state fermentation in high titers against submerged fermentation for biomass hydrolysis. It is a cheaper technology and the difficulty in purification of metabolites produced by SSF is not a limitation for the above application as the purity of enzyme doesn't matter. Besides, the enzymes produced are often less complex than bacterial glycoside hydrolases and can readily be used in developing recombinant organisms (Maki *et al.*, 2009).

Recently, bacteria are being exploited more because of their high growth rate and ability to produce multi-enzyme complex with increased function and synergy. Most of the β-glucosidase producing bacteria were isolated from Mediterranean soils. BGL producing bacteria belongs to the phylum *Proteobacteria* (Moreno *et al.*, 2013). On the other hand, use of yeasts producing sufficient amounts of β-glucosidase enzymes opens the door for consolidated bioprocessing that can lead to low cost cellulosic ethanol production (Liu *et al.*, 2012).

8. Metagenomics Approach in Exploring Natural β-glucosidase

Since, less than 1 per cent of the microorganisms are only culturable, metagenomic approach is employed in understanding β-glucosidase diversity in composite environment samples (Cristóbal *et al.*, 2009). Few β-glucosidases producing fungus have been isolated from marine environment (Fischer *et al.*, 1996; Park *et al.*, 2005). Using metagenomic approach, Li *et al.* (2013) discovered that different microbial community in soil release β-glucosidase in different concentration, together thereby favouring the composting process. β-glucosidase from insects and plants share few similar properties. β-glucosidase have been reported in the gut of several insects (Robinson 1956), mainly involved in the digestion of plant wall. In plants, β-glucosidase is involved in plant defence and synthesis of lignin precursors (Chapelle *et al.*, 2012). Hence congregated mining of the genetic information of β-glucosidase was so far explored across different environments including hydrothermal spring (Schroder *et al.*, 2014), wetlands (Nam *et al.*, 2008), sludge (Jiang *et al.*, 2011), pulp wastewater (Yang *et al.*, 2013), gut and rumen (Wang *et al.*, 2012), anaerobic digesters (Healy *et al.*, 1995) and compost (Uchiyama *et al.*, 2013).

It is interesting to know that metagenome-derived β-glucosidases are bestowed with unique characteristics. For instance, the β-glucosidase *Bgl1D* derived from alkaline polluted soil is remarkably stable across a broad range of pH (5.5 to 10.5) and is not affected by high concentration of metal ions (Jiang *et al.*, 2011). Similarly, a thermostable GH1 family β-glucosidase from archaeal origin, isolated from hydrothermal spring, was active even at 105°C. It also performed well in the presence of different reagents and solvents (Schroder *et al.*, 2014). Another Functional screening of a metagenomic library from termite gut yielded a thermostable β-glucosidase which showed its highest activity at 90°C (Wang *et al.*, 2012). β-glucosidase with very high glucose tolerance ability of upto 1000 mM was obtained from compost and it showed high transglycosylation activity by generating sophorose, laminaribiose, cellobiose, and gentiobiose (Uchiyama *et al.*, 2013). Marine metagenome derived β-glucosidase also possess similar glucose tolerance by retaining 50 per cent of its activity even at 1000 mM of glucose (Fang *et al.*, 2010). This enzyme is also active at high NaCl concentration signifying its origin from marine environment. These β-glucosidase proteins show very less protein similarity with the sequences available in database. These new entries from unculturable microbial systems incorporate novel insights into the structure- function relationship and its substrate recognition.

9. Improvement of β-glucosidases for Biorefineries

Enzymes commercially available in markets are mainly from fungi and are the most important source of β-glucosidase (Singhania *et al.*, 2013). The cellulase widely used in biomass saccharification is obtained from *Trichoderma reesi*. Although 12 β-glucosidase genes are present in the fungi, it is mostly intracellular and exhibit low β-glucosidase activity leading to incomplete saccharification (Lynd *et al.*, 2002). In order to increase the saccharification efficiency and bioethanol production several strategies has been employed.

9.1. β-glucosidases from Thermophillic Microbes

β-Glucosidases produced by thermophillic organism is gaining importance because of its higher enzyme stability, which is favourable for industrial purposes. Compared to N188BG (Novozyme 188), thermostable β-glucosidases from *Acremonium thermophilum* (*At*BG3) and *Thermoascus aurantiacus* (*Ta*BG3), showed higher cellulose hydrolysis (Teugjas and Väljamäe 2013). Thermoacidophillic β-glucosidases was produced extracellularly in higher amount by *Tolypocladium cylindrosporum* syzx4, isolated from rotten corn stover. Supplementation of β-glucosidase from *Tolypocladium cylindrosporum* syzx4, to *T. reesei* cellulocast increased the saccharification yield to 88.4 per cent (Zhang *et al.*, 2011). Enzyme cocktail preparation of thermotolerant β-glucosidase with commercial cellulase increased the saccharification efficiency greatly (Krisch *et al.*, 2010). The characteristic property of β-glucosidases produced by thermophilic microorganism was summarized in Table 11.2.

9.2. Overexpression of β-glucosidases by Transgenic Approaches

The genes responsible for expression of efficient and versatile β-glucosidase can be transferred to more industrially tractable and robust organisms to enhance the saccharification of the complex lignocellulosic biomass. Many β-Glucosidases genes have been isolated from several organisms like bacteria, yeast, fungi, plant and animal sources (Collins *et al.*, 2007; Chang *et al.*, 2011; Aftab *et al.*, 2012). The phyto-β-glucosidase (clover) gene structure and β-glucosidase of rumen microbe have also been identified (Saibi 2007). These genes can be overexpressed in expression system for its large production and application. *Pichia pastoris, Trichoderma reesei*

Table 11.2: Properties of β-glucosidases Produced by Thermophilic Microorganisms

Strain	MW^a (kDa)	pH	T (°C)	Stability	K_m (mM)	v_{max}	Reference
Humicola insolens	55	6–6.5	60	50°C, pH 6	0.51 (cellobiose) 0.16 (p-NPG^b)	86 U/mg protein (cellobiose) 18.1 U/mg protein (p-NPG)	Souza *et al.* (2010)
Scytalidium thermophilum	40	6.5	60	50°C	0.29 (cellobiose) 1.61 (p-NPG)	13.27 U/mg protein (cellobiose) 4.12 U/mg protein (p-NPG)	Zanoelo *et al.* (2004)
Penicillium citrinum YS40-5	72	5	70	60°C	32.17 (cellobiose) 17.59 (p-NPG)	72.49 U/mg protein (cellobiose) 85.93 U/mg protein (p-NPG)	Ng *et al.* (2010)
Termitomyces clypeatus	116	5	45	60°C, pH 6–7	0.148 (p-NPG)	0.077 U/mg protein (p-NPG)	Pal *et al.* (2010)
Thermotoga maritima	81	5	85		0.0039 (p-NPG)	—	Goyal *et al.* (2001)
Thermoascus aurantiacus	120	4.5	80	70°C, pH 5	0.1137 (p-NPG)	—	Parry *et al.* (2001)
Anaerobic bacterium	43	6.2	75	—	0.73 (cellobiose) 0.15 (p-NPG)	35.67 U/mg protein (p-NPG)	Patchett *et al.* (1987)
Humicola grisea var. *thermoidea*	82	4–4.5	60	60°C	0.316 (p-NPG)	0.459 IU/mL (p-NPG)	Ferreira-Filho (1996)

a: Molecular mass; b: p-Nitrophenyl-β-glucopyranoside.

†*Source:* Chandel *et al.*, 2013.

and *E. coli* species were used as expression host for β-glucosidases (Harhangi *et al.*, 2002; Murray *et al.*, 2004; Chang *et al.*, 2011). β-Glucosidases genes from various organism, their overexpression host and properties of the recombinant β-glucosidase were listed in Table 11.3.

Through transgenic approach, microbial strain can also be developed for simultaneous saccharification and fermentation with enhanced yield at reduced costs. For example, Lee *et al.* (2013) engineered a *Saccharomyces cerevisiae* strain expressing a cellodextrin transporter and an intracellular β-glucosidase from *Neurospora crassa*.

9.4. Differential Expression of beta-glucosidase Isoforms

Several filamentous fungi exhibit the property of expressing different isoforms of BGL depending on the culture conditions or carbon sources (Singhania *et al.*, 2011). Various isoforms of endoglucanase and beta-glucosidase are reported to be expressed in response to carbon sources in *Aspergillus terreus* (Nazir *et al.*, 2010). The sequential induction of isoforms has been associated with the presence of distinct metabolites. The induction of cellulases is mediated either by low molecular weight soluble oligosaccharides that are released from complex substrates as a result of hydrolysis by constitutive enzymes or by the products (positional isomers) of transglycosylation reactions mediated by constituent beta-glucosidase, xylanases, etc. (Badhan *et al.*, 2007). These metabolites enter the cell and signal the presence of extracellular substrates and provide the stimulus for the accelerated synthesis of constituent enzymes of cellulase complex.

However, this process is complex in view of the fact that many fungi and bacteria are known to express functionally multiple cellulases/hemicellulases in the presence of different carbon sources. This multiplicity may be the result of genetic redundancy, differential mRNA processing or post translational modification such as glycosylation, autoaggregation or/and proteolytic digestion (Collins *et al.*, 2005). The regulation of expression of these multiple isoforms is still not clear which necessitates further research regarding the sequential and differential expression of the isoforms.

9.5. Enzyme Immobilisation

Enzyme immobilisation technique is used for activity enhancement, efficient recovery and reuse of enzymes (Illanes 2008). Several inorganic compounds and organic polymers like chelated magnetic metal ion nanoparticles (Chen *et al.*, 2014), magnetic chitosan (Zheng *et al.*, 2013), alginate (Keerti *et al.*, 2014), polyacrylamide gel (Ortega *et al.*, 1998), agarose (Silva *et al.*, 2014) and silica (Reshmi and Sugunan 2013) were used for enzyme immobilisation. The disadvantages of both physical adsorption and covalent modification are enzyme leakage and reduced activity, respectively. In the present day, nanoparticles for enzyme immobilisation are gaining importance due to high surface area to volume ratio that favours higher enzyme loading and enhanced biocatalytic efficiency for industrial application (Verma *et al.*, 2013). Nanofiber polymer coated with β-glucosidase from *Miscanthus sinensis* feeding cattle rumen microbe was more efficient in bioethanol production (Lee

Table 11.3: Characteristics of Heterogeneously Expressed β-glucosidases Isolated from different Types of Organisms

Type of Organism	Source Organism	Expression Host	Mol. Mass (kDa)	No. of Amino Acids	pH Optima	Temp. Optima (LC)	GH Family	References
Bacteria	Acetobacter xylinum	E. coli	79	739	–	–	GH3	Tajima et al. (2001)
	Vibrio cholera	E. coli	65	574	6.0–6.5	37–42	GH9	Park et al. (2002)
	Thermus flavus	E. coli	49	431	5.0–6.0	80–90	GH1	Kang et al. (2005)
	Thermus thermophiles	E. coli	48.7	431	8.5	90	GH1	Gu et al. (2009)
	Exiguobacterium oxidotolerans	E. coli	51.6	448	7	35	GH1	Chen et al. (2010a)
	Cellulomonas biazotea	E. coli	48	447	–	–	GH1	Chan et al. (2011)
	Exiguobacterium sp.	E. coli	52	450	7.0	45	GH1	Chang et al. (2011)
	Micrococcus antarcticus	E. coli	48	472	6.5	25	GH1	Fan et al. (2011b)
	Sphingomonas sp.	E. coli	49.3	447	5.0	37	GH1	Wang et al. (2011)
	Bacillus licheniformis	E. coli	53	466	6	50	GH1	Zahoor et al. (2011)
	Bacillus licheniformis	E. coli	52.2	–	6.0	60	GH16	Aftab et al. (2012)
	Fervidobacterium islandicum	E. coli	53.41	459	6.0–7.0	90	GH1	Jabbour et al. (2012)
Yeast	Candida wickerhamii	E. coli	72	–	7	37–40	GH1	Skory and Freer (1995)
	Pichia etchellsii	E. coli	50	–	6.5–7.0	50	–	Pandey and Mishra (1997)
	Pichia etchellsii	E. coli	52.1	504	7–9	45	GH1	Roy et al. (2005)
Fungi	Piromyces sp.	Pichia pastoris	75.8	–	6.0	39	GH1	Harhangi et al. (2002)
	Uromyces fabae	–	92.4	843	–	–	GH3	Haerter and Voegele (2004)
	Sclerotinia sclerotiorum	–	–	–	5.0	55–60	–	Issam et al. (2004)
	Talalaromyces emersonii	Trichoderma reesei	90.59	857	4.02	71.5	GH3	Murray et al. (2004)
	Thermoascus aurantiacus	Pichia pastoris	–	–	5.0	70	GH3	Hong et al. (2007)

Contd...

Table 11.3–Contd...

Type of Organism	Source Organism	Expression Host	Mol. Mass (kDa)	No. of Amino Acids	pH Optima	Temp. Optima (LC)	GH Family	References
	Aspergillus fumigatus	Pichia pastoris	91.47	–	6.0	60	GH3	Liu et al. (2012a)
	Neosartorya fischeri	E. coli	–	529	6.0	40	GH1	Ramachandrana et al. (2012)
Animal/Insects	Human liver	Pichia pastoris	53	496	6.5	50	GH1	Berrin et al. (2002)
	Bombyx mori		57	491	6.0	35	GH1	Byeon et al. (2005)
	Reticulitermes santonensis	E. coli	–	–	6.0	40	GH1	Matteotti et al. (2011)
	Macrotermes barneyi	E. coli	54	493	5.0	50	GH1	Wu et al. (2012)
	Miscanthus sinensis	E. coli	85	779	5.0	38	GH3	Li et al. (2014)
Plant	Pinus contorta	E. coli	–	513	–	–	GH1	Dharmawardhana et al. (1999)
	Rauvolfia serpentine	E. coli	61	540	5.0	28	GH1	Warzecha et al. (2000)
	Glycine max	E. coli	58		7.0	30	GH1	Suzuki et al. (2006)

†Source: Singh et al., 2016.

et al., 2010). β-glucosidase enzyme immobilised on magnetic Fe_3O_4 nanoparticles retained 90 per cent of enzyme activity after 15 successive cycles (Chen *et al.*, 2014).

Supplementation of super paramagnetic magnetite immobilised *T. reesei* β-glucosidase enzyme to cellulases increased saccharification efficiency by 10 per cent compared to free β- glucosidase enzyme supplementation (Vaenzuela *et al.*, 2014). When immobilised β-glucosidases were pretreated with cellobiose and glucose, the activity of the enzyme increased but the cost of bioethanol production also increased (Jung *et al.*, 2012).

9.5. Inhibitor Tolerant β-glucosidases

Presence of excess glucose in the medium causes feedback inhibition of β-glucosidase activity leading to poor saccharification of biomass. This can be overcome by developing glucose tolerant β-glucosidase. Supplementation of glucose tolerant β-glucosidase increases the hydrolysis and thereby ethanol production (Teugjas and Väljamäe 2013). Glucose tolerant β-glucosidases are produced by few species of *Aspergillus* genus and their gene have been isolated and characterised (Riou *et al.*, 1998; Gunata and Vallier 1999). β-Glucosidases isoforms with lower molecular weight are more glucose tolerant (Rajasree *et al.*, 2013). Glucose tolerance can also be improved by changing the carbon source. For example, Gonde *et al.* (1985) reported that, when lactose was used as the carbon source, the glucose tolerance of the enzyme increased. In addition to glucose tolerance, the β-glucosidase enzyme should also possess ethanol tolerance. *Dekkera intermedia* (Blondin *et al.*, 1983), *Candida molischiana* (Gonde *et al.*, 1985) produced high amount of β-glucosidase under low ethanol concentration. The media composition also affects the β-glucosidase production. Addition of inorganic ions like $MnCl_2$ resulted in higher β-glucosidase production in *Phaffia rhodozyma* (Pera *et al.*, 1999). Apart from cellobiose and glucose, xylooligomers are also power inhibitors of saccharification process (Qing and Wyman 2011). Hence GHs with bifunctional activity or enzyme cocktail with bifunctional β-glucosidase will greatly increase the saccharification efficiency. *Penicillium piceum* produces novel bifunctional glycoside hydrolase enzymes with both β- glucosidase and β-xylosidase that can even act on xylotriose to produce xylobiose and D- xylose (Gao *et al.*, 2013). Preparation of enzyme cocktails with compatible enzyme had also shown a significant increase in saccharification and ethanol production. Simultaneous saccharification and fermentation of pretreated spruce was achieved using an enzyme cocktail containing *T. reesei* cellulase and extracellular β-glucosidase of mutant strain *T. atroviride* (Kovacs *et al.*, 2009).

During pretreatment of lignocellulosic biomass, inhibitors like, furfural and hydroxyl methyl furfural are produced. These inhibitor resistant yeast strains like *Clavispora* NRRL Y-50464 need to be identified (Liu *et al.*, 2012b). In the presence of 15 mM each of furfural and HMF, *Clavispora* NRRL Y-50464 converted furfural into furan methanol in less than 12 h and HMF into furan-2,5-dimethanol within 24h. Without exogenous β-glucosidase supplementation the ethanol production was 23 gl^{-1} by this strain (Liu *et al.*, 2012b). Compared to normal method of ethanol production by yeast, ethanol producing bacteria have gained importance because of its higher growth rate and economical production of ethanol. *Zymomonas mobilis* an ethanol producing bacterium with higher growth rate but narrow spectrum of fermentable carbohydrates reduced its use for fuel ethanol production. Genetic manipulation in β-glucosidase gene to use wide range of carbohydrate substrates had led to the production of 0.49 g ethanol/g cellobiose by recombinant strain. The microbial β-glucosidase and its use in biofuel production from cellulosic waste are listed in Table 11.4.

Table 11.4: Microbial β-glucosidase for Ethanol Production from Cellulosic Materials

Microorganism Type	Name of Organisms	Total Ethanol Production	References
Bacteria	*Exiguobacterium oxidotolerans*	–	Chen *et al.* (2010a)
	Penicillium decumbens	–	Chen *et al.* (2010b)
	Cellulomonas biazotea	–	Chan *et al.* (2011)
	Clostridium phytofermentas	25 mM	Tolonen *et al.* (2011)
	Clostridium thermocellum	1.80 g/l	Kim *et al.* (2013)
Yeast	*Saccharomycopsis fibuligera*	9.15 g/l	Jeon *et al.* (2009a)
	Saccharomycopsis fibuligera	–	Jeon *et al.* (2009b)
	Issatchenkia orientalis	29 g/l	Kitagawa *et al.* (2010)
	Saccharomyces cerevisiae	45 g/l	Ha *et al.* (2011)
	Clavispora NRRL Y-50464	23 g/l	Liu *et al.* (2012b)
	Saccharomyces cerevisiae	8.5 g/l	Tang *et al.* (2013)
Other Fungi	*Aspergillus oryzae*	21.6 g/l	Kotaka *et al.* (2008)
	Aspergillus niger; Trichoderma reesei	–	Chauve *et al.* (2010)
	Penicillium decumbens	–	Ma *et al.* (2011)
	Agaricus arvensis	–	Singh *et al.* (2011)
	Neocallimastix patriciarum	–	Chen *et al.* (2012)
	Periconia sp.	–	Dashtban and Qin (2012)
	Penicillium simplicissimum H-11	–	Bai *et al.* (2013)
	White rot fungi	–	Mfombep *et al.* (2013)
	Acremonium thermophilum and *Thermoascus aurantiacus*	–	Teugjas and Valjamae (2013)
	Aspergillus nidulans, Aspergillus fumigatus, and *Neurospora crassa*	–	Bauer *et al.* (2006)

†*Source*: Singh *et al.*, 2016.

9.6. Optimized Enzyme Production using Statistical Tools

It is well established that optimization of culture conditions helps in significant improvement in enzyme yield, time saving and cost-effective. Various extrinsic factors like temperature, pH, incubation time, inoculum size, moisture content and substrate concentration have been found to affect β-glucosidase production. Nowadays Response Surface Methodology (RSM) is widely used to optimize multi factors at one time especially during fermentation. It has been successfully employed for optimizing physiological conditions of fermentation involved in production of enzymes like *Taq*1endonuclease (Nikerel *et al.*, 2005), polygalacturonase (Tari *et al.*, 2007), cellulase (Dave *et al.*, 2013) and also maximizing β-glucosidase enzyme production. For BGL production, a number of factors like pH, temperature, incubation time, surfactant, casamino acid, methanol, type and amount of carbon and nitrogen source were chosen of which methanol plays a pronounced effect (Batra *et al.*, 2014).

Further optimization is required at the hydrolysis stage where sugar rich complex substrates and different combinations of hydrolytic enzymes are used. Tools like Artificial Neural Network (ANN) and genetic algorithm are also available that can be used for process optimization effectively in case of large scale production of BGL in automatic fermentors.

10. Conclusions

Beta-glucosidase (BGL) is a key enzyme involved in sugar-enzyme platform for bioethanol production from lignocellulosic biomass. Most of the bioconversion processes used today does not allow complete saccharification of biomass. Hydrolysis of biomass can be enhanced by several approaches, one of which is by supplementation of cellulase complex with accessory enzymes (β-glucosidase). Development of glucose, ethanol, inhibitor tolerant beta-glucosidase and enhancement in β-glucosidase activity by enzyme immobilisation can greatly increase the hydrolysis. Traditional approach focuses on isolating hyper β-glucosidase producing microorganisms, separating the enzymes and then supplementing it to commercial preparations. An alternate and more futuristic strategy is to engineer microbes for producing all major enzymes involved in hydrolysis of cellulose in optimum ratio, which would also decrease the expenditure greatly. Heterologously expressed beta-glucosidase by genetically modifying host organism could also be an effective tool in developing enzyme with desired properties. Enzyme recycling can also help in reducing the cost of ethanol production to greater extent.

At present a complex, coherent and cost economic enzyme cocktails to unlock and saccharify polysaccharides from the lignocellulose complex to fermentable sugars is the need of the hour for increasing biobased economy. Development of an efficient cellulase cocktail for biomass based biorefineries will combat the crisis of petroleum products and makes our nation 'energy sustainable' by reducing total carbon credits.

References

Aftab S, Aftab MN, Ikram-Ul-Haq MMJ, Zafar A, Iqbal I (2012). Cloning and expression of endo-1, 4–β-glucanase gene from *Bacillus licheniformis* ATCC 14580 into Escherichia coli BL21 (DE 3). *Afr J Biotechnol* 11:2846–2854.

Ait N. N, Creuzet, Cattaneo J(1982). Properties of β-glucosidase purified from *Clostridium thermocellum*. *J Gen Microbiol* 128: 569-577.

Alef K, Nannipieri, P (1995). Methods in Applied Soil Microbiology and Biochemistry: Enzyme Activities. Academic Press, London.

Amouri B. Gargouri A (2006). Characterization of a novel beta-glucosidase from a *Stachybotrys strain*. *Biochem. Eng. J.* 32: 191–197.

Bai H, Wang H, Sun J, Irfan M, Han M, Huang Y, Han X, Yang Q (2013). Production, purification and characterization of novel beta-glucosidase from newly isolated *Penicillium simplicissimum* H-11 in submergence fermentation. EXCLI J 12: 528–540.

Bao L, Huang Q, Chang L, Sun Q, Zhou J, *et al.* (2012). Cloning and characterization of two β-glucosidase/xylosidase enzymes from yak rumen metagenome. *Appl Biochem Biotechnol* 166: 72-86.

Batra J, Beri D, Mishra S (2014). Response surface methodology based optimization of β-glucosidase production from *Pichia pastoris*. *Appl Biochem Biotechnol* 172: 380-393.

Bauer S, Vasu P, Persson S, Mort A.J, Somerville C.R (2006). Development and application of a suite of polysaccharide degrading enzymes for analyzing plant cell walls. *Proc Natl Acad Sci* 103: 11417–11422.

Beguin P (1990). Molecular biology of cellulose degradation. *Annual Review of Microbiology* 44: 219–248.

Behera S, Arora R, Nandhagopal N, Kumar S (2014). Importance of chemical pretreatment for bioconversion of lignocellulosic biomass. *Renew Sust Energy Rev* 36:91–106.

Berrin J.G, McLauchlan W.R, Needs P, Williamson G, Puigserver A, Kroon P.A, Juge N (2002). Functional expression of human liver cytosolic β-glucosidase in *Pichia pastoris*: insights into its role in the metabolism of dietary glucosides. *Eur J Biochem* 269:249–258.

Blondin B, Ratomahenina R, Arnaud A, Galzy P (1983). Purification and properties of the β-glucosidase of a yeast capable of fermenting cellobiose to ethanol: *Dekkera intermedia* Van Der Walt. *European J Appl Microbiol Biotechnol* 17: 1-6.

Byeon G.M, Lee K.S, Gui Z.Z, Kim I, Kang P.D, Lee S.M, Sohn H.D, Jin B.R (2005). A digestive beta-glucosidase from the silkworm, *Bombyx mori*: cDNA cloning, expression and enzymatic characterization. *Comp Biochem Physiol B Biochem Mol Biol* 141(4):418–427.

Cairns J.R.K, Esen A (2010). β-Glucosidase. *Cell Mol Life Sci* 67:3389–3405.

Cantarel B.L, Coutinho P.M, Rancurel C, Bernard T, Lombard V, Henrissat B (2009). The Carbohydrate-Active EnZymes database (CAZy): an expert resource for glycogenomics. *Nucleic Acids Res* 37:D233–D238.

Chan A.K.N, Wang Y.Y, Ng K, Fu Z, Wong W (2011). Cloning and characterization of a novel cellobiase gene, cba3, encoding the first known β-glucosidase of glycoside hydrolase family 1 of *Cellulomonas biazotea*. *Gene* 493:52–61.

Chandel A.K, Giese E.C, Singh O.V, rio da Silva S.S (2013). Sustainable role of thermophiles in the second generation of ethanol production. In: *Extremophiles:*

Sustainable Resources and Biotechnological Implications, First Edition. Wiley-Blackwell. Published by John Wiley and Sons, Inc.

Chang J, Park I.H, Lee Y.S, Ahn S.C, Zhou Y, Choi Y.L (2011). Cloning, expression, and characterization of β-glucosidase from *Exiguobacterium sp.* DAU5 and transglycosylation activity. *Biotechnol Bioprocess Eng* 16: 97–106.

Chapelle A, *et al.* (2012). Impact of the absence of stem-specific β-glucosidases on lignin and monolignols. *Plant physiology* 160(3):1204-1217.

Chauve M, Mathis H, Huc D, Casanave D, Monot F, Ferreira N.L (2010). Comparative kinetic analysis of two fungal β-glucosidases. *Biotechnol Biofuels* 3: 1–8.

Chen H.L, Chen Y.C, Lu M.Y.J, Chang J.J, Wang H.T.C *et al.* (2012). A highly efficient beta-glucosidase from a buffalo rumen fungus *Neocallimastix patriciarum* W5. *Biotechnol Biofuels* 5(24): 1–10.

Chen M, Qin Y, Liu Z, Liu K, Wang F, Qu Y (2010a). Isolation and characterization of a b-glucosidase from *Penicillium decumbens* and improving hydrolysis of corncob residue by using it as cellulase supplementation. *Enzyme Microbial Technol* 46:444–449.

Chen S, Hong Y, Shao Z, Liu Z (2010b). A cold-active β-Glucosidase (Bgl1C) from a sea bacteria *Exiguobacterium oxidotolerans* A011. *World J Microbiol Biotechnol* 26:1427–1435.

Chen T, Yang W, Guo Y, Yuan R, Xu L, *et al.* (2014). Enhancing catalytic performance of β-glucosidase via immobilization on metal ions chelated magnetic nanoparticles. *Enzyme Microb Technol* 63: 50-57.

Christakopoulos P, Goodenough P. W, Kckos D, Macris B. J, Claeysscns M, Bhat M.K (1994). Purification and characterization of an extracellular β-glucosidase with transglycosylation and exo-glucosidase activities from *Fusarium oxysporum*. *Eur J Biochem* 224: 379-385.

Collins C.M, Murray P.G, Denman S, Morrissey J.P, Byrnes L, Teeri T.T, Tuohy M.G (2007). Molecular cloning and expression analysis of two distinct b-glucosidase genes, bg1 and aven1 with very different biological roles from the thermophilic, saprophytic fungus *Talaromyces emersonii. Mycol Res* 111:840–849.

Cristóbal H.A, Schmidt A, Kothe E, Breccia J, Abate C.M (2009). Characterization of inducible cold-active β-glucosidases from the psychrotolerant bacterium *Shewanella* sp. G5 isolated from a sub-Antarctic ecosystem. *Enzym. Microb. Technol.* 45, 498–506.

Dashtban M, Qin W (2012). Overexpression of an exotic thermotolerant β-glucosidase in *Trichoderma reesei* and its significant increase in cellulolytic activity and saccharification of barley straw. *Microb Cell Fact* 11(63): 1–15.

Dave B.R, Sudhir A.P, Parmar P, Pathak S, Raykundaliya D.P, *et al.* (2013). Enhancement of cellulase activity by a new strain of *Thermoascus aurantiacus*: Optimisation by statistical design response surface methodology. *Biocat Agri Biotechnol* 2: 108-115.

Davies G, Henrissat B (1995). Structures and mechanisms of glycosyl hydrolases. *Structure* 3:853–859.

Decker C.H, Visser J, Schreier P (200 I). β-Glucosidase multiplicity from *Aspergillus tubingensis* CBS 643.92: purification and characterization of four ll-glucosidases and their differentiation with respect to substrate specificiry glucose inhibition and acid tolerance. *Appl Microbiol Biotechnol* 55: 157-163.

Deswal D, Khasa Y.P, Kuhad R.C (2011). Optimization of cellulase production by a brown rot fungus *Fomitopsis sp*. RCK2010 under solid state fermentation. *Bioresour. Technol.* 102: 6065–6072.

Dharmawardhana D.P, Ellis B.E, Carlson J.E (1999). cDNA cloning and heterologous expression of coniferin β -glucosidase. *Plant Mol Biol* 40:365–372.

Fan H.X, Miao L.L, Liu Y, Liu H.C, Liu Z.P (2011b). Gene cloning and characterization of a cold-adapted β -glucosidase belonging to glycosyl hydrolase family 1 from a psychrotolerant bacterium *Micrococcus antarcticus*. *Enzyme Microbial Technol* 49: 94–99.

Fang Z, Fang W, Liu J, Hong Y, Peng H, *et al.* (2010). Cloning and characterization of a beta-glucosidase from marine microbial metagenome with excellent glucose tolerance. *J Microbiol Biotechnol* 20: 1351-1358.

Faure D, Desair J, Keijers V, Bekri M.A, Proost P, Henrissat B, Vanderleyden J (1999). Growth of *Azospirillum irakense* KBC1 on the aryl beta-glucoside salicin requires either salA or salB. *J Bacteriol* 18: 3003-3009.

Fischer L, Bromann R, Kengen S.W.M, Vos W.M.D, Wagner F (1996). Catalytical potency of beta-Glucosidase from the extremophile *Pyrococcus furiosus* in glucoconjugate synthesis. *Bio'Technology* 14: 88-91.

Gao L, Gao F, Zhang D, Zhang C, Wu G, *et al.* (2013). Purification and characterization of a new β-glucosidase from *Penicillium piceum* and its application in enzymatic degradation of delignified corn stover. *Bioresour Technol* 147: 658-661.

Gonde P, Ratomahenina R, Arnaud A, Galzy P (1985). Purification and properties of an exocellular β-glucosidase of *Candida molischiana* (Zikes) Meyer Yarrow capable of hydrolyzing soluble cellodextrins. *Can J Biochem Cell Biol* 63: 1160-1166.

Gonzalez-Candelas L, Raman D, Polaina J (1990). Sequences and homology analysis of two genes encoding beta-glucosidases from *Bacillus polymyxa*. *Gene* 95(1): 31-38.

Gu N.Y, Kim J.L, Kim H.J, You D.J, Kim H.W, Jeon S.J (2009). Gene cloning and enzymatic properties of hyperthermostable b-glycosidase from *Thermus thermophilus* HJ6. *J Biosci Bioeng* 107:21–26.

Gueguen Y, Chemardin P, Arnaud A, Galzy P (1995). Purification and characterization of an intracellular β-glucosidasesfrom *Botrytis cinerea*. *Enzyme Microb Technol* 78: 900-906.

Gunata Z, Vallier M, Sapis J, Baumes R, Bayonove C (1994). Enzymatic synthesis of monoterpenyl β-D-glucosides by various β-glucosidases. *Enzyme Microbial Technol* 16:1055–1058.

Gunata, Z., Vallier, M.J., 1999. Production of a highly glucose-tolerant extracellular beta-glucosidase by three Aspergillus strains. *Biotechnol. Lett.* 21, 219–223.

Ha S.J, Galazka J.M, Kim S.R, Choi J.H, Yang X, Seo J.H, Glass N.L, Cate J.H.D, Jin Y.S (2011). Engineered *Saccharomyces cerevisiae* capable of simultaneous cellobiose and xylose fermentation. *Proc Natl Acad Sci* 108(2): 2504–2509.

Haerter A.C, Voegele R.T (2004). A novel β-glucosidase in *Uromyces fabae*: feast or fight? *Curr Genet* 45:96–103.

Hang H.T (1995). Decolorization of anthocyan ins by fungal enzymes. *Agric Food Chem* 3:141-146.

Harhangi H.R, Steenbakkers P,J.M, Akhmanova A, Jetten M.S.M, van der Drift C, Op den Camp H.J.M (2002). A highly expressed family 1 β-glucosidase with transglycosylation capacity from the anaerobic fungus *Piromyces* sp. E2. B*iochim Biophys Acta* 1574: 293–303.

Harnpicharnchai P, Champreda V, Sornlake W, Eurwilaichitr L (2009). A thermotolerant beta-glucosidase isolated from an endophytic fungi *Periconia* sp. with a possible use for biomass conversion to sugars. *Protein Expr. Purif.* 67: 61–69.

Healy F.G, Ray R.M, Aldrich H.C, Wilkie A.C, Ingram L.O, *et al.* (1995). Direct isolation of functional genes encoding cellulases from the microbial consortia in a thermophilic, anaerobic digester maintained on lignocelluloses. *Appl Microbiol Biotechnol* 43: 667-674.

Henrissat B, Davies G (1997). Structural and sequence-based classification of glycoside hydrolases. *Curr Opin Struct Biol* 7:637–644.

Henrissat B, Driguez H, Viet C, *et al.* (1985). Synergism of cellulases from *Trichoderma reesei* in the degradation of cellulose. *Biotechnology* 3: 722–726.

Her S.H.S, Lee S.J, Choi S.W, Choi H.J, Choi S.S, Yoon, Oh D.H (1999). Cloning and sequencing of β--l 4-endoglucanase gene (*celA*) from *Pseudomonas* sp. YD-IS. Left *Appl Microbiol* 29: 389-395.

Himmel M.E, Ding S.Y, Johnson D.K, Adney W.S, Nimlos M.R, Brady J.W, *et al.* (2007). Biomass recalcitrance: engineering plants and enzymes for biofuels production. Science 315:804–7.

Hirofumi S, Kenro T, Akikazu A, Takaaki F, Makoto S, Yoshimichi D, Tsunco Y (1991). Enzymatic synthesis of useful alkyl β-glucosidase. *Agric Biol Chem Techno/*55:1679-1681.

Hong J, Tamaki H, Kumagai H (2007). Cloning and functional expression of thermostable β-glucosidase gene from *Thermoascus aurantiacus*. *Appl Microbiol Biotechnol* 73: 1331–1339.

Illanes A (2008). Enzyme Biocatalysis: Principles and Applications, Springer Science, New York, NY, USA.

Issam S.M, Mohamed G, Dominique L.M, Thierry M, Farid L, Nejib M (2004). A β-glucosidase from *Sclerotinia sclerotiorum*. *Appl Biochem Biotechnol* 112:63–77.

Izumi, T., Piskula, M.K., Osawa, S., Obata, A., Tobe, K., Saito, M., Kataoka, S., Kubota, Y., Kikuchi, M., 2000. Soy isoflavone aglycones are absorbed faster and in higher amounts than their glucosides in humans. *J. Nutr.* 130, 1695–1699.

Jabbour D, Klippel B, Antranikian G (2012). A novel thermostable and glucose-tolerant β -glucosidase from *Fervidobacterium islandicum*. *Appl Microbiol Biotechnol* 93:1947–1956.

Jeon E, Hyeon J.E, Eun L.S, Park B.S, Kim S.W, Lee J, Han S.O (2009a). Cellulosic alcoholic fermentation using recombinant *Saccharomyces cerevisiae* engineered for the production of *Clostridium cellulovorans* endoglucanase and *Saccharomycopsis fibuligera* β-glucosidase. *FEMS Microbiol Lett* 301: 130–136.

Jeon E, Hyeon J.E, Suh D.J, Suh Y.W, Kim S.W, Song K.H, Han S.O (2009b). Production of cellulosic ethanol in *Saccharomyces cerevisiae* heterologous expressing *Clostridium thermocellum* endoglucanase and *Saccharomycopsis fibuligera* β-glucosidase *Genes*. *Mol Cells* 28: 369–373.

Jiang C, Li SX, Luo FF, Jin K, Wang Q, *et al.* (2011). Biochemical characterization of two novel β-glucosidase genes by metagenome expression cloning. *Bioresour Technol* 102: 3272-3278.

Joo A.R, Jeya M, Lee K.M, Lee K.M, Moon H.J, Kim Y.S, Lee J.K (2010). Production and characterization of beta-1-4-glucosidase from a strain of *Penicillium pinophilum*. *Process Biochem.* 45: 851–858.

Jørgensen H, Kristensen J.B, Felby C (2007). Enzymatic conversion of lignocellulose into fermentable sugars: challenges and opportunities. Biofuels, *Bioprod. Bioref.* 1:119–134.

Jung Y.R, Shin H.Y, Song Y.S, Kim S.B, Kim S.W (2012). Enhancement of immobilized enzyme activity by pretreatment of β-glucosidase with cellobiose and glucose. *Journal of Industrial and Engineering Chemistry* 18: 702-706.

Kang S.K, Cho K.K, Ahn J.K, Kang S.H, Lee S.H, Lee H.G, Choi Y.J (2005). Cloning, expression, and enzyme characterization of thermostable b-glycosidase from *Thermus flavus* AT-62. *Enzyme Microbial Technol* 37: 655–662.

Kawaguchi T, Enoki T, Tsurumaki S, Sumitani J, Ueda M, Ooi T, Arai M (1996). Cloning and sequencing of the cDNA encoding β-glucosidase 1 from *Aspergillus aculeatus*. *Gene* 173: 287-288.

Keerti, Gupta A, Kumar V, Dubey A, Verma A.K (2014). Kinetic Characterization and Effect of Immobilized Thermostable β-Glucosidase in Alginate Gel Beads on Sugarcane Juice. *ISRN Biochemistry* 2014: 1-8.

Kempton J.B, Withers S.G (1992). Mechanism of *Agrobacterium* beta-glucosidase: kinetic studies. *Biochemistry* 31: 9961-9969.

Kim S, Baek S.H, Lee K, Hahn J.S (2013). Cellulosic ethanol production using a yeast consortium displaying a minicellulosome and β-glucosidase. *Microb Cell Fact* 12 (14):1–7.

Kitagawa T, Tokuhiro K, Sugiyama H, Kohda K, Isono N, Hisamatsu M, Takahashi H, Imaeda T (2010). Construction of a β-glucosidase expression system using the multistress-tolerant yeast Issatchenkia orientalis. *Appl Microbiol Biotechnol* 87:1841–1853.

Kohchi C, Toh E (1985). Nucleotide sequence of *Candida pelliculosa* beta-glucosidase gene. *Nucleic Acids Res* 13: 6273- 6282.

Kotaka A, Bando H, Kaya M, Kato-Murai M, Kuroda K, Sahara H, Hata Y, Kondo A, Ueda M (2008). Dirtct ethanol production from barley β-glucan by sake yeast *Aspergillus oryzae* β-glucosidase and endoglucanase. *J Biosci Bioeng* 105(6): 622–627.

Kovacs K, Megyeri L, Szakacs G, Kubicek C.P, Galbe M, Zacchi G (2008). *Trichoderma atroviride* mutants with enhanced production of cellulase and betaglucosidase on pretreated willow. *Enzyme Microb. Technol.* 43: 48–55.

Kovács K, Szakacs G, Zacchi G (2009). Comparative enzymatic hydrolysis of pretreated spruce by supernatants, whole fermentation broths and washed mycelia of *Trichoderma reesei* and *Trichoderma atroviride*. *Biores Technol* 100:1350–1357

Krisch J, Tako M, Papp T, Vagvolgyi C (2010). Characteristics and potential use of b-glucosidases from Zygomycetes. In: Mendez- Vilas A (ed) Current Research, Technology and Education, *Topics in Applied Microbiology and Microbial Biotechnology*, pp 891–896.

Krogh, K.B.R., Harris, P.V., Olsen, C.L., Johansen, K.S., Hojer-Pedersen, J., Borjesson, J.,Olsson, L., 2010. Characterization and kinetic analysis of a thermostable GH3 beta-glucosidase from Penicillium brasilianum. *Appl. Microbiol. Biotechnol.* 86,143–154.

Lauro B.D, Mose Rossi, Moracci M (2006). Characterization of a β-glycosidase from the thermoacidophilic bacterium *Alicyclobacillus acidocaldarius*. *Extremophiles* 10: 301-310.

Lee S.M, Jin L.H, Kim J.H, Han S.O, Na H.B, Hyeon T, Koo Y.M, Kim J, Lee J.H (2010). β-Glucosidase coating on polymer nanofibers for improved cellulosic ethanol production. *Bioprocess Biosyst Eng* 33:141–147

Li X, Pei I, Wu G, Shao W (2005). Expression purification and characterization of a recombinant β-glucosidase from *Volvariella volvacea*. *Biotech Lett* 27: 1369-1373.

Li H, Xu X, Chen H, Zhang, Xu J, Wang J, Lu X (2013). Molecular analyses of the functional microbial community in composting by PCR-DGGE targeting the genes of the β-glucosidase. *Bioresource Technology* 134: 51–58

Li Y, Liu N, Yang H, Zhao F, Yu Y, Tian Y, Lu X (2014). Cloning and characterization of a new β-Glucosidase from a metagenomic library of Rumen of cattle feeding with *Miscanthus sinensis*. *BMC Biotechnol* 14:85–94.

Lieberman, R.L., Wustman, B.A., Huertas, P., Powe Jr, A.C., Pine, C.W., Khanna, R.,Schlossmacher, M.G., Ringe, D., Petsko, G.A., 2007. Structure of acid betaglucosidase with pharmacological chaperone provides insight into Gaucher disease. *Nat. Chem. Biol.* 3, 101–107.

Lin J, Pillay B, Singh S (1999). Purification and biochemical characteristics of 5-D-glucosidase from a thermophilic fungus, *Thermomyces lanuginosus*-SSBP. *Biotechnology and Applied Biochemistry* 30(1): 81–87.

Litzinger S, Fischer S, Polzer P, Diederichs K, Welte W, Mayer C (2010). Structural and kinetic analysis of *Bacillus subtilis* N-acetylglucosaminidase reveals a unique Asp-His dyad mechanism. *J Biol Chem* 285(46):35675–35684.

Liu D, Zhang R, Yang X, Zhang Z, Song S, Miao Y, Shen Q (2012a). Characterization of a thermostable beta-glucosidase from *Aspergillus fumigatus* Z5, and its functional expression in *Pichia pastoris* X33. *Microb Cell Fact* 11: 25–29.

Liu Z, Weber S, Cotta M, Li S (2012). A new β-glucosidase producing yeast for lower-cost cellulosic ethanol production from xylose-extracted corncob residues by simultaneous saccharification and fermentation. *Bioresour Technol.*, 104: 410-441.

Liu Z.L, Weber S.A, Cotta M.A, Li S.Z (2012b). A new β-glucosidase producing yeast for lower-cost cellulosic ethanol production from xylose-extracted corncob residues by simultaneous saccharification and fermentation. *Bioresour Technol* 104: 410–416.

Lymer E. S. V. Renganathan (1995). Purification and characterization of a cellutose binding β-glucosidase from cellulose-degrading cultures of *Phanerochaete chrysosporium*. *Appl Environ Microbiol* 61: 2976-2980.

Lynd L.R, Weimer P.J, Van Zyl W.H, Pretorius I.S (2002). Microbial cellulose utilization: undamentals and biotechnology. *Microbiol Mol Biol Rev* 66:506–577.

Ma L, Zhang J, Zou G, Wang C, Zhou Z (2011). Improvement of cellulase activity in *Trichoderma reesei* by heterologous expression of a beta-glucosidase gene from *Penicillium decumbens*. *Enzyme Microbial Technol* 49: 366–371.

Machida M, Ohtsuki I, Fukui S,Yarnashita 1 (1988). Nucleotide sequences of *Saccharomycopsis fibuligera* genes for extracellular β-glucosidases as expressed in *Saccharomyces cerevisiae*. *Appl Environ Microbial* 54: 3147·3155.

Maki M, Leung K.T, Qin W (2009). The prospects of cellulase-producing bacteria for the bioconversion of lignocellulosic biomass. *Int J Biol Sci* 5: 500-516.

Martino A, Pifferi P.G, Spagna G (1994). Production of β-glucosidase by *Aspergillus niger* using carbon sources derived from agricultural wastes. *J Chem Technol Biolechnol* 60: 247-252.

Martins L.F, Kolling D, Camassola M, Dillon P.A.J, Ramos L.P (2008). Comparison of *Penicillium echinulatum* and *Trichoderma reesei* cellulases in relation to their activity against various cellulosic substrates. *Bioresour. Technol.* 99: 1417–1424.

Matteotti C, Thonart P, Francis F, Haubruge E, Destain J, Brasseur C, Bauwens J, Pauw E.D, Portetelle D, Vandenbol M (2011). New glucosidase activities identified by functional screening of a genomic DNA library from the gut microbiota of the termite *Reticulitermes santonensis*. *Microbiol Res* 166(8):629–642.

Mfombep P.M, Senwo Z.N, Isikhuemhen O.S (2013). Enzymatic activities and kinetic properties of β-glucosidase from selected white rot fungi. *Adv Biol Chem* 3: 198–207.

Moreno B, Canizares R, Nunez R, Benitez E (2013). Genetic diversity of β-glucosidase encoding genes as a function of soil management. *Biol Ferti Soils* 49: 735-745.

Murray P, Aro N, Collins C, Grassick A, Penttila M, Saloheimo M, Tuohy M (2004). Expression in Trichoderma reesei and characterization of a thermostable family 3 b-glucosidase from the moderately thermophilic fungus *Talaromyces emersonii*. *Protein Expr Purif* 38:248–257

Murray P, Aro N, Collins C, Grassick A, Penttila M, Saloheimo M, Tuohy M (2004). Expression in *Trichoderma reesei* and characterization of a thermostable family 3 β-glucosidase from the moderately thermophilic fungus *Talaromyces emersonii*. *Protein Expr Purif* 38: 248–257.

Nam K.H, Kim S.J, Kim M.Y, Kim J.H, Yeo Y.S, *et al.* (2008). Crystal structure of engineered beta-glucosidase from a soil metagenome. *Proteins* 73: 788-793.

Nam, K.H., Sung, M.N., Hwang, K.Y., 2010. Structural insights into the substrate recognition properties of b-glucosidase. Biochem. *Biophys. Res. Commun.* 391, 1131–1135.

Ng I.S, Li C.W, Chan S.P, Chir J.L, Chen P.T, Tong Ch.G, Yu S.M, David Ho T.H (2010). High level production of a thermophilic beta-glucosidase from *Penicillium citrinum* YS40-5 by solid-state fermentation with rice bran. *Bioresour. Technol.* 101: 1310–1317.

Nijikken Y, Tsukada T, Igarashi K, Samejima M, Wakagi T, Shoun H, Fushinobu S (2007). Crystal structure of intracellular family 1β-glucosidase BGL1A from the basidiomycete *Phanerochaete chrysosporium*. *FEBS Lett* 581:1514–1520.

Nikerel E, Toksoy E, Kirdar B, Yildirim R (2005). Optimising medium composition for TaqI endonuclease production by recombinant *Escherichia coli* cells using response surface methodology. *Process Biochemistry* 40: 1633-1639.

Okamoto K, Sugita Y, Nishikori N, Nitta Y, Yanase H (2011). Characterization of two acidic beta-glucosidases and ethanol fermentation in the brown rot fungus *Fomitopsis palustris*. *Enzyme. Microb. Technol.* 48: 359–364.

Opassiri R, Hua Y, Wara-Aswapati O, Akiyama T, Svasti J, Esen A, Cairns JRK (2004). Beta-glucosidase, exo-beta-glucanase and pyridoxine transglucosylase activities of rice BGlu1. *Biochem J* 379:125–131.

Ortega N, Busto M.D, Parez- Mateos M (1998). Optimisation of β-glucosidase entrapment in alginate and polyacrylamide gels. *Bioresource Technology* 64: 105-111.

Ozaki H, Yamada K (1992). Isolation of *Streptomyces* sp. producing glucose-tolerant β-glucosidases and properties of the enzymes. *Agric Biol Chem* 55: 979-987.

Paavilaincn S, Hellman 1, Korpcla T (1993). Purification characterization gene cloning and sequencing of a new β-Glucosidase from *Bacillus circulans* subsp. *alkalophilus*. *Appl Environ Microbiol* 59: 927-932.

Pandey M, Mishra S (1997). Expression and characterization of *Pichia etchellsii* β-glucosidase in *Escherichia coli*. *Gene* 190:45–51.

Park J.K, Wang L.X, PateL H.V, Roseman S (2002a). Molecular Cloning and Characterization of a unique β-Glucosidase from *Vibrio cholerae*. *J Biolog Chem* 277: 29555-29560.

Park T.H, Choi K.W, Park C.S, Lee S.B, Kang H.Y, Shon K.1, Park J.S, Cha J (2005). Substratc spccificity and transglycosylation catalyzcd by a thermostable beta-glucosidase from marine hypcrthermophile *Thermotoga neapolitana*. *Appl Microbial Biotechnol* 4:411-422.

Parry N.J, Beever D.E, Owen E, Vandenberghe I, Van Beeumen J, Bhat M.K (2001). Biochemical characterization and mechanism of action of a thermostable β-glucosidase purified from *Thermoascus aurantiacus*. *Biochem J* 353: 117-127.

Pera L.M, Rubinstein L, Baigori M.D, Figueroa L, Callieri D.A (1999). Influence of manganese on cell morphology, protoplasts formation and β-D-glucosidase activity in *Phaffia rhodozyma*. *FEMS Microbiology Letters* 171: 155-160.

Pouwels J, Moracci M, Cobucci-Ponzano B, Perugino G, van der Oost J, Rossi M (2000). Activity and stability of hyperthermophilic enzymes: a comparative study on two archaeal beta-glycosidases. *Extremophiles* 4: 157-164.

Qi M, Jun H.S, Forsberg C.W (2008). Cel9D, an atypical 1, 4-β-D-glucan glucohydrolase from *Fibrobacter succinogenes*: characteristics, catalytic residues, and synergistic interactions with other cellulases. *J Bacteriol* 190:1976–1984.

Qing Q, Wyman C.E (2011). Hydrolysis of different chain length xylooliogmers by cellulase and hemicellulase. *Bioresour Technol* 102: 1359-1366.

Rajan S.S, Yang X, Collart F, Yip V.L, Withers S.G, Varrot A, Thompson J, Davies G.J, Anderson W.F (2004). Novel catalytic mechanism of glycoside hydrolysis based on the structure of an NAD+/Mn2+-dependent phospho-alpha-glucosidase from *Bacillus subtilis*. *Structure* 12:1619–1629.

Rajasree K.P, Mathew G.M, Pandey A, Sukumaran R.K (2013). Highly glucose tolerant β-glucosidase from *Aspergillus unguis*: NII 08123 for enhanced hydrolysis of biomass. *J Ind Microbiol Biotechnol* 40: 967-975.

Ramachandrana P, Tiwari M.K, Singh R.K, Haw J, Jeya M, Lee J (2012). Cloning and characterization of a putative b-glucosidase (NfBGL595) from *Neosartorya fischeri*. *Process Biochem* 47:99–105.

Reshmi R, Sugunan S (2013). Improved biochemical characteristics of crosslinked β-glucosidase on nanoporous silica foams. *Journal of Molecular Catalysis B: Enzymatic* 85-86: 111-118.

Riou C, Salmon 1.M, Vallier M.J, Gunata, Barrc P (1998). Purification characterization and substrate specificity of a novel highly glucose-tolerant β-Glucosidase from *Aspergillus oryzae*. *Appl Environ Microbol* 64: 3607-3614.

Riou C, Salmon J.M, Vallier M.J, Gunata Z, Barre P (1998). Purification, characterization, and substrate specificity of a novel highly glucose-tolerant β-glucosidase from *Aspergillus oryzae. Appl Environ Microbiol* 64:3607–3614

Robinson, D (1956). The fluorimetric determination of β-glucosidase: its occurrence in the tissues of animals, including insects. *Biochem. J.*, 63: 39-44.

Roy P, Mishra S, Chaudhary T.K (2005). Cloning, sequence analysis, and characterization of a novel β-glucosidase-like activity from *Pichia etchellsii. Biochem Biophys Res Commun* 336:299–308.

Saloheimo M, Kuja-Panula J, Ylosma Ylosmaki E, Ward M, Penttila M (2002). Enzymatic properties and intracellular localization of the novel *Trichoderma reesei* betaglucosidase BGL (CellA). *Appl Environ Microbiol* 68: 4546—4553.

Saha B.C Bothast R. J (1996). Production.purification and characterization of a highly glucose-tolerant novel β-glucosidasc from *Candida peltata. Appl Environ Biotechnol* 62: 3165-3170.

Saibi W, Amouri B, Gargouri A (2007). Purification and biochemical characterization of a transglucosilating beta-glucosidase of Stachybotrys strain. *Applied Microbiology and Biotechnology* 77:293–300.

Schröder C, Elleuche S, Blank S, Antranikian G (2014). Characterization of a heat-active archaeal β-glucosidase from a hydrothermal spring metagenome.*Enzyme Microb Technol* 57: 48-54.

Shipkowski S, Brenchley J.E (2005). Characterization of an unusual cold-active β-Glucosidase belonging to family 3 of the glycoside hydrolases from the psychrophilic isolate *Paenibacillus* sp. strain C7. *Appl Environ Microbiol* 71: 4225-4232.

Silva T.M, Pessela B.C, Silva J.C.R, Lima M.S, Jorge J.A, *et al.* (2014). Immobilization and high stability of an extracellular β-glucosidase from *Aspergillus japonicas* by ionic interactions. *Journal of Molecular Catalysis B: Enzymatic* 104: 95-100.

Singh G, Verma A.K, Kumar V (2016). Catalytic properties, functional attributes and industrial applications of β-glucosidases. 3 *Biotech* 6:3.

Singh R.K, Zhang Y.W, Nguyen N.P.T, Jeya M, Lee J.K (2011). Covalent immobilization of β-1,4-glucosidase from *Agaricus arvensis* onto functionalized silicon oxide nanoparticles. *Appl Microbiol Biotechnol* 89: 337–344.

Singhania R.R, Sukumaran R.K, Rajasree K.P, Mathew A, Gottumukkala L.D, Pandey A. (2011). Properties of a major beta-glucosidase-BGL1 from *Aspergillus niger* NII-08121 expressed differentially in response to carbon sources. *Process Biochem.* 46: 1521–1524.

Singhania R.R, Patel A.K, Sukumaran R.K, Larroche C, Pandey A (2013). Role and significance of beta-glucosidases in the hydrolysis of cellulose for bioethanol production. *Bioresource Technology* 127: 500–507.

Skory C.D, Freer S.N (1995). Cloning and characterization of a gene encoding a cell-bound, extracellular beta-glucosidase in the yeast *Candida wickerhamii*. *Appl Environ Microbiol* 61:518–525.

Souza F.H.M, Nascimento C.V, Rosa J.C, Masui D.C, Leone F.A, Jorge J.A, Furriel R.P.M (2010). Purification and biochemical characterization of a mycelia glucose- and xylose-stimulated beta-glucosidase from the thermophilic fungus *Humicola insolens*. *Process Biochem*. 45: 272–278.

Suzuki H, Takahashi S, Watanabe R, Fukushima Y, Fujita N, Noguchi A, Yokoyama R, Nishitani K, Nishino T, Nakayama T (2006). An isoflavone conjugate-hydrolyzing β-Glucosidase from the roots of soybean (*Glycine max*) seedlings: purification, gene cloning, phylogenetics and cellular localization. *J Biol Chem* 281(40):30251–30259.

Tajima K, Nakajima K, Yamashita H, Shiba T, Munekata M, Takai M (2001). Cloning and sequencing of the beta-glucosidase gene from *Acetobacter xylinum* ATCC 23769. *DNA Res* 8: 263–269.

Tang H, Hou J, Shen Y, Xu L, Yang H, Fang X, Bao X (2013). High β-glucosidase secretion in *Saccharomyces cerevisiae* improves the efficiency of cellulase hydrolysis and ethanol production in simultaneous saccharification and fermentation. *J Microbiol Biotechnol* 23(11): 1577–1585.

Teugjas H, Valjamae P (2013). Selecting β-glucosidases to support cellulases in cellulose saccharification. *Biotechnol Biofuels* 6(105): 1–13.

Tolonen A.C, Haas W, Chilaka A.C, Aach J, Gygi S.P, Church G (2011). Proteome-wide systems analysis of a cellulosic biofuelproducing microbe. *Mol Syst Biol* 7(461): 1–2.

ToshioT, Susumi O (1985). Cellotriose synthesis using D-glucose as a starting material by two step reactions of immobilized β-glucosidase. *Agric Bioi Chem Technol* 49: 1267-1273.

Tribolo S, Berrin J.G, Kroon P.A, Czjzek M, Juge N (2007). The crystal structure of human cytosolic beta glucosidase unravels the substrate aglycone specificity of a family 1 glycoside hydrolase. *J Mol Biol* 370:964–975.

Tsukada T, Igarashi K, Yoshid M, Samejima M (2006). Molecular cloning and characterization of two intracellular beta-glucosidases belonging to glycoside hydrolase family 1 from the basidiomycete *Phanerochaete chrysosporium*. *Appl. Microbiol. Biotechnol.* 73, 807–814.

Uchiyama T, Miyazaki K, Yaoi K (2013). Characterization of a novel β-glucosidase from a compost microbial metagenome with strong transglycosylation activity. *J Biol Chem* 288: 18325-18334.

Vaenzuela R, Castro JF, Parra C, Baeza J, Duran N, *et al*. (2014). β-Glucosidase immobilisation on synthetic superparamagnetic magnetite nanoparticles and their application in saccharification of wheat straw and *Eucalyptus globules* pulps. *Journal of Experimental Nanoscience* 9: 177-185.

Verma M.L, Chaudhary R, Tsuzuki T, Barrow C.J, Puri M (2013). Immobilization of β-glucosidase on a magnetic nanoparticle improves thermostability: application in cellobiose hydrolysis. *Bioresour Technol* 135: 2-6.

Villena M.A, Iranzo J.F.U, Gundllapalli S.B, Otero R.R.C. Al B.P (2006). Characterization of an exocellular β-glucosidase from *Debaryomyces pseudopolymorphus*. *Enzyme Microbial Technol* 39: 229234.

Villena M.A, Iranzo J.F.Ú, Gundllapalli S.B, Otero R.R.C, Pérez I.B (2006). Characterization of an exocellular beta-glucosidase from *Debaryomyces pseudopolymorphus*. *Enzyme Microb. Technol.* 39: 229–234.

Wang L, Liu Q.M, Sung B.H, An D.S, Lee H.G, Kim S.G, Kim S.C, Lee S.T, Im W.T (2011). Bioconversion of ginsenosides Rb1, Rb2, Rc and Rd by novel β-glucosidase hydrolyzing outer 3-O glycoside from *Sphingomonas sp*. 2F2: cloning, expression, and enzyme characterization. *J Biotechnol* 156:125–133.

Wang Q, Qian C, Zhang XZ, Liu N, Yan X, *et al.* (2012). Characterization of a novel thermostable β-glucosidase from a metagenomic library of termite gut. *Enzyme Microb Technol* 51: 319-324.

Wang Q, Trimbur D, Graham R, Warren RAJ, Withers SG (1995). Identification of the acid/base catalyst in *Agrobacterium faecalis* β-glucosidase by kinetic analysis of mutants. *Biochemistry* 34:14554–14562.

Warzecha H, Gerasimenko I, Kutchan T.M, Stockigt J (2000). Molecular cloning and functional bacterial expression of a plant glucosidase specifically involved in alkaloid biosynthesis. *Phytochemistry* 54:657–666.

Wood T.M. McCrae S. I (1980). Purification and some properties of the extracellular β-glucosidase of the cellulolytic fungus *Trichoderma koningii*. *J Gen Microbiol* 128: 2973-2982.

Wu Y, Chi S, Yun C, Shen Y, Tokuda G, Ni J (2012). Molecular cloning and characterization of an endogenous digestive β-glucosidase from the midgut of the fungus-growing termite *Macrotermes barneyi*. *Insect Mol Biol* 21(6): 604–614.

Yang C, Niu Y, Li C, Zhu D, Wang W, *et al.* (2013). Characterization of a novel metagenome-derived 6-phospho-β-glucosidase from black liquor sediment. *Appl Environ Microbiol* 79: 2121-2127.

Yang S, Wang L, Yan Q, Jiang Z, Li L (2009). Hydrolysis of soybean isoflavone glycosides by a thermostable beta-glucosidase from Paecilomyces thermophila. *Food Chem.* 115: 1247–1252.

Yazdi M.T, Khosravi A.A, Nemati M, Motlagh N.D.V (2003). Purification and characterization oftwo intracellular β-glucosidases from the *Neurospora crassa* mutant cell-I. *W J Microbiol Biotechnol* 19: 79-84.

Zahoor S, Javed M.M, Aftab M.N (2011). Cloning and expression of beta glucosidase gene from *Bacillus licheniformis* into *E. coli* BL 21 (DE3). *Biologia* 66: 213–220.

Zhang Y, Chengyu L, Jingying T, Yu X, Lu J, *et al.* (2011). Enhanced saccharification of steam explosion pretreated corn stover by the supplementation of thermoacidophilic β-glucosidase from a newly isolated train, *Tolypocladium cylindrosporum* syzx4. *Afr J Microbiol* Res 5: 2413-2421.

Zheng P, Wang J, Lu C, Xu Y, Sun Z (2013). Immobilized β-glucosidase on magnetic chitosan microspheres for hydrolysis of straw cellulose. *Process Biochemistry* 48: 683-687.

Zhu L, Wu Q, Dai J, Zhang S, Wei F (2011). Evidence of cellulose metabolism by the giant panda gut microbiome. *Proc Natl Acad Sci USA* 108: 17714- 17719.

12

Biorefinery Approach of Biomass: Synthesis and Characterization of Lignin Adhesive

T. Chirangeevi[1], Y. Varun[1], K. Ravichandra[2],
A. Umav and R.S. Prakasham[2]

[1]*Center For Biotechnology, Institute of Science and Technology, Jawaharlal*
Nehru Technological University, Hyderabad – 500 085
[2]*Bioengineering and Environmental Sciences,*
CSIR-Indian Institute of Chemical Technology, Hyderabad – 500 007

1. INTRODUCTION

A biorefinary approach was applied to synthesize bioadhesive using the lignin extracted from water hyacinth (Eichhornia crassipes). More than 75 per cent of the lignin was recovered using alkali extraction method and the same was used for adhesive synthesis. The structural and thermal properties revealed the presence of guacyl groups signifying extracted lignin potential reactivity during polymerization and high suitability for the preparation of wood adhesive. The specific gravity, viscosity and tensile strength of adhesive were noticed to be 1.29, 1530mPas and 2.4 Mpa respectively. The FT-IR spectra of lignin-incorporated adhesive showed structural similarity with control phenol-formaldehyde adhesive. Thermal characterization based on TG and DTG analyses indicated that the degradation of solid matter occurred in two stages; 50 per cent of the solid matter was degraded in wide range of temperature (60-500°C) while the rest was not degraded even at 800°C indicating high thermal stability of lignin-adhesive.

Lignocellulosic biomass plays vital role in development of sustainable products such as chemicals, fuels, etc., through biorefinery approach. This is mainly because of its inexpensive, renewable, largely available and world-wide distribution nature as well as being CO_2 neutral (Lange, 2007). The structural complexity associated with lignocellulosic structure restrict the use of biomass for industrial products production and only 3 to 3.5 per cent of the annually produced biomass (amounting 170-200 x 10^9 tons) is used for non-food applications (Kamm *et al.*, 2006). Carbohydrate polymers of the biomass are basically used for production of paper and pulp and other value added compounds. During this process, lignin stream exits as waste stream amounting more than 70 million tons per year (Kumar *et al.*, 2009).

Lignin is basically an aromatic biopolymer and has high potential to produce fission products especially platform chemicals. Moreover, the complexity associated with recalcitrance nature of lignin makes difficulty to analyze its numerous degradation products. In addition, its structural chemical and physical properties differ with extraction procedure. At present more than 98 per cent of this lignin is burnt for energy generation and remaining is mainly used for formulating dispersants, adhesives and surfactants indicating a need for intensive research efforts in process engineering and biorefinery based product development (Ghaffar and Fan, 2014). The aromatic nature of lignin along with aliphatic hydroxyl groups can be aided advantage as potential reactive sites for chemical harnesses (Pizzi, 1993). All these factors have attracted biomass researchers worldwide to explore the possibility of using lignin for the development of green adhesives. Substitution of phenol by lignin reduces the cost of adhesive preparation and apparently provides value addition to lignin. However, the practical route to incorporate lignin into phenol-formaldehyde resin is still a subject of debate. In the manufacturing of wood panels, various synthetic resins such as phenol-formaldehyde, urea-formaldehyde, melamine urea-formaldehyde resins are frequently employed as wood adhesives (Keulgen, 1969, Danielson and Simonson, 1998). The raw materials for manufacturing of these synthetic adhesives include phenol and formaldehyde are commonly derived from petrochemicals (Pizzi, 1993). From long-lasting perspective, alternative materials to these petrochemicals are certainly desired from both environmental and economical points of view. Indeed, the manufacturers of medium density fiber board (MDF), particle board (PB), plywood and oriented strand board (OSB) are under pressure to economize the raw materials and eliminate harmful formaldehyde/phenol emissions (Ibrahim *et al.*, 2013). Hence renewable resources adoptability is imperative and the better exploitation of such resources is one of the cutting-edge targets in current biomass research area of biorefinery. In view of above, an effort has been made to use the lignin extracted from water hyacinth for production of bioadhesives.

Water hyacinth is an aquatic plant and is often considered a highly problematic due to its heavy infestations in water systems instigating environmental damage. While water hyacinths do confiscate inorganic pollutants from water, the eruptions cover water surfaces and devour most of the dissolved oxygen leaving in little room for aquatic species resulting of their extinction (Tham, 2012). Moreover, massive outbreaks of water hyacinth lead to further economic destruction by blocking water conveyance. In the context of natural waste biorefinery, water hyacinth has proved to be a suitable renewable resource for biofuel production (Ahn, 2012; Lin *et al.*, 2015; Yan *et al.*, 2015) using cellulose and hemicellulose portions. The objective of this study to develop and characterize wood adhesive using lignin extracted from water hyacinth.

2. Experimental

2.1. Materials

Phenol (99 per cent), formaldehyde solution (37-41 per cent), sodium hydroxide (pellets) (98 per cent), orthophosphoric acid (88 per cent), methanol (LR grade) (99 per cent) were used for experimental work. All these chemicals were obtained from HiMedia, Mumbai, India. Water hyacinth biomass was collected from the local water ponds surrounding of Hyderabad. It was washed with distilled water and dried in a hot air oven at 105°C for overnight, grounded using a laboratory mixer and subsequently used for lignin extraction.

2.2. Compositional Analysis of Water Hyacinth

The ground water hyacinth was analyzed for various components such as cellulose, hemicellulose, and acid soluble lignin (ASL) the National Renewable Energy-Laboratory protocols (NREL-LAP) (Sluiter *et al.*, 2011). In brief, 300 mg of dried extractive free biomass sample was hydrolyzed in the presence of 1 mL of 72 per cent (w/v) sulfuric acid for 1 h at 4°C followed by autoclaving at 121°C for 30 min. The resulting hydrolysate was filtered using G3 crucible and the filtrate was used for the estimation of ASL and hydrolyzed sugars (formed from cellulose and hemicellulose). The ASL was analyzed spectrophotometrically based on the absorbance coefficients at specified wave length according to Sluiter *et al.* (2009). On the other hand, some portion of the hydrolysate was neutralized using solid $CaCO_3$ and used for the analysis of hydrolyzed sugars. The type of sugar and its concentration was analysed by HPLC (High Performance/Pressure Liquid Chromatography) (Waters, Austria) using an amino column (Luna 5 μ NH_2 column 100Å, 250 x 4.6 mm, Phenomenex) connected to RID (Refractive Index Detector). The conditions employed were, water: acetonitrile (20:80), mobile phase; 1 mLmin⁻¹, flow rate; 45°C, column temperature (external temperatures); 30°C, detector temperature (internal temperature).

Klason-Lignin (KL) Content

Klason-lignin/acid insoluble (AIL) content of water hyacinth was analysed according to method of Technical Association of Paper and Pulp Industry (TAPPI). In brief, 1g of oven dried and extractive free water hyacinth was added with 15 mL of 72 per cent (v/v) sulfuric acid and the hydrolysis was carried out at 20°C for 2 h under occasional stirring. Later on, the reaction mixture was transferred to one-neck round bottom flask (1L) and then the sulfuric acid concentration was brought down to 3 per cent (v/v) by diluting it with 650 mL of distilled water. Subsequently, the contents were refluxed for 4 h and then filtered using G3 crucible. The crucible along with the solids was kept in an oven at 105°C to a constant weight. The KL content was calculated as follows:

$$\text{Klason - lignin (per cent)} = \frac{\text{Oven dry weight of lignin}}{\text{Oven dry weight of sample}} \times 100$$

2.3. Extraction of Lignin from Water Hyacinth

The procedure for lignin extraction from water hyacinth is illustrated in Figure 12.1. The extraction was performed in a two-neck round bottom flask (2000 mL) equipped with a double-surface condenser. A 150 g of ground water hyacinth and 1.5 L of 8 per cent (w/v) sodium hydroxide solution were charged to the flask (the solid-to-liquid ratio was 1:10). The temperature was maintained at 100°C ± 5°C and the extraction was sustained for 4 h. The black liquor so obtained was separated from the suspended fibers by filtration. The solids were further subjected to hot water washes (3 times) to collect the residual lignin trapped in the fibers. Afterwards, orthophosphoric acid (85 per cent H_3PO_4) was added until reaching the pH 2 in order to precipitate lignin. Subsequently, the lignin was collected by centrifugation and then dried in an oven at 50°C for overnight. No additional purification processes were performed and the yield of lignin represents triplicate analysis.

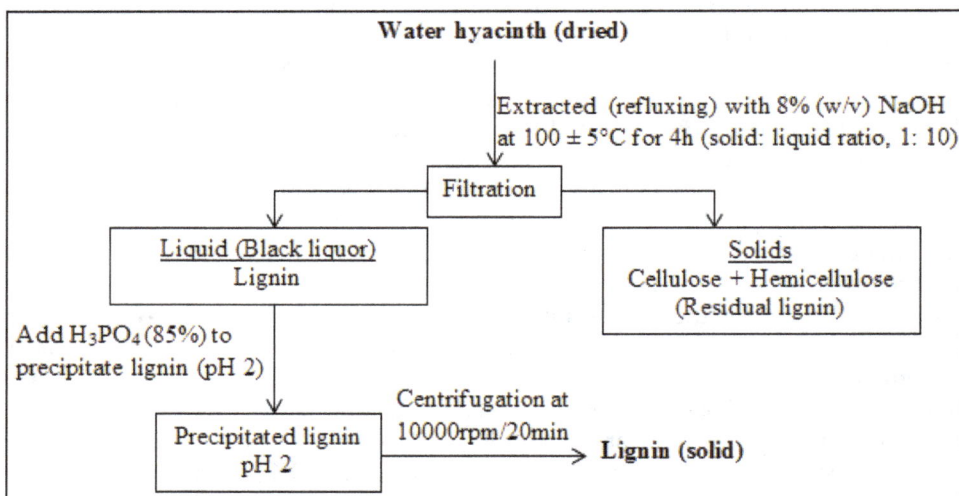

Figure 12.1: Scheme of Extraction of Lignin from Water Hyacinth.

2.4. Characterization of Lignin

a) Hydroxyl Content

The hydroxyl value of lignin was determined according to IS 548-1 (1964), wherein a known amount of lignin was acetylated under controlled environment and titrated it against alcoholic 1N KOH solution. The hydroxyl value was calculated using the following formula:

$$Hydroxyl value = \frac{56.1 \times N \times Y}{W}$$

where,

N = Normality of alc. KOH solution

Y= alc. KOH solution required to titrate blank

W = Weight of the sample taken

b) FT-IR Analysis

The dried lignin sample was mixed with IR grade KBr (Spectro Inc) at a ratio of 2:200 (lignin: KBr) to form a disc for analysis in FT-IR spectrometer (PerkinElmer). The background spectrum of control KBr was subtracted from that of the lignin spectrum. The sample was analysed over a wave number range of 400 to 4000 cm^{-1} region to characterize the extracted lignin.

c) Elemental Analysis

Carbon, Hydrogen and Oxygen analysis of lignin was carried out using a Rapid Elemental analyzer. 20 mg of lignin was subjected to oxidative combustion at 950°C using Cerium oxide catalyst. The gases so produced were analysed and the concentration of each element was calculated from the integration of the corresponding peak.

2.5. Preparation of Adhesives

Both lignin-phenol-formaldehyde (LPF) and phenol-formaldehyde (PF) (control) adhesives were prepared in a three-neck round bottom flask (250 mL) equipped with a double-surface reflux condenser, dropping funnel, magnetic bar (stirrer) and thermometer (0-250°C).

LPF

The preparation of lignin-phenol-formaldehyde adhesive was a two-step process, where in the first step involved the preparation of lignin-phenol adducts and the second step consisted of the preparation of lignin-phenol-formaldehyde adhesive. In the first step, 25 g of lignin was taken into a beaker (250mL), after which equal amount of molten phenol (25 g) was added and the mixture so formed was stirred for 2 h at 37°C to obtain a homogeneous lignin-phenol mixture (adduct contained 50 per cent moisture). While in the latter step, the lignin-phenol adduct was mixed with 22.5 g of formaldehyde, 10 g of methanol and 1.5 g of sodium hydroxide (present in 10 mL of d. H_2O) and the polymerization reaction was carried out at 80°C for 4 h under constant stirring.

PF

The reaction for the preparation of conventional phenol-formaldehyde (control) adhesive was a single step process and wherein, 10 g of molten phenol, 20 g of formaldehyde, 10 g of methanol, 1.5 g of sodium hydroxide and 10 g of distilled water were charged to the flask. The mixture was heated to 80°C under constant stirring and the reaction was continued for 4 h.

2.6. Characterization of Adhesives

Non-volatile Content

The content of solids (non-volatile content) of the adhesives was determined according to the ASTM D 4426-93. Briefly, in order to determine the solid content, the solvent was stripped off from the samples by heating them at 150°C for 30 min.

$$Non - volatile/solid\ content\ (per\ cent) = \frac{Weight\ of\ the\ solid\ resin}{Weight\ of\ the\ solution} \times 100$$

pH

The apparent pH of the adhesive samples was measured by using pH indicator strip.

i) *Specific gravity*: The specific gravity of the adhesives was determined by using the standard specific gravity bottles.

ii) *Tensile strength analysis*: The strength of the adhesive was determined according to international standard method of IS 851-1978. Two teak-wooden blocks (600 x100 mm) having ~7 per cent moisture were taken and the adhesive was applied on both sides of interfaces. The glued specimens were then pressed with clamps under the pressure of ~50N. After conditioning for about 48 h, the tensile strength of the specimens was determined by using a tensile testing machine.

iii) *Thermal stability*: Thermal stability of lignin-phenol-formaldehyde and phenol-formaldehyde adhesives was determined using thermo gravimetric analyzer (TG, DTA, PerkinElmer Instruments). The test and control adhesive samples were heated from ambient temperature to 900°C at a heating rate of 10°Cmin⁻¹under nitrogen atmosphere.

3. Results and discussion

3.1. Chemical Composition of Water Hyacinth

The chemical composition of water hyacinth was analyzed by following the NREL-LAP (Sluiter *et al.*, 2009) and TAPPI (TAPPI) methods and compared with that of reported values (Table 12.1). The glucan (cellulose) and hemicellulose (xylan + arabinan + galactan + mannan) contents were 31±1.6 per cent and 20±0.9 per cent, respectively while the lignin was 17±2 per cent. Cellulose, hemicellulose and lignin contents of water hyacinth reported in the literature vary remarkably, which indicates that the chemical composition of this biomass was affected by its growing environment. The considerable lignin content in water hyacinth is advantageous to use it in the development of valuable products such as wood adhesives, resins, *etc.*

Table 12.1: Chemical Composition of Water Hyacinth and its Comparison with Reported Values

Cellulose (per cent)	Hemicellulose (per cent) (Xyl + Ara + Gal + Man)*	Lignin (per cent)	Reference
31 ± 1.6	20 ± 0.9	17 ± 2.0	Present study
31.81	25.64	3.55	Yan *et al.*, 2015
23.3	27	10	Cheng *et al.*, 2013
23.3 ± 0.3	18.3 ± 0.5	17.7 ± 0.1	Gao *et al.*, 2013
34.19	17.66	12.22	Ahn *et al.*, 2012
25 ± 1.4	18.1 ± 0.2	13.3 ± 1.6	Mishima *et al.*, 2006
36.5	22	15.2	Mukherjee *et al.*, 2004
33.97	18	26.36	Chanakya *et al.*, 1996

*Xyl: xylan; Ara: Arabinan; Gal: Galactan; Man: Mannan.

3.2. Yield of Extracted Lignin

In general, alkaline extraction process is an effective route to extract lignin from lignocelluloses. Sodium hydroxide, one of the most reactive and frequent alkaline agents, has been employed for delignification of various agro-residues in lignocellulosic biorefinery (Liu and Wu, 2013). In the present study, the lignin from water hyacinth was extracted with 8 per cent (w/v) sodium hydroxide at 100±5°C for 4 h. Yield of lignin was 13 g and it was calculated based on the dry weight of the recovered sample. This yield was observed to be two fold higher than the reported values (Ahn *et al.*, 2012; Yan *et al.*, 2015). The sodium hydroxide (8 per cent w/v) extraction followed by hot water treatment, totally released 76 per cent of the original lignin from water hyacinth. The cellulose and hemicellulose obtained in this process could be used for various valuable bioproducts such as biofuel production.

3.3. Characteristics of Extracted Lignin

The phenolic hydroxyl content indicates the reactivity of any lignin and in fact, these chemical groups participates the polymerization process which is the basic reaction mechanism involves in the synthesis of phenolic wood adhesives. The total phenolic hydroxyl value of the present water hyacinth lignin was 242. In addition, the isolated lignin was also analyzed for carbon, hydrogen and oxygen, which were determined to be 45.42 per cent, 4.39 per cent and 50.19 per cent, respectively. The high oxygen and low hydrogen content signifies that the present lignin was highly unsaturated.

Fourier transform infrared spectra of lignin displays its chemical fingerprint as well as the purity (Figure 12.2). The wide band in range of 3200-3530 cm⁻¹ indicated the stretching vibration of phenolic and aliphatic –OH groups. As compared to the reported standard lignin FT-IR spectrums (Kline *et al.*, 2010), the lignin derived from the present process lacks a spectral band around 1700 cm⁻¹, which is usually attributed to the carbonyl (aldehyde/ketone) moieties, signifying

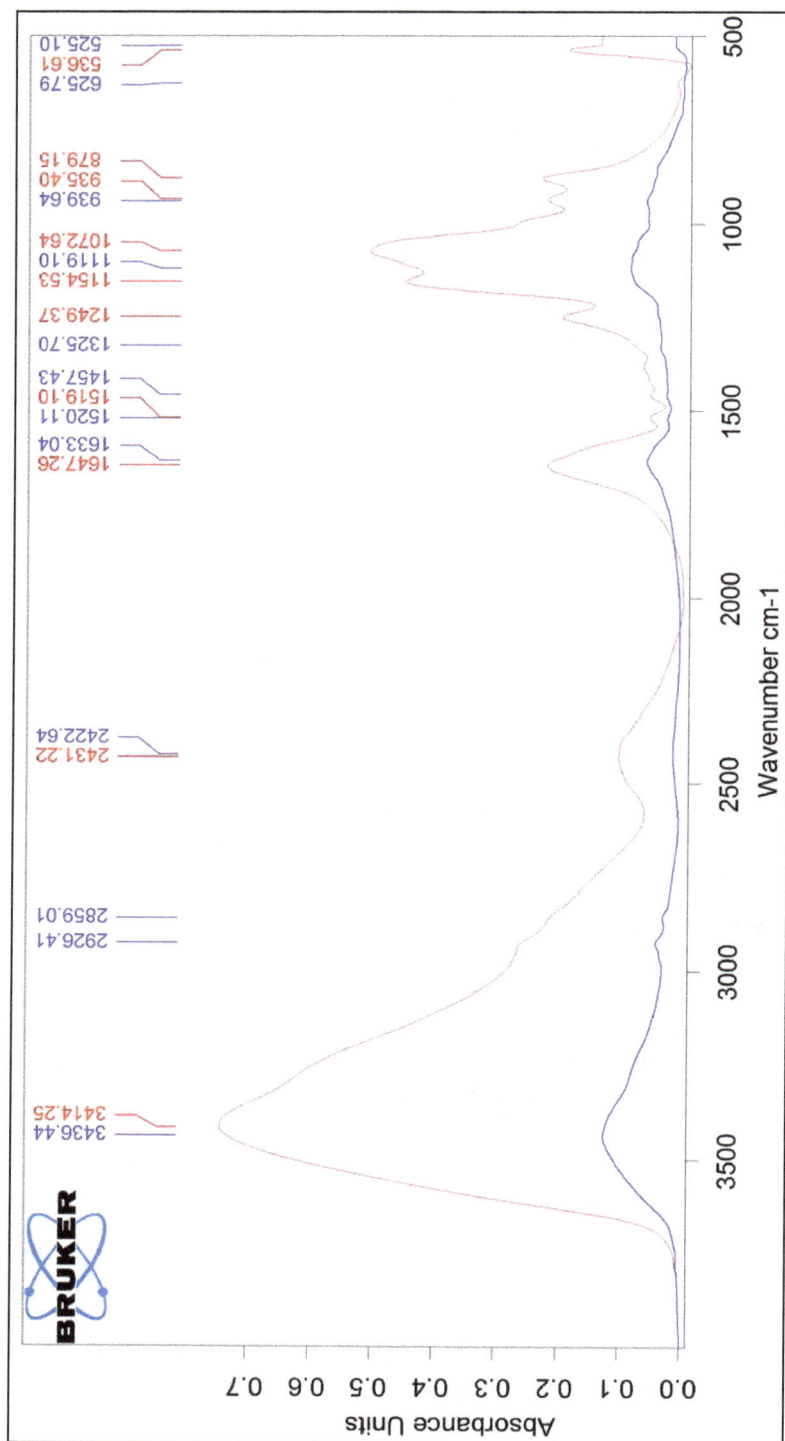

Figure 12.2: FTIR Spectra of Sodium Hydroxide (Alkaline) Extracted Water Hyacinth Lignin.

the chemical modification of most of the carbonyl groups under alkaline (high pH) conditions (Orton *et al.,* 2004). The strong band in the range of 1618-1638 cm^{-1} reflects C-C stretching of the aromatic skeleton in lignin while the band at ~1425 reflects in-plane asymmetric vibration in phenolic OH group (Δ_{ip} O-H) (Moubarik *et al.,* 2013). Further, the bands in range of 1110-1130 cm^{-1} signifies the presence of syringyl (S) and guacyl (G) lignin, while a sharp band around 1165 cm^{-1} indicates the presence of phenyl propane (H) in the lignin (Zhao., *et al.,* 2009). The presence of guacyl groups signifies that the water hyacinth lignin had potential reactive sites for polymerization reaction. Moreover, it revealed that this lignin could react with formaldehyde and could be cross-linked in the same way as in the phenol-formaldehyde polymerization. The presence of a signal band around 1458 cm^{-1} (assigned to CH deformation/asymmetric in methyl, methylene and methoxyl groups), confirms that the aromatic structures of lignin did not change significantly during the alkaline extraction process.

The TG and DTG curves of lignin extracted from water hyacinth were displayed in Figure 12.3. As it can be seen, the decomposition of water hyacinth lignin covered wide range of temperature from ambient to 900°C. Lignin degradation could be seen in two distinct stages. The initial weight loss occurred at 60-130°C due to the evaporation of absorbed water (moisture). After that, lignin was degraded over a wide range of temperature, 150-900°C. The degradation of lignin generates volatile products and most of them were phenolics, acids, alcohols, aldehydes, which further get converted into various gaseous products including CO, CO_2 and CH_4 (Liu *et al.,* 2008). The TG curve also indicates that thermal degradation of lignin began to

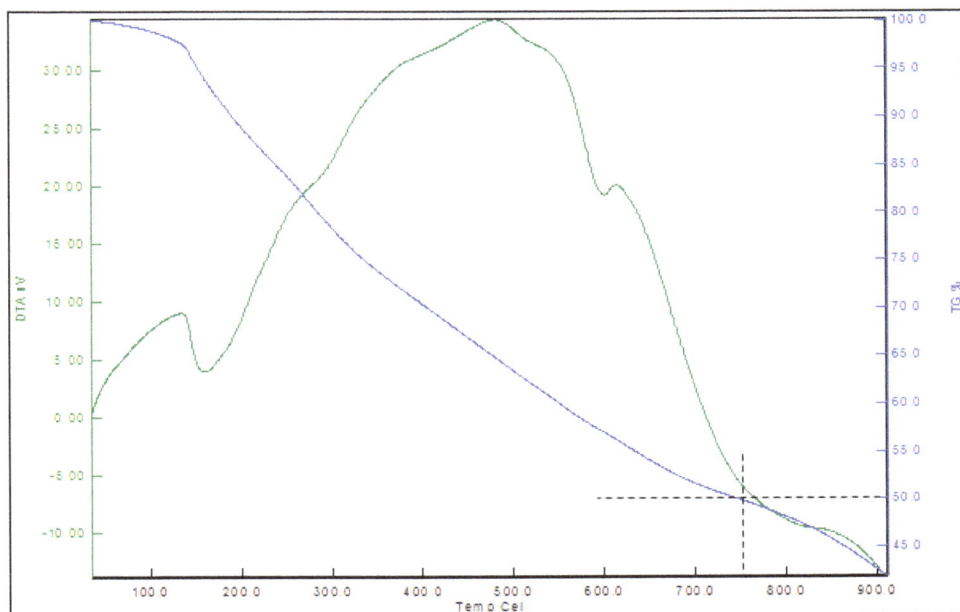

Figure 12.3: TGA and DTG Curves of the Water Hyacinth Lignin,
$\beta= 10°C\ min^{-1}$ **in Nitrogen Atmosphere.**

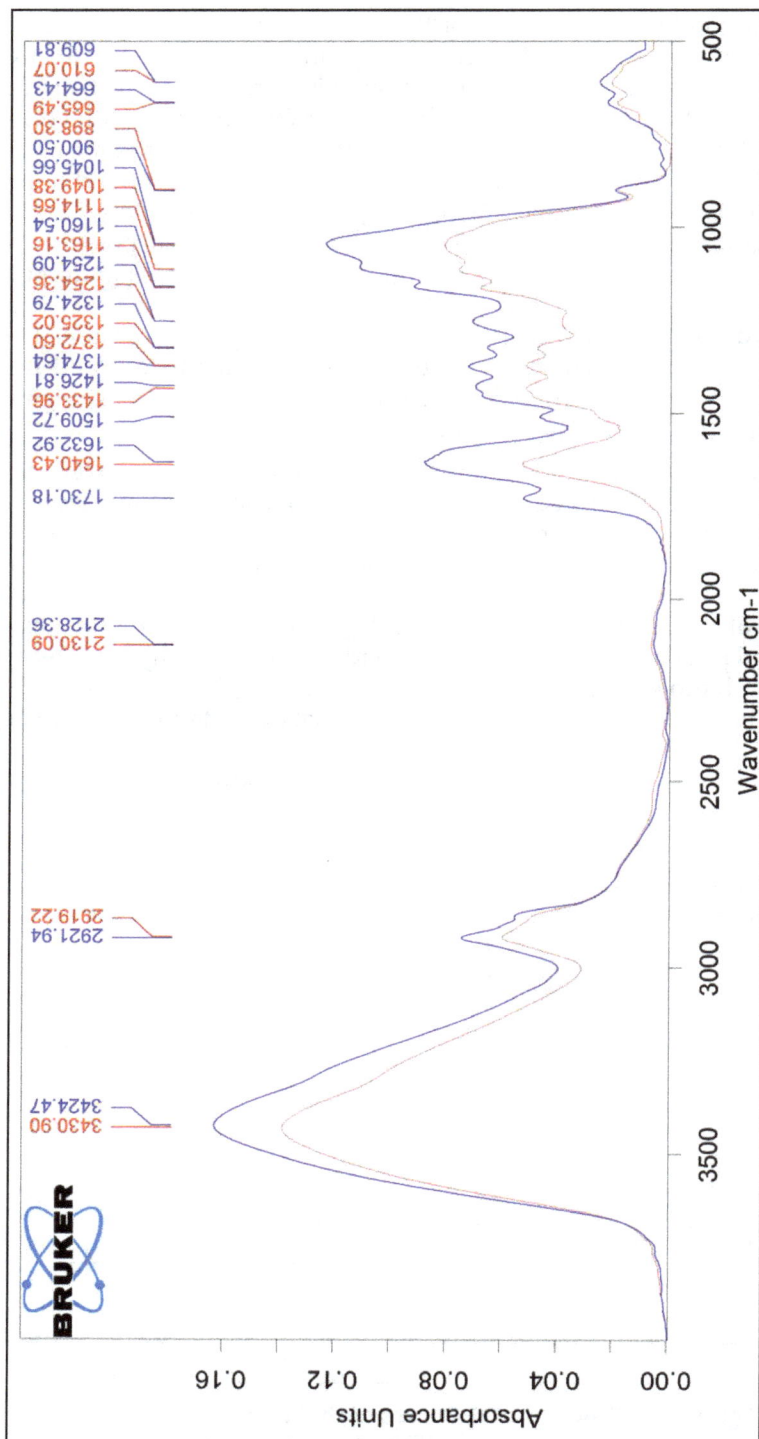

Figure 12.4: FTIR Spectra of Lignin-Phenol-Formaldehyde (Blue) and Phenol-Formaldehyde (Red) Adhesive.

take place only after the sample has absorbed certain amount of heat energy. The absorbed heat initiates the degradation process and the disrupting of molecular structure. At 750°C, about 50 per cent of the non-volatile matter sill remained (Figure 12.3), which revealed that water hyacinth lignin was stable at higher temperatures.

3.4. Characterization of Adhesives

Water hyacinth lignin was substituted for phenol in the conventional phenol-formaldehyde adhesive. The solid content (non-volatile content) of the lignin-based adhesive was determined as 49.19 per cent and it had a pH of 10.1. The specific gravity and viscosity of this resin were 1.29 and 1530 mPas, respectively. Moreover, the tensile strengths of the lignin-PF and control (PF) adhesives were calculated to 2.4 and 2.9 Mpa. Although, the observed tensile strength of lignin-based adhesive was low when compared to the control adhesive, this could be ameliorated by further optimizing the critical reaction parameters.

The FT-IR spectrums of the test lignin-phenol-formaldehyde and control phenol-formaldehyde resins were shown in Figure 12.4. As it can be seen, there was a characteristic broad O-H stretching band at 3200-3530cm^{-1} and an intense C-H band at ~2921cm^{-1}. The bands that observed at 1509 cm^{-1}, 1436 cm^{-1} and 1433 cm^{-1} in both samples could be attributed to the C=C stretching vibration for aromatic

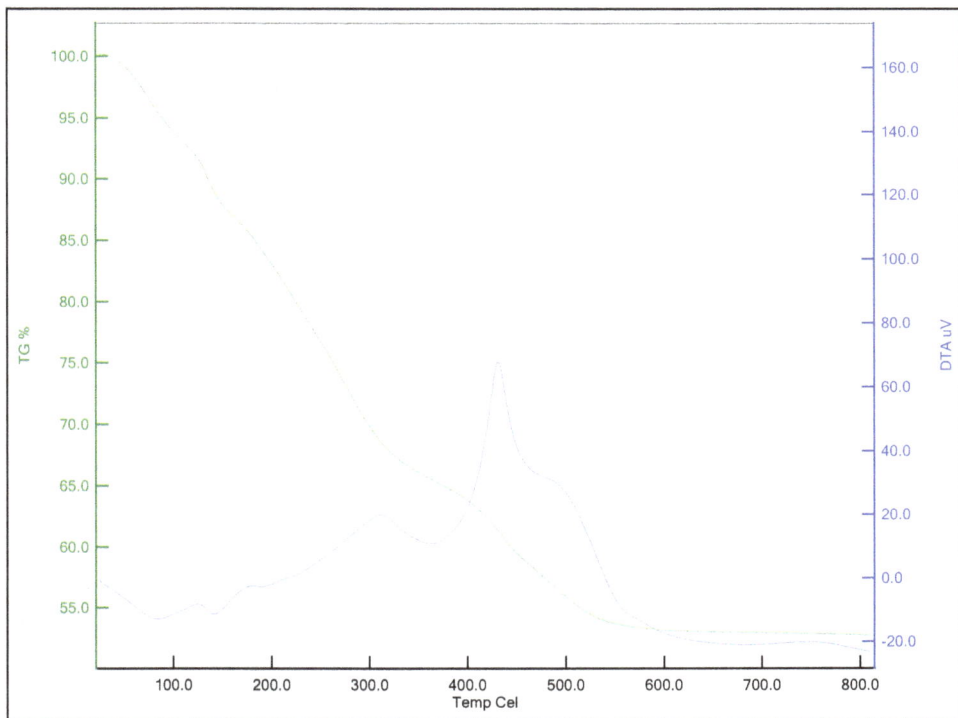

Figure 12.5a: TGA and DTG Curves of Lignin-Phenol-Formaldehyde Adhesive, β= 10°C min^{-1} in Nitrogen Atmosphere

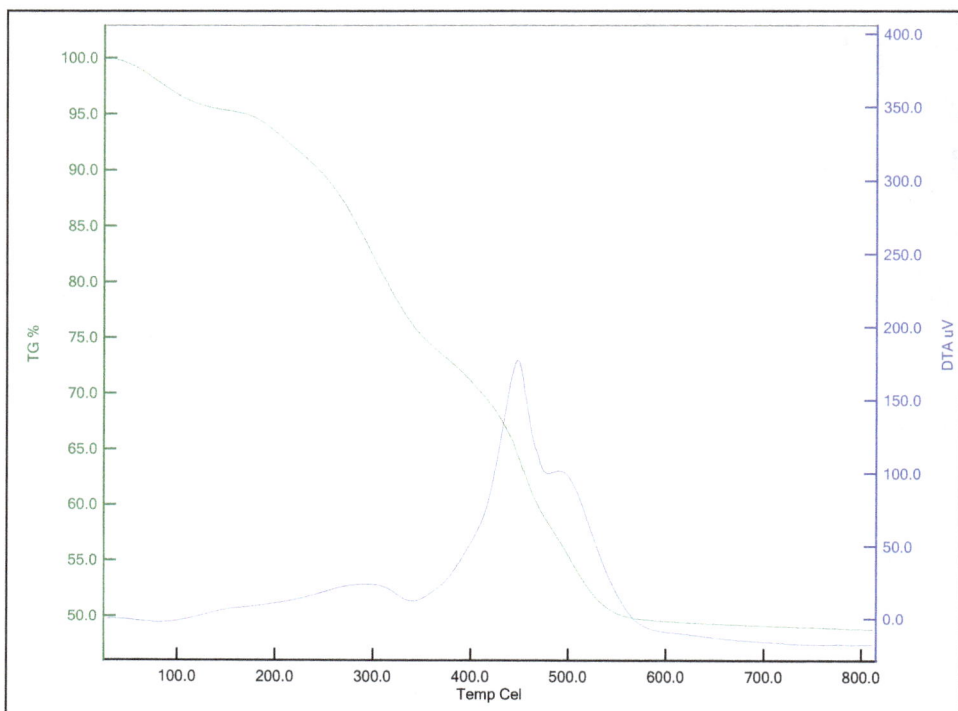

**Figure 12.5b: TGA and DTG Curves of Phenol-Formaldehyde Adhesive,
β= 10°C min⁻¹ in Nitrogen Atmosphere.**

rings. Similarly, the absorption at 1730 cm⁻¹ could be attributed to the carbonyl stretching of carboxylic and ester groups. A band was observed at 898 cm⁻¹ in the spectrum of lignin-phenol-formaldehyde resin that resulted from the reaction between phenols and (α)-hydroxyl groups of present in the side chains of lignin (Alonso *et al.*, 2005). The spectral difference between these two resins was observed ~900 cm⁻¹, which was attributed to C-H stretching vibration of vinyl group in the phenol-formaldehyde resin.

TG and DTG curves of the lignin-PF and PF resins were presented in Figure 12.5. Thermal stability of lignin-based adhesive (Figure 12.5a) was comparable with the control (phenol-formaldehyde) adhesive (Figure 12.5b). In both cases, the degradation of solid matter occurred in two stages *i.e.*, up to 50 per cent of the solid matter was degraded in wide range of temperature, 60-500°C while the rest was not degraded even at 800°C, which reveals the stability of the present phenol substituted lignin-based adhesive. Thermal stability profiles of the adhesives prepared in this study are comparable with the profiles reported by other researchers for various lignin-based adhesives (Khan *et al.*, 2004; Gosselink *et al.*, 2004; Sun *et al.*, 2004; Moubarik *et al.*, 2013).

Conclusion

In this work, the exploitation of lignin isolated from water hyacinth in the preparation of phenol substituted lignin-PF adhesive was investigated. The chemical composition reveals that water hyacinth contains comparable amount of lignin than present in other lignocellulosic materials. Sodium hydroxide catalyzed extraction process resulted in considerable yield of lignin. FT-IR spectroscopy reveals that the isolated lignin composed of guaiacyl and syringyl units with free hydroxyls and absence of carbohydrate bands signifies the purity of lignin. Both structural and thermal properties suggest that water hyacinth lignin could be a better substitute in making of lignin-based wood adhesives. As evidence, lignin-PF wood adhesive was prepared by replacing 50 per cent phenol and resulted adhesive had similar structural and thermal properties with that of control adhesive. However, it exhibited less tensile strength which could be improvedthrough optimizationof reaction parameters.

Acknowledgements

We acknowledge Indo-US JCERDC, Bilateral Project for the financial support.

References

Ahn, DJ., Kim, SK., Yun, HS, 2012. Optimization of pretreatment and saccharification for the production of bioethanol from water hyacinth by *Saccharomyces cerevisiae*. *Bioprocess Biosyst Eng.*, 35, 35–41.

Alonso, MV., Oliet, M., Rodriguez, F., Garcia, J., Gilarranz, MA., Rodriguez, JJ, 2005. Modification of ammonium lignosulfonate by phenolation for use in phenolic resins. *Biores. Technol.*, 96(9), 1013-1018.

Danielson, B., Simonson, R, 1998. Kraft lignin in PF resin. Part 1. Partial replacement of phenol by kraft lignin in PF adhesives for plywood. *J. Adhes. Sci. Technol.*, 12, 923–939.

Ghaffar, SH., Fan, M, 2014. Lignin in straw and its applications as an adhesive. *International Journal of Adhesion and Adhesives*, 48, 92–101.

Gosselink, RJA., Snijder, MHB., Kranenbarg, A., Keijsers, ERP., de Jong, E., Stigsson, LL, 2004. Characterisation and application of NovaFiber lignin. *Industrial Crops and Products*, 20, 191–203.

Ibrahim, V., Mamo, G., Gustafsson, PJ., Hatti-Kaul, R, 2013. Production and properties of adhesives formulated from laccase modified Kraft lignin. *Industrial Crops and Products*, 45, 343– 348.

IS 548-1, 1964. Methods of sampling and test for oils and fats, Part I: Methods of sampling, physical and chemical tests [FAD 13: Food and Agriculture].

Kamm, B., Kamm, M., Gruber, PR., Kromus, S, 2006. Biorefineries-Industrial Processes and Products Status Quo and Future Directions, Wiley-VCH, 1.

Keulgen, WA. Encyclopedia of Polymer Science and Technology; Wiley: New York, 1969; Vol. 10, p. 1.

Khan, MA., Ashraf, SM., Malhotra, VP, 2004. Eucalyptus bark lignin substituted phenol formaldehyde adhesives: A study on optimization of reaction parameters and characterization, *Journal of Applied polymer science*, 92, 3514-3523.

Kline, LM., Hayes, DG, womac, AR., Labbe, N, 210. Lignin determination in IL. *Bioresources*, 5(3), 1366-1383.

Kumar, MNS., Mohanty, AK., Erickson, L., Misra, M, 2009. Lignin and its applications with polymers. *Journal of Biobased Materials and Bioenergy*, 3 (1), 1–24.

Lange, JP, 2007. Lignocellulose conversion: an introduction to chemistry, process and economics. *Biofuels, Bioproducts and biorefining*, 1(1), 39–48.

Lin, R., Cheng, J., Song, W., Ding, L., Xie, B., Zhou, J., Cen, K, 2015. Characterisation of water hyacinth with microwave-heated alkali pretreatment for enhanced enzymatic digestibility and hydrogen/methane fermentation. *Biores. Technology.*, 182, 1–7.

Liu, Q., Wang, SR., Zheng, Y., Luo, ZY., Cen, KF, 2008. Mechanism study of wood lignin pyrolysis by using TG-FTIR analysis. *J.of anal.and app pyrol.*, 82(1), 170-177.

Liu, S., Li, J., Wu, Y, 2013. Pre-treatment by sodium hydroxide for hydrolysis of wheat straw and analysis of the product. *Sci. Technol. Food Ind.*, 4, 180-183.

Moubarik, A., Grimi, N., Boussetta, N., Pizzi, A, 2013. Isolation and characterization of lignin from Moroccan sugar cane bagasse: Production of lignin–phenol-formaldehyde wood adhesive. *Industrial Crops and Products*, 45, 296– 302.

Orton, CR., Parkinson, DY., Evans, PD., Owen, NL, 2004. Fourier transform infrared studies of heterogeneity, photodegradation, and lignin/hemicellulose ratios within hardwoods and softwoods. *Appl. Spectrosc.* 58(11), 1265-1271.

Pizzi, A., 1993. *Wood Adhesives Chemistry and Technology*, vol. 1. Marcel Dekker, New York.

Polymer Science, Vol. 92, 3514–3523 (2004). Sun R. Lignin. In: Lu F, Ralph J, editors. Cereal straw as a resource for sustainable biomaterials and biofuels. Amsterdam, the Netherlands: Elsevier; 2010.p.169–207.

Sluiter, A., Hames, B., Ruiz, R., Scarlata, C., Sluiter, J., Templeton, D., *et al.*, 2011. Determination of structural carbohydrates and lignin in biomass. National Renewable Energy Laboratory (NREL), Analytical Procedure (LAP), http://www.nrel.gov/biomass/analytical_procedures.html.

Tham, HT, 2012. Water Hyacinth (*Eichhornia crassipes*) – Biomass Production, Ensilability and Feeding Value to Growing Cattle. Doctoral Thesis, Swedish University of Agricultural Sciences, Uppsala.

Yan, J., Wei, Z., Wang, Q., He, M., Li, S., Irbis, Ch, 2015. Bioethanol production from sodium hydroxide/hydrogen peroxidepretreated water hyacinth via simultaneous saccharification and fermentation with a newly isolated thermotolerant *Kluyveromyces marxianu* strain *Bioresource Technology*, 193, 103–109.

Youn, HD., Hah, YC., Kang, SO, 1995. Role of laccase in lignin degradation by white rot fungi. *FEMS Microbiol. Lett.*,132, 183–188.

Zhao, H., Jones, CIL., Baker, GA., Xia, S., Olubajo, O., Person, VN, 2009. Regenerating cellulose from ionic liquids for an accelerated enzymatic hydrolysis. *J. Biotechnol.*, 139, 47-54.

13

The Perspectives of Chemocatalytic Biomass Valorization into Bioproducts (Platform Molecules, Fuels) and Energy

Nikolay V. Gromov[1,2] Oxana P. Taran[1,2],
Valentin N. Parmon[1,3] and S. Uthandi[4]

[1]Boreskov Institute of Catalysis, Russia, 630090, Novosibirsk, Lavrentiev av., 5
[2]Novosibirsk State Technical University, Russia, 630037, Novosibirsk, Karl Marx av., 20
[3]Novosibirsk State University, Russia, 630090, Novosibirsk, Ulitsa Pirogova, 2
[4]Tamil Nadu Agricultural University, Coimbatore – 641 003, India

1. Introduction

In recent years, valorization of biomass aimed in production valuable chemicals, fuels or just energy has become a one of the main fields of research. The interest to biomass as source for industry has started in XIX century. That time industrial scale processes have been developed for producing cellulose esters (nitrate and acetate), alcohols, organic acids, sugars, *etc.* [1, 2] and the amount of investigations in this point have been steel permanently increased. The reason that causes such interest to biomass is a clear understanding of running out fossil fuels (oil, coal, gas) [3]. Also utilization of fossil fuels in petrochemical and chemical processes becomes a serious environmental problem due to emission of big amount of CO_2 [4-8]. This reduces to serious climate changes accompanied by economic troubles [3, 9]. To overcome this problems, alternative solutions are sought and biomass utilization is one of them. Plant materials seem to be very perspective as they are renewable, or even inexhaustible. Biomass is produced from CO_2 and H_2O under the Sun radiation about 200 billion tons of biomass are annually generated [10]. The main component of plant materials is lignocellulose which could take up to 99 per cent of biomass [1]. It should be also noted that overwhelming amounts of lignocellulose main components (40–50 per cent cellulose, 16–33 per cent hemicellulose and 15–30 per cent lignin [11-13]) are inedible and they are not used in the industry. A lot of unused biomass is available from industrial wastes, paper industry and agriculture first of all. The following sources of biomass for valorization could be mentioned: corn-cobs, straw, rise and sunflower peels produced in agriculture, and energy crops (miscantus, cane, millet), and wastes of wood appeared in paper industry. Utilization of such biomass let to the competition between lignocellulose valorisation and human food supply and improve the economy and environmental safety of biomass valorization methods.

One of the main goals of transforming the renewable resources is to develop a series of novel chemical processes based on renewable feedstocks, typically biomass and biomass-derived chemicals [14]. Catalytic methods of biomass transformation appear to be most effective for producing biofuels and valuable chemicals. Development of such techniques is a serious challenge, due to robust and very stable structure of biomass main components - cellulose and lignin, first of all. Of particular interest is the search for catalytic one-pot technologies for synthesis of valuable chemicals or biofuels directly from biomass or it's main components to eliminate technological stages of semiproduct isolation and purification. In this way it is possible to produce biofuels or platform molecules, for example levulinic acid, 5-hydroxymetilfurfural, furfural and furfuric alcohol, ethylen glycol, γ-valerolactone, polyols (sorbitol, mannitol, etc.) from polysaccharides namely cellulose and hemicelluloses [7, 15-24]. Lignin could be a renewable source of aromatics [25, 26].

This chapter is a brief overview of the biomass transformation methods and of a progress in these techniques. It is devoted to catalytic thermochemical (gasification, liquefaction, pyrolysis) and aqueous thermal (hydrolysis, hydrogenatio, oxidation, liquid-phase reforming) methods of biomass transformation.

2. Lignocellulose Composition

As mentioned above, lignocellulose contains mainly cellulose, hemicellulose, lignin. The ratio of the three main components strongly depends on a plant species. It may also depend upon climatic conditions of the natural growth zone and the season (drought, excessive rainfall, etc.). Generally, the amount of cellulose could be up to 50-60 per cent, hemmicelluloses and lignin upto 30 per cent [11-13, 27, 28][29, 30].

Cellulose is a polysaccharide which is consist of monomer units (anhydro-β-D-glucopyranose or glucose residues) being connected to each other by 1,4-glycosidic bonds. Cellulose is the main component of lignocellulosic biomass. General chemical formula of cellulose can be written as $(C_6H_{10}O_5)_n$ or $(C_6H_7O_2(OH)_3)_n$. The origination of the term "cellulose" refers to the first half of the 19th century (see. designation in [12]). The main purpose of natural cellulose is structural one. This polymer provides a shape of plant cells, and also determines the rigidity of plant tissues. The highest content of cellulose is in the wood, the lowest one is in the leaves of annual plants [27, 31]. The degree of polymerization of natural cellulose depends on the plant species.

The large number of hydroxyl groups in close-packed cellulose polymer chains results in a significant number of hydrogen bonds. They determine the physical and chemical properties of cellulose including ability of cellulose molecules to form supramolecular structures (microfibrils) made of molecular beams possessing different crystallinity degrees [32]. According to XRD, microfibrils consist of alternate crystalline and non-crystalline (amorphous) regions, in which the order of arrangement of molecules is maintained by hydrogen interactions [33]. In amorphous parts long-range order is absent, and only a general longitudinal direction of the chains is conserved. Sizes highly ordered domains (crystallites) vary significantly depending on the type of cellulose. This very robust crystal structure of cellulose being in pure state or as part of lignocellulosic materials makes further processing into valuable compounds difficult. Over that, it necessary to conduct a preliminary stage of cellulose transformation - it's activation. Activation significantly changes the physico-chemical properties of raw materials and increases its' reactivity because of the destroying the supramolecular cellulose structure and reducing the crystallinity degree [34, 35].

Hemicelluloses is a wide range of polysaccharides of different structures. The exact composition of hemicellulose strongly depends on plan species. Hemicelluloses are made of hexoses and pentoses such as glucose, xylose, mannose, galactose, arabinose, etc. [36]. The most widespread in nature are arabinogalactane (polymer of arabinose and galactose), galactoglucomannan or galactomannan (polymers of galactose, glucose and mannose) and xylane (polymer of xylose) [36, 37]. In spite of cellulose, hemicelluloses are soluble in water what makes then more available for transformations. Processing hemicelluloses does not need applying harsh conditions: concentrated bases or acids, high temperature, pressure.

Lignin is an aromatic oxygen-containing polymer. It consists of three main fragments: p-coumaryl alcohol, coniferyl alcohol, and sinapyl alcohol residues [38]. The main sources of lignin are wastes of pulp and paper industries and plant materials. Lignin transformations seems to be very challenging. Even revealing lignin exact structure is a serious task [39-42].

3. Catalytic Methods of Biomass Transformation

Nowadays a very wide range of biomass transformation methods have been explored. These techniques make it possible to produce different products from fine chemicals to fuels and energy. Lignocellulose transformation methods could be divided into two main groups depending on the process temperature [43]. It is possible to mention thermochemical and aqueous thermal transformation techniques (Figure 13.1).

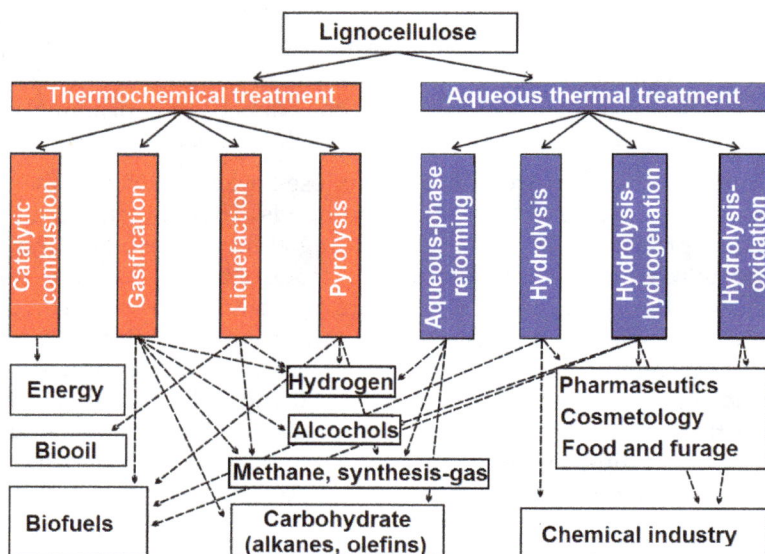

Figure 13.1: Catalytical Ways of Lignocellulose Biomass Transformation into Chemicals.

Therrnochemical methods (catalytic combustion, gasification, liquefaction, and pyrolysis) are characterized by caring out experiments under very high temperatures (Figure 13.1). Catalytic combustion of lignocellulose is normally made to produce energy in fixed bed reactors under temperatures up to 1000°C. Gasification is carried out under oxidative atmosphere of air, oxygen, or water steam at 800–1300°C. It yields hydrogen, carbon dioxide, methane and small amounts of light hydrocarbons. This gas mixture (synthesis gas) is later used as a fuel or industrial recourse [30]. Lignocellulose liquefaction produces a mixture of hydrocarbon and there derivatives so call biooil. It is made over catalysts under reductive atmosphere of carbon oxide or hydrogen at 450 – 500°C and pressure 200 – 700 bar. Biooil produced in this way is transformed to fuels or for production of phenol compounds. Pyrolysis is processed at 300 – 500°C [44]. The products of pyrolysis are gases (hydrogen, methane, *etc.*), liquids and solid biochar.

The second group of catalytic ways of lignocellulosic biomass conversion is aqueous thermal methods. Such transformations are usually realized in the presence of catalysts in water medium. Temperature and pressure are not very high (up to 250°C, and 70 bar). Here, it is possible to mention hydrolysis, hydrolysis-hydrogenation, hydrolysis-oxidation and aqueous-phase reforming. As lignocellulose is mainly not soluble in water, preliminary activation is necessary for aqueous thermal treatment [32, 45, 46]. Chemical [45-49], physical [45, 46, 49, 50] and biological [45, 46] ways of the pretreatment of the raw materials could be divided [32]. In the result of aqueous thermal methods it is possible to produce from lignocellulose different types of chemicals: monosaccharides, polyols, organic acids, furan and phenol derivatives, hydrocarbons. The reason which makes these methods very attractive is a possibility of one-pot realizing transformations of plant materials.

3.1. Cellulose and Hemicellulose Hydrolysis-Dehydration

Transformation of polysaccharide components of biomass namely cellulose and hemicelluloses is under very intensive scientific investigations. Plenty of research papers and reviews have been published since the pioneer researches in this field made in XIX century [12]. Nowadays, hydrolysis is a one of the main reactions used for cleavage of glycoside bonds in polysaccharides [12, 51]. Standard reaction conditions of biomass hydrolysis are temperature in the range of 97–297°C. The reaction is usually processed in the presence of acid or base catalysts. Acid hydrolysis is the most useful. It consists in cleavage of $C-O-C$ bonds which connect monosaccharides. Acid catalysts also initiate the following dehydration reactions of sugars produced in hydrolysis. Furfural, 5-hydroxymethylfurfural, levulinic acid and some other chemicals could be produced. For example, hydrolysis of cellulose makes possible the production of glucose, 5-hydroxymethylfurfural, and levulinic acid, and transformation of xylane yields xylose, furfural [61]. All this products are well-known as platform molecules and could be applied in industry [23, 30, 62, 63].

Hydrotic transformations of biomass is possible under biocatalysis or chemocatalysis. Biocatalytic transformation of plant materials is realized in the presence of enzyme catalysts. Processing cellulose is performed in the presence of cellulase enzyme. The yields of glucose equal to 80 per cent could be reached in this way [52-54]. However, application of biocatalysis demands activation of plant biomass. Pretreatment increases availability of substrates for enzymes, removes reaction inhibitors such as lignin (delignification procedure) [55]. Delignification could also been carried out via biocatalysis by laccase enzymes [56]. Nevertheless, some disadvantages of biocatalysis coulb be mentioned, among them low reaction rates, high costs of enzymes and consumption of sugars for feeding enzyme-producing microorganisms [3, 55, 57, 58].

All biocatalysis drawbacks mentioned above causes a grate interest in chemocatalytic hydrolysis techniques [1, 3, 9, 59, 60].

Soluble acid or base catalysts are able to be applied for biomass transformations, including cellulose hydrolysis. Polysaccharide hydrolysis in acid solutions has a long history. Such methods have been developed and instilled into industry at the first part of XX century [64-67]. Commonly used are sulfuric acid [68] and its derivatives [65] and sodium hydroxide [69]. Hydrochloric acid [70, 71], nitric acid [72], phosphoric acid [68, 72] and organic acids [65, 73] also have catalytic activity in this process. Among examined mineral acids H_2SO_4 is a most active cellulose hydrolysis catalyst [74]. Industrial methods developed let to produce glucose with a vary high yields up to 80 per cent [65]. However, a strong corrosive impuct of concentrated acid prevents the widespread use of these techniques [46]. More promising method is cellulose depolymerization by diluted solutions of acids [65]. Such techniques produce upto 80-90 per cent of glucose under heating up to 200 °C [68, 75]. Mineral acid solutions catalyze not only cellulose hydrolysis reactions. They could be used for producing such compounds as 5-HMF or levulinic acid [76, 77]. However such feature of mineral acids is undesirable if target product is glucose.

Organic acids are also usefull for biomass transformations. Investigations on acetic, formic, malic, oxalic acids have been made [65, 67, 78, 79]. Among them the most interesting is oleic acid [78]. However the yields of the main products are lower compared to mineral acids [78, 79].

Another perspective soluble hydrolysis catalyst is a heteropoly acid [80, 81]. These sysmens could be more effective then mineral acids [71] due to stronger acidity [80, 81]. In cellulose hydrolysis glucose yield could achieve 60-75 per cent [71, 82].

Another perspective approach to overcoming the task of biomass depolymerization is a developing solid catalysts. This method of attack has some advantages including simple separation of solid catalyst from soluble products. Also application of solid catalysts reduces the risk of corrosion impact on equipment. A wide range of catalysts such as insoluble heteropolyacids [83], carbons with different structure [84], supported noble metals [85], polymer resins [86], oxides of aluminium and silicon [86-88], zeolites [88, 89] could be used for biomass transformations.

Catalytic systems based on P-W HPA are usually studed as HPAs of such composition have demonstrated the best activity in cellulose hydrolysis among [83]. Cesium HPA salts are only investigated as a solid acid catalysts of cellulose hydrolysis. For the first time, they were applied in 2010 in [71]. Such catalysts let tot produce glucose with the yield up to 30 per cent only [71, 90]. The activity of the Cesium HPA salts depends on the catalyst acidity (amount of H in the structures).

Another type of catalysts is carbon materials. These catalysts are called as the most promising for cellulose transformations [84, 86, 91, 92]. Different carbons (e.g. activated carbons, pyrolytic carbons derived from monosaccharides (glucose, sucrose), lignin, cellulose or phenol-containing organic waste) could be used for the creation acid catalysts [86, 91, 93-95]. Before used carbon materials are usually treated by sulfuric acid [84, 86, 91, 92, 96-100], by solutions of HNO_3 [84, 101], H_2O_2 [101, 102], oxidation by air [101], wet air [103], NaOCl [103] or by combination of these methods [102] for increasing catalyst acidity and, hence, activity.

Carbon catalysts have shown a high activity in cellulose hydrolysis. For example, Onda et al., compared a wide range of solid acid catalysts (oxides, zeolites, sulfonated carbon, Amberlyst resin). The catalytic activity of the sulfonated carbon was several times greater than the activity of the other acid catalysts. The yield of glucose over the carbon reached 40 per cent [86, 92]. In the paper [84] the authors reached the glucose yield of 62 per cent in mildly conditions at 150°C and 24 hours of reaction between mechanically activated microcrystalline cellulose. Studies of solid carbon acid catalysts in several reaction cycles demonstrate high stability of the catalyst systems [94, 104-106]. Yang et al., in [104] showed that catalyst Fe_3O_4@C-SO_3H exhibit a slight decrease in catalytic activity over 5 reaction cycles.

To facilitate the removal of a carbon catalyst from the reaction mixture containing saccharides and their dehydration products as well as unreacted solid particles of cellulose, catalysts are developed, which contain a magnetic Fe_3O_4 particles along with carbonaceous materials [104, 107]. Such catalysts could produce glucose form cellulose with yield up to 50 per cent [104]. Carbon catalysts may be used in a combination with other catalytic systems. Thus, using a carbon catalyst in a combination with hydrochloric or sulfuric acid produces glucose with yields up to 35 per cent [108]. With adding carbon catalyst in cellulose mechanical activation step the yields of desired products increases significantly. Thus, the system in the presence of carbon-sulfuric acid and carbon-hydrochloric acid glucose yield was 69 and 88 per cent while the conversion of the substrate was 95 and 98 per cent, respectively.

Highly dispersed metal particles supported on carbon supports may also be catalysts of polysaccharide hydrolysis [85, 88]. Ru-containing catalysts have been found to have the highest activity compared to other noble metals (Pt, Pd) and more then 30 per cent of glucose yield could be reached [85]. The activity of Ru/C increases with metal loading [85]. Ru/C also demonstrates high stability under the aqueous thermal reaction conditions [109, 110]. The high activity in the hydrolysis of cellulose has been demonstrated by polystyrene resins. Thus, using of a polystyrene resin containing -Cl and -SO$_3$H groups on the surface makes it possiblw to obtain glucose from cellulose with the yield of 95 per cent after 10 hours of the reaction under very mild conditions at 120 ° C [111]. Testing the catalyst in several reaction cycles showed reproducibility of its catalytic activity. Polymeric ion exchange resin Amberlist®-15 representing a phenylsulfo acid based on polyvinyl(styrene) is a catalyst of a large number of organic synthesis processes including hydrolysis [112]. However, an information about Amberlist activity in the cellulose hydrolysis reaction is mutually contradictory. Thus, the authors of [92] observed a significant activity of this catalyst, but in [113] the activity was negligible. Lower glucose yield in the second case may be explained by the lower hydrolysis temperature (130 vs. 150 °C, respectively) and 2 times less contact time.

Oxide catalysts (oxides of aluminum, silicon, niobium, and others) and zeolites with different composition and structure may also be of interest to produce valuable products from cellulose due to the high acid properties. Oxides are characterized by high mechanical and thermal stabilities what is very important for aqueous thermal chemistry [65]. The number of studies aimed at the study of the catalytic activity of oxide or zeolite systems in the hydrolysis-dehydration of cellulose is insignificant [114]. Unfortunately, such systems usually show low activity in one-pot transformation of cellululose.

Zeolites demonstrates low activity in the hydrolysis of cellulose to glucose. They are less active then any other type of solid catalysts [86, 93]. Using ionic liquids being capable of dissolving cellulose in place of water makes zeolite catalysts much more effective [93, 115]. However, in [89] the possibility of a direct one-pot conversion of cellulose into a valuable product 5-HMF over bimodal zeolite H-Z-5 (ratio Si/Al equal to 30.15) was demonstrated. This is the only one example of producing 5-HMF with so high yield (a. 49 per cent) from biomass polysaccharide under aqueous thermal conditions over zeolite catalyst. Silicate MCM-41 type structure representing mesoporous materials with a large specific surface area and mesopores also is not active in cellulose hydrolysis in water at temperature of 190-230 ° C [116]. Only 2-7 per cent of 5-HMF yield were achieved from cellulose.

Compared to zeolites, oxide catalysts seem to be much more perspective for hydrolysis of polysaccharides. Most often tested catalysts based on oxides of silicon (IV), zirconium (IV), niobium (V), titanium (IV), tungsten (VI). However, the activity of oxides is also not very significant. Niobium oxide is a promising catalyst for a number of processes of the hydrolysis [117] and dehydration [93] due to the high content of acid sites on the surface. Catalysts based on niobium oxide (Nb_2O_5), tantalum (Ta_2O_5) and zirconia (ZrO_2) have been very effective in the process of hydrolysis of soluble polysaccharides, as well as dehydration of fructose and glucose into 5-HMF [118-121]. However when using in one-pot cellulose hydrolysis-dehydration Nb-containing catalysts are not so effective and yields of glucose lower 8 per cent could be evaluated [113, 122]. ZrO_2, TiO_2, Al_2O_3 also demonstrate higher catalytic activity in the hydrolysis-dehydration of celluloses [86, 87, 106, 123]. Oxides ZrO_2 and Al_2O_3 demonstrates activity close to blank experiment in pure water in some cases [86]. Low activity of oxides is usually attributed by the authors either insufficient pore size of catalysts either insufficient force acidic groups or their low concentration. As in the case of zeolites, oxide catalyst activity and reaction yields of the target products during the hydrolysis rise in ionic liquids [93]. Glucose yields over ZrO_2, Al_2O_3 are normally close to 10-20 per cent [106, 120, 124]. Greater efficiency titanium (IV) oxide has. Over this catalyst 5-HMF or furfural could be produced from cellulose and xylane with the yields up to 15 per cent [120]. On the activity of SiO_2 in the process of hydrolysis-dehydration of cellulose are conflicting data. The studies [106, 113, 125] showed that pure silica does not show activity in the hydrolysis-dehydration of cellulose, which in this case can be explained by the extremely low acidity of catalyst. On the other hand, in [87] glucose yield was 50 per cent at 73 per cent substrate conversion in the presence of SiO_2 catalyst.

The low content of acidic groups on the surface of oxide catalysts compared to, for example, with the sulfonated coals (according to [84, 123, 126, 127] for oxide systems 0.1-1 mmol • g^{-1}, and for the sulfonated coals 1.5-3.5 mmol • g^{-1}) makes it necessary to promotion of acid catalysts. By analogy with the sulfonated coals can be assumed that the treatment with sulfuric acid of transition metal oxides can also enhance the catalytic activity of the materials in comparison with untreated oxides. Most often investigated sulfated system SO_4^{2-}/Al_2O_3, SO_4^{2-}/TiO_2, SO_4^{2-}/ZrO_2, SO_4^{2-}/SnO_2 ᵒ SO_4^{2-}/V_2O_5 [65]. Promotion by sulfuric acid ZrO_2-SiO_2 mixed oxide catalysts with different ratios of oxide forms obtained by simultaneous hydrolysis of the precursor that increased activity of to hydrolyze cellobiose by 4-5 times, allowing glucose to obtain with a yield of up to 50 per cent [128]. Sulfation of SiO_2 also significantly increased the activity of the catalyst, thus yielding glucose from cellulose with a 50 per cent yield [106].

The work of Onda *et al.*, in the presence of sulphated zirconium oxide (IV) was able to receive the glucose yield 14.2 per cent, which is 3.5 times less than in the presence of sulfonated carbon. At the same time the total yield of reducing sugars for ZrO_2-SO_3H are only slightly lower than in the experience with coal, which indicates the low selectivity of the developed sulfated oxide catalyst. Yield 24 per cent of glucose may be achieved by using as catalyst sulfated silica(IV), coated on the ferromagnetic particles Fe_3O_4 [124]. As the additive promoter may be not only mineral sulfuric acid but also organic sulfonic acid [126]. Supported on SiO_2 7.5 and 15 per cent arensulfonic and 15 per cent propilsulfonic acid catalyzed hydrolysis of cellobiose to glucose, providing deep conversion (over 80 per cent) for half an hour of the reaction [126]. Carboxylic acid supported on an oxide carrier not show activity comparable with sulfonic acids [126]. A significant drawback of sulfated catalysts washability of active ingredient is proven. So test the catalysts SiO_2 and ZrO_2 in a few cycles of the reaction showed a slight decrease in their activity [87, 123], while the activity of the sulfated ZrO_2 samples gradually decreases from cycle to cycle [123]. Sulfonic groups washed from the catalyst, not only during the reaction, but even at aqueous thermal tests. For example, in [128] aqueous thermal test sulfated ZrO_2-SiO_2 at 160 ° C leads to almost complete (90 per cent) of acid leaching of the active groups have 30 minutes. Within one hour sulfo washed away completely.

The most widespread method of hemicellulose hydrolysis is a catalysis by diluted solutions of mineral acids. Sulfuric acid is the most often used [129, 130]. Also there are examples of utilizing HCl [129], H_3PO_4 [130], trifluoroacetic [130], maleic [131], oxalic [132], acitic [133] acids and there mixtures [129]. Hemicellulose hydrolysis proceeds under much softer conditions compared to transformation of cellulose and biomass delignification. Optimal temperature range is 100 - 220°C [129]. Removal of hemicellulose is possible directly from real biomass samples, activation could be avoided [130]. In this case cellulose and lignin rest almost untreated [129].

Xylane which is a polymer of xylose undergoes hydrolysis under softer conditions compared to ones for transforming arabinogalactane or mannane [134]. It is possible to convert xylane at 90°C and reaction time 15 min. A ratio of xylane hydrolytic transformations is 90 - 160°C. Hydrolysis of manne becomes possible in the range of 180 - 230°C and of arabinogalactane at 100°C [129].

In contrast to soluble catalysts, the amount of works where solid catalysts are applied is much lower.

In [135] inuline hydrolysis to fructose was investigated in the presence of zeolite catalyst LZ-M-8 at 100°C. The main product yield was 96 per cent at almost 100 per cent substrate conversion. Formation of 5-HMF was not observed under the reaction conditions applied. Using polystyrene resins let to hydrolyze starch to glucose with the main product yield 100 per cent at 120 °C [111]. Zeolite catalysts are also able to convert starch and glucose into 5-HMF with low yields. Thus, zeolites H-Beta, H-Y, H-ZSM-5 produce 5-HMF with the yields 9 per cent, 9 per cent and 12 per cent, respectively [136]. Comparable investigation of Zr(IV) and Ti(IV) oxides as well as sulfated ZrO_2 in xylane depolymerization demonstrates that titanium oxide (IV) possess the highest activity among catalysts tested. Furfural was produced with the yield 25 per cent in this case [123].

Transformations of polysaccharides in the mixtures of water and organic solvents have also been investigated. Biphasic reaction systems stabilize reaction products and prevent there decomposition at high temperature via extraction [137]. Plenty of papers are devoted to synthesis of 5-HMF from monosaccharides in biphasic water-organic system for extraction of 5-HMF from water reaction medium [7]. The results achieved so far demonstrate 90 per cent of 5-HMF yield from fructose when using butanol-2, methylisobutylketone (MIBK) and dimethyl sulfoxide (DMSO) [7, 138]. 50 per cent 5-HMF yield is possible when glucose is a substrate and organic co-solvent is butanol-2 [119] or tetrahydrofuran [139]. The first works dealing with conversion of polysaccharides to 5-HMF in biphasic media have been carried out in 80s of XX century. And this field becomes more and more attractive for investigations. Attempts have been made to produce 5-HMF from inulin, starch, raffinose, cellobiose, sucrose [7]. Thus, the possibility of producing 70 per cent of 5-HMF from inulin in water-MIBK system at 78°C has been demonstrated. In 2011 5-HMF was synthesized from inulin with 87 per cent yield in water-butanol-2 mixture Ta_2O_5 catalyst [118]. The authors of [140] have revealed optimal temperature for transforming easy-hydrolysable and hard-hydrolysable polysaccharides which are a. 180°C and 270°C, respectively, in flow system over TiO_2 in water-MIBK system. Under the optimal conditions developed 5-HMF yields were 15 per cent and 35 per cent when using starch and cellulose as a substrates. Found in [140], the yield of 5-HMF produced from cellulose is the best one which is available now in the literature.

3.2. Cellulose and Hemicellulose Hydrolysis-Hydrogenation

Hydrolysis-hydrogenation is combined one-pot process. It consists of two stages: first, hydrolysis of polysaccharides and hydrogenation of hydrolysis products (monosaccharides) into corresponding polyols. Thus, for carrying out this process acid-base and reducing active sites are needed. Nowadays, two main approaches are offered for one-pot hydrolysis-hydrogenation of biomass produced polysaccharides. The first one is simultaneous using of two catalysts. One is for hydrolysis and another one is for hydrogenation. Both catalysts are placed together in one reactor. In this case, mineral acids or heteropolyacids in combination with noble metal nanoparticles (Ru, Pt, etc.) are usually applied for hydrolysis and hydrogenation, respectively. Preparing a bifunctional catalyst which brings both hydrolysis and hydrogenation sites is a second point of view on transformation of polysaccharides into polyols. Such catalysts are prepared via immobilization of noble metal nanostructures on a support with acid and/or base surface groups.

Hydrogen gas is used for hydrolysis-hydrogenation processes. It is the most widespread reducing agent in such reactions. However, H_2 gas has some drawbacks. The reaction should be carried out under high gas pressure. Nevertheless, it is possible to apply propanol-2 as a reducing agent [141]. During the reaction it is oxidized to acetone, high pressures are not needed.

Thus, the reaction path of cellulose transformation is follows. Cellulose is hydrolyzed to glucose and aldehyde group of monosaccharide is reduced to alcohol one. Sorbitol is formed in this way [21].

As a whole, among noble metals Ru, Pt, Au, Rh, Ir, Pd ruthenium and platinum are most perspective [142]. As it mentioned above, solid high dispersed metal catalysts of Ru, Pt are able to catalyze hydrolysis of polysaccharides [43][48]. Carrying out the same process of cellulose transformation under hydrogen pressure produces sorbitol and mannitol as main products. The most active catalyst is 2.5 per cent $Pt/\gamma-Al_2O_3$. The highest yields of sorbitol and mannitol are 25 per cent and

6 per cent, respectively [48]. Modification of Ru/C catalysts by phosphoric acid increases the catalyst activity in 2 times [143].

Ni-containing catalysts could also be applied for hydrolysis-hydrogenation [144]. Sorbitol and mannitol yields equal to 50 and 6 per cent could be produced over 3 per cent Ni/C under aqueous thermal conditions at 190°C H_2 pressure 60 atm. Ni supported on carbon nanotubes produces 64 per cent of sorbitol and 7 per cent of mannitol [145]. Ni catalysts are quiet stable under reaction conditions. In contrast to Ru catalysts, Ni/C promoted by phosphorous is unstable, although, sorbitol yield 48-60 per cent in the first reaction cycle is achievable [146, 147]. Bimetalic Ir-Ni catalysts supported on mesopore carbon material produces sorbitol with the yield about 58 per cent [145]. Iridium is shown to improve catalyst stability.

Ru/C in combination with mineral acids hydrolyzes cellulose to sorbitol under H_2 pressure 70 bar. The product yields could reach 39 per cent [148, 149]. Ru supported on carbon nanotubes produces 73 per cent of sorbitol in solution of mineral acids [146]. Mixing Ru/C with soluble heteropolyacid $H_4SiW_{12}O_{40}$ increases a rate of cellulose hydrolysis-hydrogenation. Sorbitol yield equal to 85 per cent has been reached [150].

Another one-pot technique of cellulose hydrolysis-hydrogenation is combined using solid cesium salts of heteropolyacids (CsHPA) with Ru/C. At 190 °C and hydrogen pressure 50 atm the highest yields of sorbitol and mannitol equal to 59 per cent and 12 per cent, respectively, could be produced in the presence of mechanical mixture $Cs_{2.5}H_{0.5}PW_{12}O_{40}$ – Ru/C. It should be noted that the combination of two solid catalysts is in 3 times more effective then a system of Ru/C with solution of HPA [151]. Bifunctional catalysts has also been applied for the process under investigation. 1 per cent $Ru/Cs_2HPW_{12}O_{40}$ and 1 per cent $Ru/Cs_3PW_{12}O_{40}$ are able to produce sorbitol from cellulose with the yield upto 40 per cent at 160 °C [152]. Using bifunctional catalyst increases the selectivity of sorbitol formation compared to mixtures of catalysts.

Investigations on hemicellulose transformation into respective polyols are also made. Thus, it is possible to convert arabinogalactane to arabitol and galacitol over Ru/MCM-48 at 185°C and hydrogen pressure 20 bar in 24 h. The total yield of polyols reached via this method is 25 per cent [37].

In case of using higher temperatures and pressures of hydrogen hydrogenation process is accompanied by hydrogenolysis reaction which forms such products as propylene glycol and ethylene glycol. This products have important industrial applications and are usually produced from a natural gas. Nowadays, investigations are made to find a method of propylene glycol and ethylene glycol directly from plant biomass [153-155]. Hydrolysis-hydrogenolysis process is carried out in water under H_2 pressure 60-100 bar and temperature 230-250 °C [156]. Reaction time is usually in the range of 0.5 to 4 h. It is determined by substrate amount and process temperature. Most effective catalysts of the cellulose transformation to glycols appear to be W-containing structures [153, 156]. Tungsten carbide supported on carbon produces 73 per cent of ethlene glycol from cellulose [157]. Nanoparticles of platinum group metals and Ni supported on tungsten oxides (WO_3, H_2WO_4) and

W-containing heteropolyacids are also under investigations [153, 157, 158]. Ni seems to be most active among them. Ru is also very perspective while Pt, Pd, Rh, Ir have low activity [153, 157, 158]. So far, it is possible to produce ethylene glycol with a yield 50 per cent from cellulose 3 per cent $Ru/NbOPO_4$ catalyst [155].

3.3. Cellulose Hydrolysis-Oxidation

Hydrolysis-oxidation of biomass and particular cellulose is also a perspective way of plant material valorization [8, 159]. This method of transformations let to produce such valuable chemicals as gluconic acid [3, 160], furandicarbonic acid [161], formic acid [162]. Chemicals produced via biomass oxidation are perspective as substances of chemical synthesis, polymer industry [163], textile and chemical industries [164]. Formic acid can be used as reducing agent or hydrogen source [165-169].

High dispersed supported metal particles of noble metals Au, Pt, Pd are well known as a catalysts of oxidizing saccharides into corresponding acids, including oxidation of glucose to gluconic acid [170-172]. One-pot synthesis of gluconic acid from cellulose over biphunctional acid-oxidative catalysts seems very interesting [159]. Thus, cellobiose hydrolysis-oxidation to gluconic acid has been studied in the presence of nanoparticles of gold and platinum supported on sulphonated activated carbon [171] and carbon nanotubes pretreated with nitric acid [170]. Yield of gluconic acid has reached 80 per cent. Onda and co-authors found a synergetic effect of using bifunctional catalyst $Au/AC-SO_3H$ compared to mechanical mixture of two catalysts. For example, the yield of gluconic acid produced from starch was 40 per cent over bifunctional catalyst and only 10 per cent in the presence of mechanical mixture of Au/C and $AC-SO_3H$ [171]. Another type of polysaccharide hydrolysis-oxidation catalysts is noble metal nonoparticles supported on solid HPAs [173, 174]. This system makes possible the production of gluconic acid from cellobiose with the yield closed to 95 per cent.

Mo-V-P heteropolyacids which have both acid and oxidative properties seem to be very perspective for one-pot producing formic acid from biomass polysaccharides under soft conditions [162, 166, 175, 176]. The first investigations on cellulose hydrolysis-oxidation over Mo-V-P HPA have been made in [177, 178]. Formic acid could be formed from cellulose with high yields up to 66 per cent [162]. Influence of HPA composition on formic acid yield in cellulose hydrolysis-oxidation process is presented in our previous work [162]. The following HPAs with different composition (different amount of acid and vanadium contents) have been investigated $H_5PMo_{10}V_2O_{40}$ (HPA-2), $Co_{0.6}H_{3.8}PMo_{10}V_2O_{40}$ (Co-HPA-2), $H_{11}P_3Mo_{16}V_6O_{76}$ (HPA-6′) ▯ $H_{17}P_3Mo_{16}V_{10}O_{89}$ (HPA-10′). The experiments have been made at 130-180 °C and air pressure 10 - 50 bar. The highest reaction rate has been demonstrated in the most acidic solutions. Linear dependence of the initial reaction rate on H^+ concentration is observed. Both formic acid yield and initial reaction rate decreases as follows: HPA-2 > Co-HPA-2 > HPA-6′ > HPA-10′. [162].

3.4. Aqueous-Phase Reforming

The most traditional way for converting biomass into ▯$_2$ is a high temperature pyrolysis [13]. Aqueous-phase reforming (APR) is a fine way for producing H_2 containing gas mixture (H_2, CO and CO_2) which could be transformed into pure gas and be used for energy applications or be used in Fischer–Tropsch process [182]. A lot of efforts are made for APR process of biomass [180]. Catalysts play a crucial role in APR of polysaccharides H_2 yield is strongly depend on a catalyst type. A ratio between gases depends on a catalyst used. Pt, Ni or Sn containing catalysts are usually applied for the process under the temperature a 230°C [179-181]. Pt–containing catalysts provides high selectivity of H_2 formation during APR [180]. On the other hand, Ni is deactivated very fast due to high yields of methane which causes зауглероживанию. Combination of Ni and Sn let to improve characteristics of Ni-containing systems and makes nickel catalysts close to Pt ones [181]. Pt/C is most often used catalyst in APR [183]. For metal particles supported on SiO_2 it was shown that the activity decreases in the following manner: Pt ~ Ni > Ru > Rh ~ Pd > Ir in the temperature range 180 – 225°C. The activity of Pd catalyst could be improved by promoting system with Fe (Pd_1Fe_9/Al_2O_3 as an eample). Pt and Pd are more selective for hydrogen production while, a Ni, Ru and Rh to C_1–C_{15} products [184].

3.5 Biomass to Biofuels via Gazification and Pyrolysis

Biomass transformation to fuels (biooil and biogas) takes place under very high temperatures. Two main processes are in thermochemical conversion of biomass are gasification and pyrolysis. Biomass gasification yields synthesis gas. It is used in Fischer–Tropsch synthesis to produce fuel or in production of pure hydrogen. Biooil is a liquid product of the biomass pyrolysis and can be converted catalytically to fuels or fuel additives. The advantage of the thermochemical approach compared to aqueous thermal is that all the biomass is converted, including the lignin. Several reviews [179] [185-189] have discussed different aspects of biomass thermochemical valorization via gasification and pyrolysis.

Pyrolysis of biomass is performed under temperature in the range of 400-500 °C [190]. The lower and upper values of temperature are typical for slow and fast pyrolysis processes, respectively. These processes produce three type of products: char, gas and liquid. The last one can be then transformed into fuel components. From this point of view, the fast pyrolysis is the most perspective and can produce up to 75 per cent of oil [190].

A lot of different catalysts have been applied to facilitate pyrolysis reaction. The most often investigated are zeolites [186-188, 191, 192]. Especial interest to them is caused by there thermal stability, high selectivity to hydrocarbons and limitied deactivation [193]. It should be also noted that zeolites are well-known catalysts of oil cracking. Pyrolysis conducted in the presence of zeolite catalysts yields a crude oil which could contains up to 30 per cent of aromatic products [187]. Also examples of using MCM-41 and metals supported on alumina for pyrolysis exist [194, 195]. However, zeolites are most promising as an organic liquid mixture with large proportion of aromatics produced over zeolites could be used as a fuel.

Gasification of biomass produces synthesis gas which is very useful in industry. This process is usually performed in the presence of water steam and small amounts of oxygen and under high temperature (over 700°C) and in the presence of supported metal catalysts [196, 197]. The catalysts for gasification should effectively remove tars, be capable of reforming methane, be resistant to deactivation, be easy to regenerate [198]. For processing biomass gasification it is possible to apply supported metals (Fe, Ni mainly), alkali metal salts, alumina and some natural minerals (dolomites, olivines) [179, 198]. Among them Ni seems to be the most perspective for biomass gasification, but the optimisation of the metal content, particle size, as well as the catalyst support is still needed.

6. Conclusion

In conclusion, the conversion of lignocellulosic biomass to fuel and valuable chemicals has sharply developed with the application of heterogeneous catalysis allowing scientists to increase the selectivity and yields of useful biobased chemicals. The main important approaches appeared to be (i) aqueous thermal catalytic methods in the presence solid acid and bifunctional catalysts; (ii) thermochemical catalytic treatment based on flash catalytic pyrolisys to produce bio-oil and its catalytic hydrotreatment to reduce the content of oxygen. In recent years, many heterogeneous catalysts were proposed for the discussed processes. Meanwhile, rational processes design and their optimization clearly require optimization of catalysts activity and improvement their stability. In this regards, future investigations will be addressed to solving mentioned problems.

References

1. Gallezot, P. Conversion of biomass to selected chemical products//Chemical Society Reviews. – 2012. – V. 41. – N 4. – P. 1538-1558.

2. Gallezot, P. Catalytic Conversion of Biomass: Challenges and Issues// ChemSusChem. – 2008. – V. 1. – N 8-9. – P. 734-737.

3. Geboers, J., Van de Vyver, S., Ooms, R., *et al.*, Chemocatalytic conversion of cellulose: opportunities, advances and pitfalls//Catalysis Science and Technology. – 2011. – V. 1. – N – P. 714-726.

4. Bhaumik, P.,Dhepe, P. L. Solid acid catalyzed synthesis of furans from carbohydrates//Catalysis Reviews. – 2016. – V. 58. – N 1. – P. 36-112.

5. Murzin, D.,Salmi, T. Catalysis for Lignocellulosic Biomass Processing: Methodological Aspects//Catalysis Letters. – 2012. – V. 142. – N 6. – P. 676-689.

6. Van de Vyver, S., Geboers, J., Jacobs, P. A.,*et al.*, Recent Advances in the Catalytic Conversion of Cellulose//Chem. Cat. Chem. – 2011. – V. 3. – N – P. 82-94.

7. van Putten, R.-J., van der Waal, J. C., de Jong, E.,*et al.*, Hydroxymethylfurfural, A Versatile Platform Chemical Made from Renewable Resources//Chemical Reviews. – 2013. – V. 113. – N 3. – P. 1499-1597.

8. Besson, M., Gallezot, P.,Pinel, C. Conversion of Biomass into Chemicals over Metal Catalysts//Chemical Reviews. – 2014. – V. 114. – N 3. – P. 1827-1870.

9. Dutta, S.,Pal, S. Promises in direct conversion of cellulose and lignocellulosic biomass to chemicals and fuels: Combined solvent–nanocatalysis approach for biorefinary//Biomass and Bioenergy. – 2014. – V. 62. – N 0. – P. 182-197.

10. Imhoff, M. L., Bounoua, L., Ricketts, T., *et al.,* Global patterns in human consumption of net primary production//Nature. – 2004. – V. 429. – N 6994. – P. 870-873.

11. Deutschmann, R.,Dekker, R. F. H. From plant biomass to bio-based chemicals: Latest developments in xylan research//Biotechnology Advances. – 2012. – V. 30. – N 6. – P. 1627–1640.

12. Heinze, T. Chemical Functionalization of Cellulose//Polysaccharides. Structural diversity and functional versatility. Second edition/Severian Dumitriu. - New York: Marcel Dekker, 2005. - с. 551.

13. Ruppert, A. M., Weinberg, K.,Palkovits, R. Hydrogenolysis Goes Bio: From Carbohydrates and Sugar Alcohols to Platform Chemicals//Angew. Chem. Int. Ed. – 2012. – V. 51. – N 11. – P. 2564 – 2601.

14. Zhou, C.-H., Xia, X., Lin, C.-X., *et al.,* Catalytic conversion of lignocellulosic biomass to fine chemicals and fuels//Chemical Society Reviews. – 2011. – V. 40. – N 11. – P.

15. Gallezot, P.,Kiennemann, A. Conversion of Biomass on Solid Catalysts// Handbook of Heterogeneous Catalysis/Ertl. - Wiley-VCH Verlag GmbH and Co. KGaA, 2008. - с. 2447-2476.

16. Bruggink, A., Schoevaart, R.,Kieboom, T. Concepts of Nature in Organic Synthesis: Cascade Catalysis and Multistep Conversions in Concert//Organic Process Research and Development. – 2003. – V. 7. – N 5. – P. 622-640.

17. Thananatthanachon, T.,Rauchfuss, T. B. Efficient Production of the Liquid Fuel 2,5-Dimethylfuran from Fructose Using Formic Acid as a Reagent//Angew. Chem. Int. Ed. – 2010. – V. 49. – N 37. – P. 6616-6618.

18. Пат. 2363698 С1 Российская Федерация, МПК C07D307/46. Способ получения 5-гидрокисметилфурфурола/Тарабанько, В. Е., Смирнова, М. А.,Черняк, М. Ю.; заявитель и патентообладатель Институт химии и химической технологии СО РАН. – опубл. 10.08.2009.

19.//Новый справочник химика и технолога. Сырье и продукты промышленности органических и неорганических веществ. Ч. II./Ю.В. Поконова; В.И.Стархова. - С.-Пб.: АНО НПО «Профессионал», 2005. - с. 882-891 с.

20. Пат. 4339387 США, МПК C07D307/46. Process For Manufacturing 5-Hydroxymethylfurfural/Flèche, G.; заявитель и патентообладатель Roquette Frères SA. – опубл. 13.07.82.

21. Werpy, T.,Petersen, G. Results of screening for potential candidates from sugars and synthesis gas//Top Value Added Chemicals from Biomass/Todd Werpy;Gene Petersen. - U.S. Department of Energy: Energy Efficiency and Renewable Energy, 2014. - с. 63.

22. Rubin, E. M. Genomics of cellulosic biofuels//Nature. – 2008. – V. 454. – N 7206. – P. 841-845.

23. Gallezot, P. Conversion of biomass to selected chemical products//Chem. Soc. Rev. – 2012. – V. 41. – N 4. – P. 1538-1558.

24. Пат. 8772515 В2 США, МПК C07D307/48. Method to convert biomass to 5-(hydroxymethyl)-furfural (HMF) and furfural using lactones, furans, and pyrans as solvents/Dumesic, J. A., Ribeiro Gallo, J. M.,Alonso, D.; заявитель и патентообладатель Wisconsin Alumni Research Foundation. – опубл. 17.08.14.

25. van Haveren, J., Scott, E. L.,Sanders, J. Bulk chemicals from biomass//Biofuels, Bioproducts and Biorefining. – 2008. – V. 2. – N 1. – P. 41-57.

26. Zakzeski, J., Bruijnincx, P. C. A., Jongerius, A. L., *et al.*, The Catalytic Valorization of Lignin for the Production of Renewable Chemicals//Chem. Rev. – 2010. – V. 110. – N 6. – P. 3552–3599.

27. Химия древесины и ее спутников./З.А. Роговин;Н.Н. Шорыгина. - М.-Л.: Госхимиздат, 1953. - 679 с.

28. Уайз, Л. Э.,Джан, Э. С. Химия древесины/под ред. Б. Д. Богомолов. - М.: Гослесбумиздат, 1959. - 608 с.

29. Древесина. Химия, ультраструктура, реакции/Д. Фенгел, Г. ВегенерА. А. Леонович. - М.: Лесная промышленность, 1988. - 512 с.

30. Кузнецов, Б. Н. Каталитическая химия растительной биомассы//Соросовский образовательный журнал. – 1996. – V. – N 12. – P. 47-55.

31. Химия древесины и целлюлозы/В.М. Никитин, А.В. ОболенскаяВ.П. Щеголев. - М.: Лесная промышленность, 1978. - 368 с.

32. Gromov, N. V., Taran, O. P., Sorokina, K. N., *et al.*, New methods for the one-pot processing of polysaccharide components (cellulose and hemicelluloses) of lignocellulose biomass into valuable products. Part 1: Methods for biomass activation//Catalysis in Industry. – 2016. – V. 8. – N 2. – P. 176-786.

33. Химия древесины и синтетических полимеров: Учебник для вузов/В.И. Азаров, А.В. БуровА.В. Оболенская. - СПб: СПбЛТА, 1999. - 682 с.

34. Bioconversion of Forest and Agricultural Plant Wastes/J. N. Saddler, L. P. RamosC. Breuil. - London: C. A. B. International, 1993. - 73 с.

35. Синицын, А. П.,Клёсов, А. А. Влияние предобработки на эффективность ферментативного превращения хлопкового линта//Прикладная биохимия и микробиология. – 1981. – V. 17. – N 5. – P. 682 - 694.

36. Spiridon, I. Hemicelluloses: structure and properties//Polysaccharides. Structural diversity and functional versatility. Second edition/Severian Dimitriu. - New York: Marcel Dekker, 2005. - с. 475 - 487.

37. Kusema, B. T., Faba, L., Kumar, N., *et al.*, Hydrolytic hydrogenation of hemicellulose over metal modified mesoporous catalyst//Catalysis Today. – 2012. – V. 196. – N 1. – P. 26-33.

38. Menon, V.,Rao, M. Trends in bioconversion of lignocellulose: Biofuels, platform chemicals and biorefinery concept//Progress in Energy and Combustion Science. – 2012. – V. 38. – N 4. – P. 522-550.

39. Химия лигнина: Учебное пособие./А. В. Оболенская. - Спб.: ЛТА, 1993. - 80 с.

40. Пат. 2009037281 А3 EP, МПК C10G1/08. Method for hydrogenating separation of lignin using transition metal carbides/Kotrel, S., Emmeluth, M.,BENÖHR, A.; заявитель и патентообладатель Basf Se. – опубл. 25.03.2010.

41. C.-L., C., Y., L. S.,W., D. C./- Berlin/Heidelberg: Eds.; Springer-Verlag, 1992. - 350 с.

42. Kadla, J. F., Chang, H. M.,Jameel, H. Reactions of lignins with high temperature hydrogen peroxide//Holzforschung. – 1999. – V. 53. – N 3. – P. 277-284.

43. Hick, S. M., Griebel, C., Restrepo, D. T., *et al.*, Mechanocatalysis for biomass-derived chemicals and fuels//Green Chemistry. – 2010. – V. 12. – N 3. – P. 468-474.

44. Elliott, D. C., Beckman, D., Bridgwater, A. V., *et al.*,//Energy Fuels. – 1991. – V. 5. – N – P. 399.

45. Singh, R., Shukla, A., Tiwari, S., *et al.*, A review on delignification of lignocellulosic biomass for enhancement of ethanol production potential// Renewable Sustainable Energy Rev. – 2014. – V. 32. – N – P. 713-728.

46. Zheng, Y., Zhao, J., Xu, F., *et al.*, Pretreatment of lignocellulosic biomass for enhanced biogas production//Progress in Energy and Combustion Science. – 2014. – V. 42. – N – P. 35-53.

47. Kim, J. S., Lee, Y. Y.,Kim, T. H. A review on alkaline pretreatment technology for bioconversion of lignocellulosic biomass//Bioresour. Technol. – 2016. – V. 199. – N – P. 42-48.

48. Mosier, N., Wyman, C., Dale, B., *et al.*, Features of promising technologies for pretreatment of lignocellulosic biomass//Bioresour. Technol. – 2005. – V. 96. – N 6. – P. 673-686.

49. Pretreatment of Biomass/Ashok Pandey, Sangeeta Negi, Parameswaran Binod, *et al.* - Amsterdam: Elsevier, 2015. - 264 с.

50. Singh, R., Krishna, B. B., Kumar, J., *et al.*, Opportunities for utilization of non-conventional energy sources for biomass pretreatment//Bioresour. Technol. – 2016. – V. 199. – N – P. 398–407.

51. Chheda, J. N., Huber, G. W.,Dumesic, J. A. Liquid-Phase Catalytic Processing of Biomass-Derived Oxygenated Hydrocarbons to Fuels and Chemicals// Angewandte Chemie International Edition. – 2007. – V. 46. – N 38. – P. 7164-7183.

52. Saha, B. C., Iten, L. B., Cotta, M. A., *et al.*, Dilute acid pretreatment, enzymatic saccharification and fermentation of wheat straw to ethanol//Process Biochem. – 2005. – V. 40. – N 12. – P. 3693-3700.

53. Kivaisi, A. K.,Eliapenda, S. Pretreatment of bagasse and coconut fibres for enhanced anaerobic degradation by rumen microorganisms//Renewable Energy. – 1994. – V. 5. – N 5–8. – P. 791-795.

54. Monlau, F., Barakat, A., Steyer, J. P., *et al.*, Comparison of seven types of thermo-chemical pretreatments on the structural features and anaerobic digestion of sunflower stalks//Bioresour. Technol. – 2012. – V. 120. – N – P. 241-247.

55. Yao, W.,Nokes, S. E. Phanerochaete chrysosporium pretreatment of biomass to enhance solvent production in subsequent bacterial solid-substrate cultivation//Biomass and Bioenergy. – 2014. – V. 62. – N 0. – P. 100-107.

56. Mot, A. C.,Silaghi-Dumitrescu, R. Laccases: complex architectures for one-electron oxidations//Biochemistry (Mosc). – 2012. – V. 77. – N 12. – P. 1395-1407.

57. Scott, G. M., Akhtar, M., Swaney, R. E., *et al.*, Recent developments in biopulping technology at Madison, WI//Progress in Biotechnology/L. ViikariR. Lantto. - Elsevier, 2002. - c. 61-71.

58. Moilanen, U., Kellock, M., Galkin, S., *et al.*, The laccase-catalyzed modification of lignin for enzymatic hydrolysis//Enzyme and Microbial Technology. – 2011. – V. 49. – N 6–7. – P. 492-498.

59. Delidovich, I., Leonhard, K.,Palkovits, R. Cellulose and hemicellulose valorisation: an integrated challenge of catalysis and reaction engineering// Energy and Environmental Science. – 2014. – V. 7. – N 9. – P. 2803-2830.

60. Dhepe, P.,Fukuoka, A. Cracking of Cellulose over Supported Metal Catalysts// Catalysis Surveys from Asia. – 2007. – V. 11. – N 4. – P. 186-191.

61. Zeitsch, K. J.//Production of sugar series. The chemistry and technology of furfural and its many by-products, vol. 13./- New York:: Elsevier Science, 2000. - c.

62. Новый справочник химика и технолога. Сырье и продукты промышленности органических и неорганических веществ. Часть 2/В. А. Столярова, С. А. Апостолов, С. Е. Бабаш, *et al.* - СПб: НПО Профессионал, 2002. - 1142 c.

63. Serrano-Ruiz, J. C., Luque, R.,Sepulveda-Escribano, A. Transformations of biomass-derived platform molecules: from high added-value chemicals to fuels via aqueous-phase processing//Chemical Society Reviews. – 2011. – V. 40. – N 11. – P. 5266–5281.

64. Saeman, J. F. Kinetics of Wood Saccharification - Hydrolysis of Cellulose and Decomposition of Sugars in Dilute Acid at High Temperature//Industrial and Engineering Chemistry. – 1945. – V. 37. – N 1. – P. 43-52.

65. Amarasekara, A. S. Acid Hydrolysis of Cellulose and Hemicellulose// Handbook of Cellulosic Ethanol/Ananda S. Amarasekara. - Beverly (USA): John Wiley and Sons, 2013. - c. 247-281.

66. Immergut, E. A.,Ranby, B. G. Heterogeneous Acid Hydrolysis of Native Cellulose Fibers//Industrial and Engineering Chemistry. – 1956. – V. 48. – N 7. – P. 1183-1189.

67. Kupiainen, L., Ahola, J.,Tanskanen, J. Distinct Effect of Formic and Sulfuric Acids on Cellulose Hydrolysis at High Temperature//Industrial and Engineering Chemistry Research. – 2012. – V. 51. – N 8. – P. 3295-3300.

68. Lenihan, P., Orozco, A., O'Neill, E., *et al.,* Dilute acid hydrolysis of lignocellulosic biomass//Chemical Engineering Journal. – 2010. – V. 156. – N 2. – P. 395-403.

69. Soldi, V. Stability and degradation of polysaccharides//Polysaccharides. Structural diversity and functional versatility. Second edition/Severian Dimitriu. - New York: Marcel Dekker, 2005. - c. 395 – 406.

70. Bustos, G., Ramírez, J., Garrote, G., *et al.,* Modeling of the hydrolysis of sugar cane bagasse with hydrochloric acid//Applied Biochemistry and Biotechnology. – 2003. – V. 104. – N 1. – P. 51-68.

71. Tian, J., Wang, J., Zhao, S., *et al.,* Hydrolysis of cellulose by the heteropoly acid $H_3PW_{12}O_{40}$//Cellulose Journal. – 2010. – V. 17. – N – P. 587-594.

72. Abeer, M. A., Zeinab, H. A. E. W., Atef, A. I., *et al.,* Characterization of microcrystalline cellulose prepared from lignocellulosic materials. Part I. Acid catalyzed hydrolysis//Bioresource Technology. – 2010. – V. 101 –N12. – P. 4446-4455.

73. Amarasekara, A. S.,Wiredu, B. Aryl sulfonic acid catalyzed hydrolysis of cellulose in water//Applied Catalysis A: General. – 2012. – V. 417–418. – N 0. – P. 259-262.

74. Laopaiboon, P., Thani, A., Leelavatcharamas, V., *et al.,* Acid hydrolysis of sugarcane bagasse for lactic acid production.//Bioresour Technology. – 2010. – V. 101. – N 3. – P. 1036-1043.

75. Mok, W. S., Antal, M. J.,Varhegyi, G. Productive and parasitic pathways in dilute acid-catalyzed hydrolysis of cellulose//Industrial and Engineering Chemistry Research. – 1992. – V. 31. – N 1. – P. 94-100.

76. Mukherjee, A., Dumont, M.-J., Raghavan, V. Review: Sustainable production of hydroxymethylfurfural and levulinic acid: Challenges and opportunities// Biomass and Bioenergy. – 2015. – V. 72. – N – P. 143-183.

77. van Dam, H. E., Kieboom, A. P. G.,van Bekkum, H. The Conversion of Fructose and Glucose in Acidic Media: Formation of Hydroxymethylfurfural//Starch - Stärke. – 1986. – V. 38. – N 3. – P. 95-101.

78. Mosier, N. S., Sarikaya, A., Ladisch, C. M., *et al.,* Characterization of Dicarboxylic Acids for Cellulose Hydrolysis//Biotechnology progress. – 2001. – V. 17. – N – P. 474-480.

79. vom Stein, T., Grande, P., Sibilla, F., *et al.,* Salt-assisted organic-acid-catalyzed depolymerization of cellulose//Green Chemistry. – 2010. – V. 12. – N 10. – P. 1844-1849.

80. Izumi, Y., Ono, M., Kitagawa, M., *et al.,* Silica-included heteropoly compounds as solid acid catalysts//Microporous Materials. – 1995. – V. 5. – N 4. – P. 255-262.

81. Hill, C. L.,Prosser-McCartha, C. M. Homogeneous catalysis by transition metal oxygen anion clusters//Coordination Chemistry Reviews. – 1995. – V. 143. – N C. – P. 407-455.

82. Ogasawara, Y., Itagaki, S., Yamaguchi, K., *et al.*, Saccharification of Natural Lignocellulose Biomass and Polysaccharides by Highly Negatively Charged Heteropolyacids in Concentrated Aqueous Solution//ChemSusChem. – 2011. – V. 4. – N 4. – P. 519-525.

83. Shimizu, K.-I., Furukawa, H., Kobayashi, N., *et al.*, Effects of Bronsted and Lewis acidities on activity and selectivity of heteropolyacid-based catalysts for hydrolysis of cellobiose and cellulose//Green Chemistry. – 2009. – V. 11. – N 10. – P. 1627-1632.

84. Pang, J., Wang, A., Zheng, M., *et al.*, Hydrolysis of cellulose into glucose over carbons sulfonated at elevated temperatures//Chemical Communications. – 2010. – V. 46. – N – P. 6935-6937.

85. Kobayashi, H., Komanoya, T., Hara, K., *et al.*, Water-tolerant mesoporous-carbon-supported ruthenium catalysts for the hydrolysis of cellulose to glucose//ChemSusChem. – 2010. – V. 3. – N 4. – P. 440-443.

86. Onda, A., Ochi, T.,Yanagisawa, K. Selective hydrolysis of cellulose into glucose over solid acid catalysts//Green Chemistry. – 2008. – V. 10. – N 10. – P. 1033-1037.

87. Wang, H., Zhang, C., He, H., *et al.*, Glucose production from hydrolysis of cellulose over a novel silica catalyst under hydrothermal conditions//Journal of Environmental Sciences. – 2012. – V. 24. – N 3. – P. 473-478.

88. Dhepe, P. L.,Fukuoka, A. Cellulose Conversion under Heterogeneous Catalysis//ChemSusChem. – 2008. – V. 1. – N 12. – P. 969-975.

89. Nandiwale, K. Y., Galande, N. D., Thakur, P., *et al.*, One-Pot Synthesis of 5-Hydroxymethylfurfural by Cellulose Hydrolysis over Highly Active Bimodal Micro/Mesoporous H-ZSM-5 Catalyst//ACS Sustainable Chemistry and Engineering. – 2014. – V. 2. – N 7. – P. 1928-1932.

90. Cheng, M., Shi, T., Guan, H., *et al.*, Clean production of glucose from polysaccharides using a micellar heteropolyacid as a heterogeneous catalyst// Applied Catalysis B: Environmental. – 2011. – V. 107. – N 1вБ"2. – P. 104-109.

91. Suganuma, S., Nakajima, K., Kitano, M., *et al.*, Hydrolysis of Cellulose by Amorphous Carbon Bearing SO3H, COOH, and OH Groups//Journal of American Chemical Society. – 2008. – V. 130. – N 5. – P. 12787-12793.

92. Onda, A., Ochi, T.,Yanagisawa, K. Hydrolysis of Cellulose Selectively into Glucose Over Sulfonated Activated-Carbon Catalyst Under Hydrothermal Conditions//Topics in Catalysis. – 2009. – V. 52. – N 6. – P. 801-807.

93. Guo, F., Fang, Z.,Xu, C. C.,*et al.*, Solid acid mediated hydrolysis of biomass for producing biofuels//Progress in Energy and Combustion Science. – 2012. – V. 38. – N 5. – P. 672-690.

94. Toda, M., Takagaki, A., Okamura, M., *et al.*, Green chemistry: Biodiesel made with sugar catalyst//Nature. – 2005. – V. 438. – N 7065. – P. 178-178.

95. Shen, S., Wang, C., Han, Y., *et al.*, Influence of reaction conditions on heterogeneous hydrolysis of cellulose over phenolic residue-derived solid acid//Fuel. – 2014. – V. 134. – N 0. – P. 573-578.

96. Guo, H., Qi, X., Li, L., *et al.*, Hydrolysis of cellulose over functionalized glucose-derived carbon catalyst in ionic liquid//Bioresource Technology. – 2012. – V. 116. – N – P. 355-359.

97. Dora, S., Bhaskar, T., Singh, R., *et al.*, Effective catalytic conversion of cellulose into high yields of methyl glucosides over sulfonated carbon based catalyst//Bioresource Technology. – 2012. – V. 120. – N – P. 318-321.

98. Li, S., Gu, Z., Bjornson, B. E., *et al.*, Biochar based solid acid catalyst hydrolyze biomass//Journal of Environmental Chemical Engineering. – 2013. – V. 1. – N – P. 1174-1181.

99. Kitano, M., Yamaguchi, D., Suganuma, S., *et al.*, Adsorption-Enhanced Hydrolysis of OI-1,4-Glucan on Graphene-Based Amorphous Carbon Bearing SO3H, COOH, and OH Groups//Langmuir. – 2009. – V. 25. – N 9. – P. 5068-5075.

100. Nakajima, K.,Hara, M. Amorphous Carbon with SO3H Groups as a Solid Brensted Acid Catalyst//ACS Catalysis. – 2012. – V. 2. – N 7. – P. 1296-1304.

101. Van Pelt, A. H., Simakova, O. A., Schimming, S. M., *et al.*, Stability of functionalized activated carbon in hot liquid water//Carbon. – 2014. – V. 77. – N – P. 143-154.

102. Foo, G. S.,Sievers, C. Synergistic Effect between Defect Sites and Functional Groups on the Hydrolysis of Cellulose over Activated Carbon//ChemSusChem. – 2015. – V. 8. – N 3. – P. 534-543.

103. Taran, O. P., Polyanskaya, E. M., Ogorodnikova, O. L., *et al.*, Sibunit-based catalytic materials for the deep oxidation of organic ecotoxicants in aqueous solution: I. Surface properties of the oxidized sibunit samples//Catalysis in Industry. – 2011. – V. 2. – N 4. – P. 381-386.

104. Yang, Z., Huang, R., Qi, W., *et al.*, Hydrolysis of cellulose by sulfonated magnetic reduced graphene oxide//Chemical Engineering Journal. – 2015. – V. 280. – N – P. 90-98.

105. Suganuma, S., Nakajima, K., Kitano, M., *et al.*, Synthesis and acid catalysis of cellulose-derived carbon-based solid acid//Solid State Sciences. – 2010. – V. 12. – N 6. – P. 1029-1034.

106. Lai, D.-m., Deng, L., Li, J., *et al.*, Hydrolysis of Cellulose into Glucose by Magnetic Solid Acid//ChemSusChem. – 2011. – V. 4. – N 1. – P. 55-58.

107. Guo, H., Lian, Y., Yan, L., *et al.*, Cellulose-derived superparamagnetic carbonaceous solid acid catalyst for cellulose hydrolysis in an ionic liquid or

aqueous reaction system//Green Chemistry. – 2013. – V. 15. – N 8. – P. 2167-2174.

108. Kobayashi, H., Yabushita, M., Komanoya, T., *et al.,* High-Yielding One-Pot Synthesis of Glucose from Cellulose Using Simple Activated Carbons and Trace Hydrochloric Acid//ACS Catalysis. – 2014. – V. 3. – N 4. – P. 581-587.

109. Komanoya, T., Kobayashi, H., Hara, K., *et al.,* Catalysis and characterization of carbon-supported ruthenium for cellulose hydrolysis//Applied Catalysis A: General. – 2011. – V. 407. – N 1вЋ"2. – P. 188-194.

110. Kobayashi, H., Komanoya, T., Guha, S. K., *et al.,* Conversion of cellulose into renewable chemicals by supported metal catalysis//Applied Catalysis A: General. – 2011. – V. 409-410. – N – P. 13-20.

111. Shuai, L.,Pan, X. Hydrolysis of cellulose by cellulase-mimetic solid catalyst// Energy and Environmental Science. – 2012. – V. 5. – N – P. 6889-6894.

112. Pal, R., Sarkar, T.,Khasnobis, S. Amberlyst-15 in organic synthesis//ARKIVOC. – 2012. – V. 2012. – N 1. – P. 570-609

113. Takagaki, A., Tagusagawa, C.,Domen, K. Glucose production from saccharides using layered transition metal oxide and exfoliated nanosheets as a water-tolerant solid acid catalyst//Chemical Communications. – 2008. – V. – N – P. 5363-5365.

114. Hu, L., Lin, L., Wu, Z., *et al.,* Chemocatalytic hydrolysis of cellulose into glucose over solid acid catalysts//Appl. Catal. B: Environ. – 2015. – V. 174–175. – N – P. 225-243.

115. Cai, H., Li, C., Wang, A., *et al.,* Zeolite-promoted hydrolysis of cellulose in ionic liquid, insight into the mutual behavior of zeolite, cellulose and ionic liquid// Applied Catalysis B: Environmental. – 2012. – V. 123-124. – N 0. – P. 333-338.

116. Jiang, C.-W., Zhong, X.,Luo, Z.-H. An improved kinetic model for cellulose hydrolysis to 5-hydroxymethylfurfural using the solid SO42-/Ti-MCM-41 catalyst//RSC Advances. – 2014. – V. 4. – N 29. – P. 15216-15224.

117. Tanabe, K. Catalytic application of niobium compounds//Catalysis Today. – 2003. – V. 78. – N 1–4. – P. 65-77.

118. Yang, F., Liu, Q., Yue, M., *et al.,* Tantalum compounds as heterogeneous catalysts for saccharide dehydration to 5-hydroxymethylfurfural//Chemical Communications. – 2011. – V. 47. – N 15. – P. 4469–4471.

119. Yang, F., Liu,Q., Bai,X.,*et al.,* Conversion of biomass into 5-hydroxymethylfurfural using solid acid catalyst//Bioresource Technology. – 2011. – V. 102. – N 3. – P. 3424-3429.

120. Chareonlimkun, A., Champreda, V., Shotipruk, A., *et al.,* Catalytic conversion of sugarcane bagasse, rice husk and corncob in the presence of TiO2, ZrO2 and mixed-oxide TiO2–ZrO2 under hot compressed water (HCW) condition// Bioresour. Technol. – 2010. – V. 101. – N 11. – P. 4179-4186.

121. Watanabe, M., Aizawa, Y., Iida, T., *et al.,* Catalytic glucose and fructose conversions with TiO2 and ZrO2 in water at 473 K: relationship between

reactivity and acid-base property determined by TPD measurement//Appl. Catal. A. – 2005. – V. 295. – N – P. 150-156.

122. Tagusagawa, C., Takagaki, A., Iguchi, A., *et al.*, Highly active mesoporous Nb-W oxide solid-acid catalyst.//Angew. Chem. Int. – 2010. – V. 49. – N – P. 1128-1132.

123. Chareonlimkun, A., Champreda, V., Shotipruk, A., *et al.*, Reactions of C5 and C6-sugars, cellulose, and lignocellulose under hot compressed water (HCW) in the presence of heterogeneous acid catalysts//Fuel. – 2010. – V. 89. – N 10. – P. 2873-2880.

124. Lai, D.-m., Deng, L., Guo, Q.-x., *et al.*, Hydrolysis of biomass by magnetic solid acid//Energy Environ. Sci., 2011,4, 3552-3557. – 2011. – V. 4. – N 9. – P. 3552-3557

125. Takagaki, A., Nishimura, M., Nishimura, S., *et al.*, Hydrolysis of Sugars Using Magnetic Silica Nanoparticles with Sulfonic Acid Groups//Chemical Letters. – 2011. – V. 40. – N 10. – P. 1195-1197.

126. Bootsma, J. A.,Shanks, B. H. Cellobiose hydrolysis using organic–inorganic hybrid mesoporous silica catalysts//Applied Catalysis A: General. – 2007. – V. 327. – N 1. – P. 44-51.

127. Takagaki, A., Tagusagawa, C.,Domen, K. Glucose production from saccharides using layered transition metal oxide and exfoliated nanosheets as a water-tolerant solid acid catalyst//Chem. Comm. – 2008. – V. 2008. – N 42. – P. 5363-5365.

128. Degirmenci, V., Uner, D., Cinlar, B., *et al.*, Sulfated Zirconia Modified SBA-15 Catalysts for Cellobiose Hydrolysis//Catalysis Letters. – 2011. – V. 141. – N 1. – P. 33-42.

129. Mäki-Arvela, P., Salmi, T., Holmbom, B., *et al.*, Synthesis of Sugars by Hydrolysis of Hemicelluloses- A Review//Chemical Reviews. – 2011. – V. 111. – N 9. – P. 5638-5666.

130. Marzialetti, T., Valenzuela Olarte, M. B., Sievers, C., *et al.*, Dilute Acid Hydrolysis of Loblolly Pine: A Comprehensive Approach//Ind. Eng. Chem. Res. – 2008. – V. 47. – N 19. – P. 7131–7140.

131. Mosier, N. S., Ladisch, C. M.,Ladisch, M. R. Characterization of acid catalytic domains for cellulose hydrolysis and glucose degradation//Biotechnology and Bioengineering. – 2002. – V. 79. – N 6. – P. 610-618.

132. Пат. CN 101525355 В Китайская Народная республика, МПК C07H3/02, C07H1/00 Method for preparing xylose and arabinose by hydrolyzing lignocellulose/Li, S., Liu, X., Li, T. - опубл. 09.09.2009.//– 2009. – V. – N – P.

133. Blecker, C., Fougnies, C., Van Herck, J.-C., *et al.*, Kinetic Study of the Acid Hydrolysis of Various Oligofructose Samples//J. Agric. Food Chem. – 2002. – V. 50. – N 6. – P. 1602-1607.

134. Sun, H.-J., Yoshida, S., Park, N.-H., *et al.*, Preparation of (1→4)-β-d-xylooligosaccharides from an acid hydrolysate of cotton-seed xylan: suitability

of cotton-seed xylan as a starting material for the preparation of (1→4)-β-d-xylooligosaccharides//Carbohydrate Research. – 2002. – V. 337. – N 7. – P. 657–661.

135. Abasaeed, A. E.,Lee, Y. Y. Inulin hydrolysis to fructose by a novel catalyst// Chemical Engineering and Technology. – 1995. – V. 18. – N 6. – P. 440-444.

136. Yang, Y., Xiang, X., Tong, D., *et al.*, One-pot synthesis of 5-hydroxymethylfurfural directly from starch over SO_4^{2-}/ZrO_2-Al_2O_3 solid catalyst//Bioresourse Technologies. – 2012. – V. 116. – N – P. 302-306.

137. Тарабанько, В. Е., Смирнова, М. А., Челбина, Ю. В., *et al.*, Низкотемпературный синтез 5-гидроксиметилфурфурола//Химия растительного сырья. – 2011. – V. 1. – N – P. 87-92.

138. Chheda, J. N., Roman-Leshkov, Y.,Dumesic, J. A. Production of 5-hydroxymethylfurfural and furfural by dehydration of biomass-derived mono- and poly-saccharides//Green Chemistry. – 2007. – V. 9. – N 4. – P. 342-350.

139. Nikolla, E., Román-Leshkov, Y., Moliner, M., *et al.*, "One-Pot" Synthesis of 5-(Hydroxymethyl)furfural from Carbohydrates using Tin-Beta Zeolite//ACS Catalysis. – 2011. – V. 1. – N 4. – P. 408–410.

140. McNeff, C. V., Nowlan, D. T., McNeff, L. C., *et al.*, Continuous production of 5-hydroxymethylfurfural from simple and complex carbohydrates//Applied Catalysis A: General. – 2010. – V. 384. – N 1-2. – P. 65-69.

141. Komanoya, T. K., H.; Hara, K.; Chun, W.-J.; Fukuoka, A. Catalysis and characterization of carbon-supported ruthenium for cellulose hydrolysis// Applied Catalysis A: General. – 2011. – V. 407. – N 1-2. – P. 188-194.

142. Kobayashi, H., Matsuhashi, H., Komanoya, T., *et al.*, Transfer hydrogenation of cellulose to sugar alcohols over supported ruthenium catalysts//Chemical Communications. – 2011. – V. 47. – N 8. – P. 2366-2368.

143. Deng, W., Tan, X., Fang, W., *et al.*, Conversion of Cellulose into Sorbitol over Carbon Nanotube-Supported Ruthenium Catalyst//Catalysis Letters. – 2009. – V. 133. – N – P. 167-174.

144. Van de Vyver, S., Geboers, J., Dusselier, M., *et al.*, Selective Bifunctional Catalytic Conversion of Cellulose over Reshaped Ni Particles at the Tip of Carbon Nanofibers//ChemSusChem. – 2010. – V. 3. – N 6. – P. 698-701.

145. Pang, J., Wang, A., Zheng, M., *et al.*, Catalytic conversion of cellulose to hexitols with mesoporous carbon supported Ni-based bimetallic catalysts//Green Chemistry. – 2012. – V. 14. – N 3. – P. 614-617.

146. Ding, L.-N., Wang, A.-Q., Zheng, M.-Y., *et al.*, Selective Transformation of Cellulose into Sorbitol by Using a Bifunctional Nickel Phosphide Catalyst// ChemSusChem. – 2010. – V. 3. – N 7. – P. 818-821.

147. Yang, P., Kobatashi, N.,Fukuoka, A. Recent Developments in the Catalytic Conversion of Cellulose into Valuable Chemicals//Chinese Journal of Catalysis. – 2011. – V. 32. – N 5. – P. 716-722.

148. Luo, C., Wang, S.,Liu, H. Cellulose Conversion into Polyols Catalyzed by Reversibly Formed Acids and Supported Ruthenium Clusters in Hot Water// Angewandte Chemie International Edition. – 2007. – V. 46. – N 40. – P. 7636-7639.

149. Rinaldi, R.,Schuth, F. Design of solid catalysts for the conversion of biomass// Energy and Environmental Science. – 2009. – V. 2. – N 6. – P. 610-626.

150. Geboers, J., Van de Vyver, S., Carpentier, K., *et al.*, Efficient catalytic conversion of concentrated cellulose feeds to hexitols with heteropoly acids and Ru on carbon//Chemical Communications. – 2010. – V. 46. – N 20. – P. 3577-3579.

151. Geboers, J., Van de Vyver, S., Carpentier, K., *et al.*, Hydrolytic hydrogenation of cellulose with hydrotreated caesium salts of heteropoly acids and Ru/C// Green Chemistry. – 2011. – V. 13. – N – P. 2167-2174.

152. Liu, M., Deng, W., Zhang, Q., *et al.*, Polyoxometalate-supported ruthenium nanoparticles as bifunctional heterogeneous catalysts for the conversions of cellobiose and cellulose into sorbitol under mild conditions//Chemical Communications. – 2011. – V. 47. – N 34. – P. 9717-9719.

153. Ji, N., Zhang, T., Zheng, M., *et al.*, Catalytic conversion of cellulose into ethylene glycol over supported carbide catalysts//Catalysis Today. – 2009. – V. 147. – N 2. – P. 77-85.

154. Ji, N., Zhang, T., Zheng, M., *et al.*, Direct Catalytic Conversion of Cellulose into Ethylene Glycol Using Nickel-Promoted Tungsten Carbide Catalysts// Angewandte Chemie International Edition. – 2008. – V. 47. – N 44. – P. 8510-8513.

155. Xi, J., Ding, D., Shao, Y., *et al.*, Production of Ethylene Glycol and Its Monoether Derivative from Cellulose//ACS Sustainable Chemistry and Engineering. – 2014. – V. 2. – N 10. – P. 2355-2362.

156. Li, H., Fang, Z., Smith Jr, R. L., *et al.*, Efficient valorization of biomass to biofuels with bifunctional solid catalytic materials//Progress in Energy and Combustion Science. – 2016. – V. 55. – N – P. 98-194.

157. Zhang, Y., Wang, A.,Zhang, T. A new 3D mesoporous carbon replicated from commercial silica as a catalyst support for direct conversion of cellulose into ethylene glycol//Chemical Communications. – 2010. – V. 46. – N 6. – P. 862-864.

158. Zhao, X., Xu, J., Wang, A., *et al.*, Porous carbon in catalytic transformation of cellulose//Chinese Journal of Catalysis. – 2015. – V. 36. – N 9. – P. 1419-1427.

159. Deng, W., Zhang, Q.,Wang, Y. Catalytic transformations of cellulose and cellulose-derived carbohydrates into organic acids//Catalysis Today. – 2014. – V. 234. – N – P. 31-41.

160. Deng, W., Zhang, Q.,Wang, Y. Polyoxometalates as efficient catalysts for transformations of cellulose into platform chemicals//Dalton Transactions. – 2012. – V. 41. – N 33. – P. 9817-9831.

161. Davis, S. E., Ide, M. S.,Davis, R. J. Selective oxidation of alcohols and aldehydes over supported metal nanoparticles//Green Chemistry. – 2013. – V. 15. – N 1. – P. 17-45.

162. Gromov, N. V., Taran, O. P., Delidovich, I. V., *et al.*, Hydrolytic Oxidation of Cellulose to Formic Acid in the Presence of Heteropoly Acid Catalysts for Efficient Processing of Lignocellulosic Biomass//Catalysis Today. – 2016. – V. 278. – N 1. – P. 74-81.

163. Moreau, C., Belgacem, M. N.,Gandini, A. Recent Catalytic Advances in the Chemistry of Substituted Furans from Carbohydrates and in the Ensuing Polymers//Topics in Catalysis. – – V. 27. – N 1. – P. 11-30.

164. Hustede, H.-J.,Haberstroh, E. S. Gluconic Acid//Ulmann's Encyclopedia Industrial Chem. 5th Edition. Volume A12/Hans Jorgen Arpe. - Гамбург: Wiley-VCH Verlag GmbH and Co., 1989. - c. 449-456.

165. Fellay, C., Dyson, P. J.,Laurenczy, G. A Viable Hydrogen-Storage System Based On Selective Formic Acid Decomposition with a Ruthenium Catalyst// Angewandte Chemie International Edition. – 2008. – V. 47. – N 21. – P. 3966-3968.

166. Li, J., Ding, D.-J., Deng, L., *et al.*, Catalytic Air Oxidation of Biomass-Derived Carbohydrates to Formic Acid//ChemSusChem. – 2012. – V. 5. – N 7. – P. 1313-1318.

167. //–!!! INVALID CITATION !!! – V. – N – P.

168. Liu, X., Li, S., Liu, Y., *et al.*, Formic acid: A versatile renewable reagent for green and sustainable chemical synthesis//Chinese Journal of Catalysis. – 2015. – V. 36. – N 9. – P. 1461-1475.

169. Serrano-Ruiz, J. C., Braden, D. J., West, R. M., *et al.*, Conversion of cellulose to hydrocarbon fuels by progressive removal of oxygen//Applied Catalysis B: Environmental. – 2010. – V. 100. – N – P. 184–189.

170. Tan, X., Deng, W., Liu, M., *et al.*, Carbon nanotube-supported gold nanoparticles as efficient catalysts for selective oxidation of cellobiose into gluconic acid in aqueous medium//Chemical Communications. – 2009. – V. – N 46. – P. 7179-7181.

171. Onda, A., Ochi, T.,Yanagisawa, K. New direct production of gluconic acid from polysaccharides using a bifunctional catalyst in hot water//Catalysis Communications. – 2011. – V. 12. – N 6. – P. 421-425.

172. Delidovich, I. V., Moroz, B. L., Taran, O. P., *et al.*, Aerobic selective oxidation of glucose to gluconate catalyzed by Au/Al2O3 and Au/C: Impact of the mass-transfer processes on the overall kinetics//Chemical Engineering Journal. – 2013. – V. 223. – N – P. 921-931.

173. An, D., Ye, A., Deng, W., *et al.*, Selective Conversion of Cellobiose and Cellulose into Gluconic Acid in Water in the Presence of Oxygen, Catalyzed by Polyoxometalate-Supported Gold Nanoparticles//Chemistry – A European Journal. – 2012. – V. 18. – N 10. – P. 2938-2947.

174. Zhang, J., Liu, X., Hedhili, M. N., *et al.*, Highly Selective and Complete Conversion of Cellobiose to Gluconic Acid over Au/Cs2HPW12O40 Nanocomposite Catalyst//ChemCatChem. – 2011. – V. 3. – N 8. – P. 1294-1298.

175. Wolfel, R., Taccardi, N., Bosmann, A., *et al.*, Selective catalytic conversion of biobased carbohydrates to formic acid using molecular oxygen//Green Chemistry. – 2011. – V. 13. – N 10. – P. 2759-2763.

176. Albert, J., Wolfel, R., Bosmann, A., *et al.*, Selective oxidation of complex, water-insoluble biomass to formic acid using additives as reaction accelerators// Energy and Environmental Science. – 2012. – V. 5. – N 7. – P. 7956-7962.

177. Evtuguin, D. V., Pascoal Neto, C.,Pedrosa De Jesus, J. D. Bleaching of kraft pulp by oxygen in the presence of polyoxometalates//Journal of Pulp and Paper Science. – 1998. – V. 24. – N 4. – P. 133-140.

178. Shatalov, A. A., Evtuguin, D. V.,Pascoal Neto, C. Cellulose degradation in the reaction system O2/heteropolyanions of series [PMo(12–n)VnO40](3+n)–// Carbohydrate Polymers. – 2000. – V. 43. – N 1. – P. 23-32.

179. Bulushev, D. A.,Ross, J. R. H. Catalysis for conversion of biomass to fuels via pyrolysis and gasification: A review//Catalysis Today. – 2011. – V. 171. – N – P. 1-13.

180. Cortright, R. D., Davda, R. R.,Dumesic, J. A.//Nature. – 2002. – V. 418. – N – P. 964.

181. Huber, G. W., Shabaker, J. W.,Dumesic, J. A.//Science. – 2003. – V. 300. – N – P. 2075.

182. Ma, L., Wang, T., Liu, Q., *et al.*, A review of thermal-chemical conversion of lignocellulosic biomass in China//Biotechnology Advances. – 2012. – V. 30. – N 4. – P. 859-873.

183. Wen, G., Xu, Y., Xu, Z., *et al.*, Direct conversion of cellulose into hydrogen by aqueous-phase reforming process//Catalysis Communications. – 2010. – V. 11. – N 6. – P. 522-526.

184. Wen, G., Xu, Y., Ma, H., *et al.*, Production of hydrogen by aqueous-phase reforming of glycerol//International Journal of Hydrogen Energy. – 2008. – V. 33. – N 22. – P. 6657-6666.

185. Alonso, D. M., Bond, J. Q.,Dumesic, J. A. Catalytic conversion of biomass to biofuels//Green Chemistry. – 2010. – V. 12. – N 9. – P. 1493-1513.

186. Bridgwater, A. V. Catalysis in thermal biomass conversion//Applied Catalysis A: General. – 1994. – V. 116. – N 1. – P. 5-47.

187. Carlson, T. R., Tompsett, G. A., Conner, W. C., *et al.*, Aromatic Production from Catalytic Fast Pyrolysis of Biomass-Derived Feedstocks//Topics in Catalysis. – 2009. – V. 52. – N 3. – P. 241.

188. Huber, G. W.,Corma, A. Synergies between Bio- and Oil Refineries for the Production of Fuels from Biomass//Angewandte Chemie International Edition. – 2007. – V. 46. – N 38. – P. 7184-7201.

189. Balat, M. An Overview of the Properties and Applications of Biomass Pyrolysis Oils//Energy Sources, Part A: Recovery, Utilization, and Environmental Effects. – 2011. – V. 33. – N 7. – P. 674-689.

190. Bridgwater, T. Biomass for energy//Journal of the Science of Food and Agriculture. – 2006. – V. 86. – N 12. – P. 1755-1768.

191. Aho, A., Kumar, N., Eränen, K., *et al.*, Catalytic pyrolysis of woody biomass in a fluidized bed reactor: Influence of the zeolite structure//Fuel. – 2008. – V. 87. – N 12. – P. 2493-2501.

192. Samolada, M. C., Papafotica, A.,Vasalos, I. A. Catalyst Evaluation for Catalytic Biomass Pyrolysis//Energy and Fuels. – 2000. – V. 14. – N 6. – P. 1161-1167.

193. Gayubo, A. G., Aguayo, A. T., Atutxa, A., *et al.*, Deactivation of a HZSM-5 Zeolite Catalyst in the Transformation of the Aqueous Fraction of Biomass Pyrolysis Oil into Hydrocarbons//Energy and Fuels. – 2004. – V. 18. – N 6. – P. 1640-1647.

194. Iliopoulou, E. F., Antonakou, E. V., Karakoulia, S. A.,*et al.*, Catalytic conversion of biomass pyrolysis products by mesoporous materials: Effect of steam stability and acidity of Al-MCM-41 catalysts//Chemical Engineering Journal. – 2007. – V. 134. – N 1–3. – P. 51-57.

195. Wang, C., Hao, Q., Lu, D., *et al.*, Production of Light Aromatic Hydrocarbons from Biomass by Catalytic Pyrolysis//Chinese Journal of Catalysis. – 2008. – V. 29. – N 9. – P. 907-912.

196. Czernik, S., French, R., Feik, C., *et al.*, Hydrogen by Catalytic Steam Reforming of Liquid Byproducts from Biomass Thermoconversion Processes//Industrial and Engineering Chemistry Research. – 2002. – V. 41. – N 17. – P. 4209-4215.

197. Rioche, C., Kulkarni, S., Meunier, F. C., *et al.*, Steam reforming of model compounds and fast pyrolysis bio-oil on supported noble metal catalysts//Applied Catalysis B: Environmental. – 2005. – V. 61. – N 1–2. – P. 130-139.

198. Sutton, D., Kelleher, B.,Ross, J. R. H. Review of literature on catalysts for biomass gasification//Fuel Processing Technology. – 2001. – V. 73. – N 3. – P. 155-173.

www.ingramcontent.com/pod-product-compliance
Lightning Source LLC
Chambersburg PA
CBHW050513190326
41458CB00005B/1526